Contents

AS 1: Physical Geography

1A Processes and features in fluvial environments 6
1B Human interaction with the fluvial environment 22

2A The ecosystem as an open system 30
2B Plant succession 38
2C Human interaction with ecosystems 46

3A Atmospheric processes 56
3B Mid-latitude weather systems 68
3C Extreme weather events 74

AS 2: Human Geography

1A Population data 84
1B Population structure 89
1C Population distribution and resources 109

2A Challenges for rural environments 114
2B Planning issues in rural environments 126
2C Challenges for the urban environment 135

3A The nature and measurements of development 148
3B Issues of development 164

Examination Technique

Maximising your potential 186

Glossary

Physical Geography 194
Human Geography 197

AS 1 PHYSICAL GEOGRAPHY

Below: Frozen meanders and ox-bow lakes in Labrador, Canada

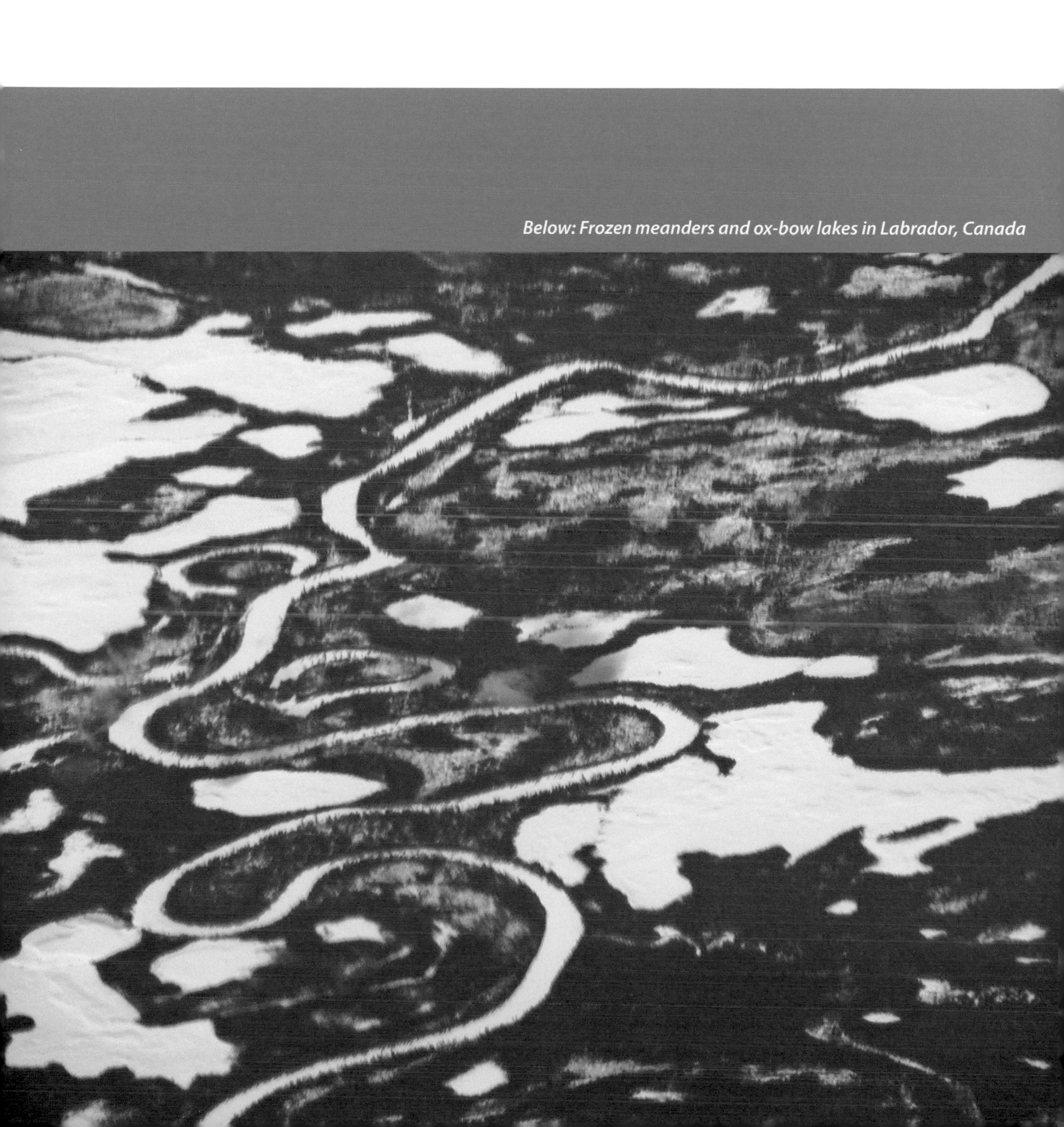

1A PROCESSES AND FEATURES IN FLUVIAL ENVIRONMENTS

"Geography; always my favourite subject"
Michael Palin

Few subjects touch the world in which we live quite like Geography. Pick up any newspaper or listen to any news bulletin and topic after topic relates to some aspect of this multidimensional subject. So all-embracing is Geography that it defies easy definition leaving some to declare: 'Geography is what Geographers do'. This textbook, like the AS specification on which it is based, attempts to illustrate the unity in diversity of the subject. While its structure reflects three aspects of both physical and human geography, the overall theme is of interaction and understanding of the world through skills.

Seeing the world geographically – a systems approach

Making sense of the world has been a universal goal across time and space. The need to create order from the night sky initially led to the identification of patterns and constellations representing earth-bound items: lions and ploughs or mythical wonders, gods and giants.

While some scientists require the study of geography to be quantitative, ie measurable and testable, other experts desire a more qualitative geography in which a sense of place is equally valid. One approach to help unify the study of geography is known as **systems theory**. In essence this is simply a pattern or framework through which aspects of the world can be viewed.

The personal computer is a good illustration of a system. In a personal computer there are objects such as the motherboard, the monitor, the hard drive and the printer. Between these flows information in the form of energy pulses. As illustrated (***Resource 1***) a system exists where a number of objects are linked by interactions between them. The objects are known as **stores** and the interactions as **transfers**. One key part in recognizing a system is to identify its boundaries. A computer has clear boundaries and even the most advanced computer needs input including power and data to operate. A system is said to be **isolated** if nothing transfers in or out across its boundaries or **closed** if only energy can enter or leave it. But most natural systems are **open** as both materials and energy, enter (**input**) and leave (**output**) them. The whole universe and its numerous parts can be regarded as systems. This can help us simplify the world and aid our understanding of how it operates.

The elements of an open system | Resource 1

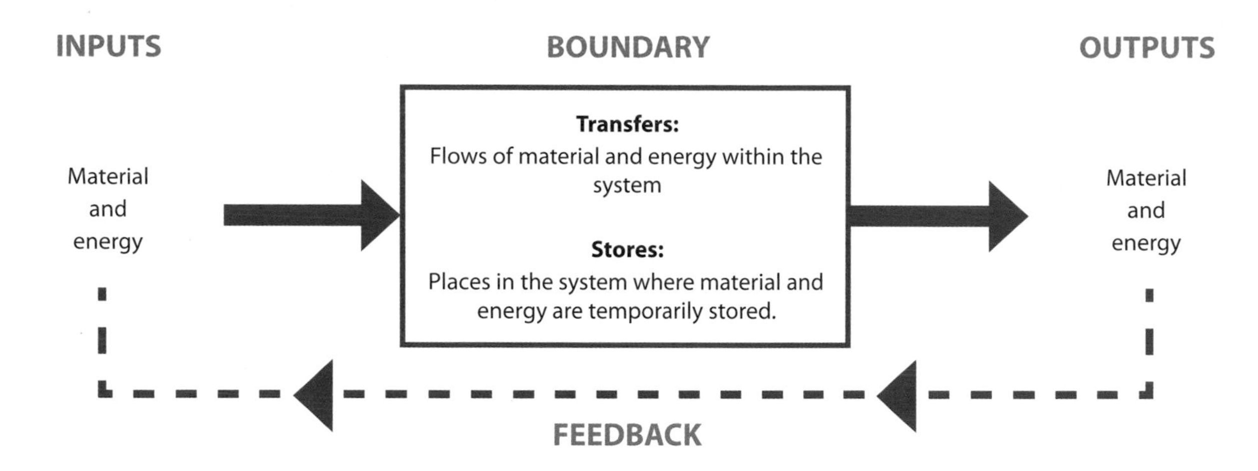

The drainage basin as an open system

Fluvial environments are those that result from the action of river processes.

The area of land that gathers water for a river is known as a **drainage basin**. The land boundary of a drainage basin is its **watershed** – an area of higher ground separating drainage down one slope from another. When **precipitation** (rain, hail or snow) falls from the atmosphere onto the land it must either return to the air (evaporate), sink into the ground or flow over the surface. The vegetation, ground surface, soil, underlying rocks and the channels are all stores in the basin. Infiltration, overland flow, through flow, and groundwater flow are some processes within the basin system. Transfers of energy and materials across the boundaries may be either **inputs** – precipitation and potential energy – or **outputs** – evapotranspiration, and runoff. Most natural river networks have many **channels** that join together as **tributaries** (contributing water) to form larger rivers and drainage basins.

The drainage basin as an open system | Resource 2

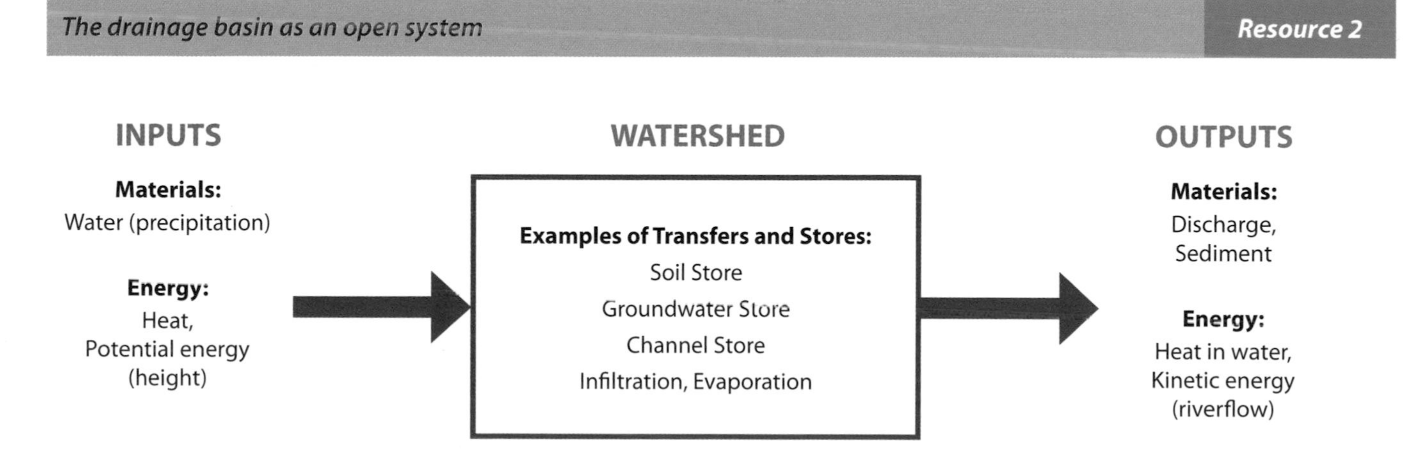

While water is the most obvious component that moves in, through and out of a drainage basin open system, there are others including sediment and energy (heat and potential energy). **Resource 3** outlines the transfers of water across the boundaries of a drainage basin and between its stores.

Resource 3 *Movement of water into, through and out of the drainage basin*

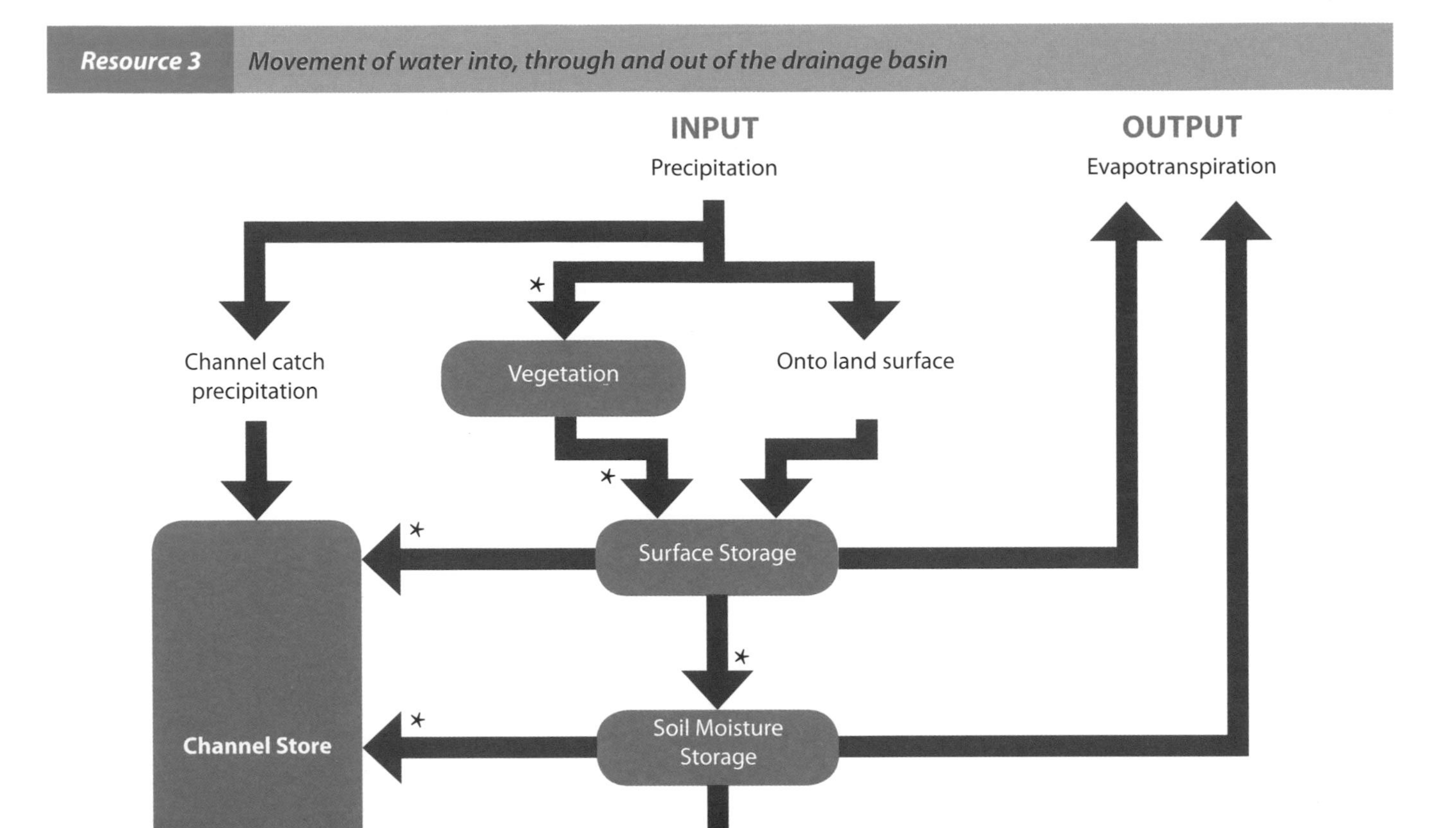

Exercise On a copy of Resource 3, add the following transfers to the relevant arrows marked with *:

- Ground water flow
- Through flow
- Overland flow
- Interception
- Stem/drip flow
- Percolation
- Infiltration

Processes in a drainage basin

A) Surface flows and outputs

Precipitation may enter a drainage basin directly into the river itself (Channel Catch), onto the land surface, or caught on the vegetation (Interception). Once it reaches the surface, water may be lost to the system by evaporation or it may move across the land (Overland flow), or into the soil (Infiltration). Once in the soil, water may be taken up by plant root systems and ultimately lost through leaves back to the air (Transpiration). The output from a drainage basin into the atmosphere is termed **evapotranspiration**. Water that flows over the land will gather in the river channels at the base of slopes.

B) Infiltration and subsurface flows

One of the key factors in how a river basin responds to rainfall is the movement of water into the soil and through the soil and groundwater stores. Soil is mostly made of inorganic particles of various sizes, from fine *clays* through medium *silt-sized* to large *sand-sized* material. The exact mixture of particles is called the **soil texture (*see Resource 59*)**.

Soil particles are separated from each other by small spaces called pores and it is through these that water can:

- infiltrate from the surface into the soil;
- percolate from the soil down into the underlying rock; or
- through-flow downslope within the soil.

Different soils have different rates of infiltration, for example sandy soils have large pores and water moves in quickly. This characteristic is called the soil's infiltration rate and this is usually measured in mm per hour.

Soils that allow water to enter rapidly are described as **permeable** whereas **impermeable** soils, often clay-rich, allow water to enter only slowly. A final key factor to note about infiltration is that the rate changes during the course of a rainfall episode (Storm). As **Resource 4** shows, no matter what soil type is found, the infiltration capacity is relatively high when rain starts. This is because the soil is dry and has pore space available, but the infiltration capacity falls to a lower, stable level as the pores fill up over time.

The movement of water from the soil into the underlying rock is termed **percolation**. Most rocks can store water either within the rock itself (porous), or within the cracks and joints they contain. This stored water is referred to as **groundwater** and it slowly makes its way into river channels as groundwater flow. The height of the groundwater level in rock is called the **water table**. This underground source of water can maintain river flow even after weeks or months without rainfall.

Typical infiltration rates for different soil types ***Resource 4***

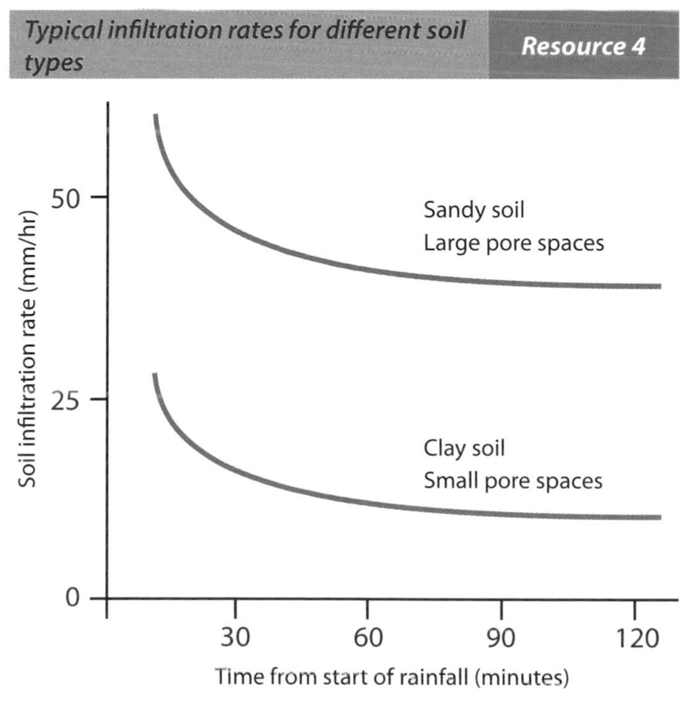

Measuring and recording river flow

Discharge is the volume of water passing any one point in a given time. It is calculated by multiplying the channel's cross-sectional area (metres squared) by the river's velocity (metres per second). The units of discharge are therefore cubic metres per second ($m^2 \times m/s = m^3/sec$) or **cumecs.**

Resource 5 *How to find the discharge of a river*

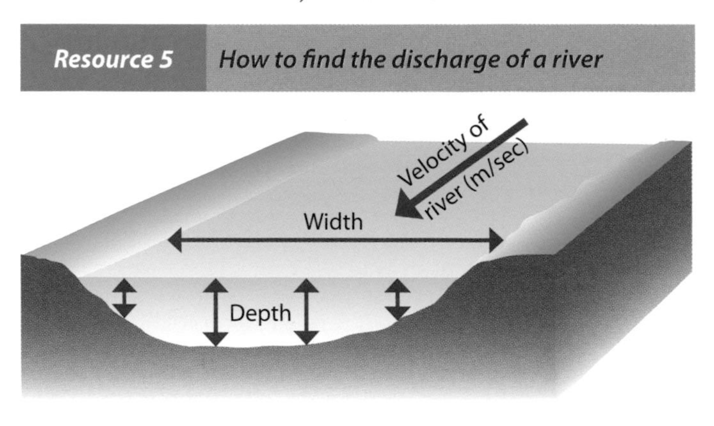

The pattern of changing discharge at any point along a river can be plotted on a time/discharge graph called a **hydrograph**. An **annual hydrograph** would show the variation in river discharge over a year whereas a **Storm** or **Flood hydrograph** records the impact of one period of precipitation on water flow. **Resource 6** shows a model hydrograph, including the terms used to describe the line of changing discharge, and the area beneath the line that represents the total volume of channel flow. The superimposed bar graph shows the amount and pattern of precipitation (storm) that causes the subsequent changes in flow.

Two sets of transfers within drainage basins impact the shape and nature of a hydrograph: the fast (surface) flow processes of channel catch and overland flow; and the slow (sub-surface) flows of throughflow (soil) and groundwater (rock) flow. A typical storm hydrograph follows a sequence. After rainfall commences, some falls directly into the channel and discharge rises, but much of the rain falls on the basin and may infiltrate into the soil. Eventually the soil becomes saturated, or if the rainfall is intense and exceeds the soil's infiltration capacity, water will flow over the surface as overland flow and quickly reach the channel. This accounts for the rapid increase in discharge shown as the **rising limb**. Even in small drainage basins it takes time for water to make its way to the channel and this delay between the **peak rainfall** (bar graph) and **peak discharge** (hydrograph) is called the **lag time,** and may be measured in minutes or days depending on basin size.

Resource 6 *Model hydrograph of storm flow*

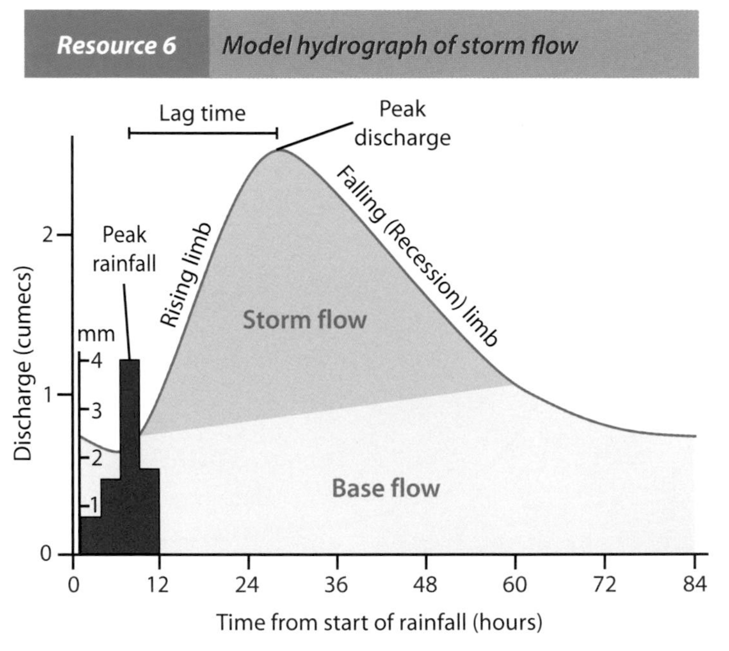

The peak discharge is a critical factor because if it exceeds the carrying capacity of the river channel, flooding will occur. Following the peak discharge is the **falling** or **recession limb** where the remaining water from the storm moves from the basin. The hydrograph levels out at a lower level but flow continues as a slow transfer from the groundwater store feeds the channel. This 'background' flow is called **base flow**, while **storm flow** refers to the discharge that resulted from the period of precipitation involved.

In general terms the shape of hydrographs varies between two extremes:

- A wide, shallow curve with gentle rising and recession limbs, and a low peak discharge following a long time lag (**flat**); and
- A steep-sided, rapidly changing graph with a high peak discharge and short lag times. This form is known as a **flashy** response and often represents the threat of river flooding.

Hydrograph shapes **Resource 7**

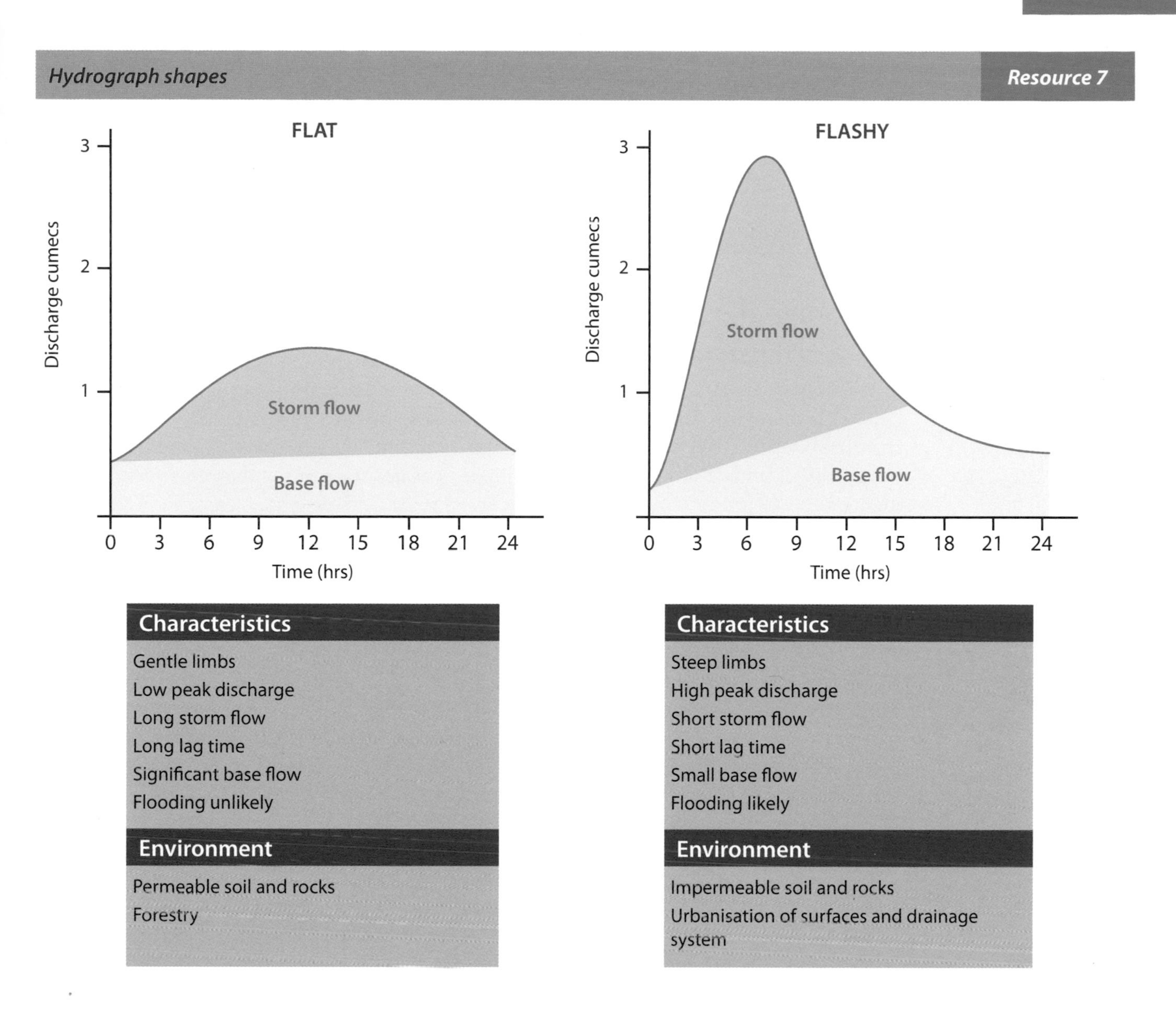

The way a drainage basin responds to rainfall and the shape of the flood hydrograph depends on many different factors; these can be classified as either the nature of the storm or the nature of the basin.

Some factors influencing basin hydrographs **Resource 8**

Nature of the storm	Nature of the basin 1 *Permanent features*	Nature of the basin 2 *Temporary change*
Total precipitation	Size of basin	Vegetation (seasons)
Type of precipitation (rain/snow)	Shape of basin	Land use (farming/urban)
Intensity of precipitation	Aspect of basin	Previous weather
Duration of storm	Slopes in basin	Soil moisture
Path of storm across basin	Drainage density	Groundwater storage
Frequency of storm	(length of channel per unit area)	
Speed of storm movement	Channel size	
	Nature of soil	
	Nature of rock	

Exercise

For any factor, discuss how it would influence water transfer in the basin and how that would look on a hydrograph.

For example:

Basin slopes – steeper slopes will encourage water to flow more rapidly, over the surface and soil, to the river channels. This will decrease the lag time, increase the rising limb curve, the peak discharge and therefore the flood risk.

Factors affecting basin discharge and the storm hydrograph

A) Nature of the storm

A period of precipitation has many variables, all of which can influence how drainage basins deal with the input. The duration of the storm, the intensity of the rainfall, the nature of the precipitation (rain, hail, or snow) and the direction and speed of movement across the basin will all impact on the hydrograph shape. It is accurate to say that no two storms are identical.

B) Nature of the basin

1. Permanent features

Basin size and shape

In a small basin of a few square kilometres, rain will be gathered and transferred quickly producing a short period of storm flow and a short lag time. In larger basins, such as the Nile or Congo, the precipitation may take days or weeks to flow through the basin. The most efficient shape for a basin would be circular with a central outflow; this is unrealistic but long narrow basins will produce longer, more even-shaped hydrographs compared to the flashy response of roughly circular basins.

Relief

Naturally steep basin slopes shed water more rapidly than gentle slopes. For this reason, mountainous drainage basins will produce more dramatic and responsive hydrographs in comparison to those found in lowland basins where gentle slopes slow down transfers.

Underlying soil and geology

The ability of soil and the underlying geology to take in and store water will have a fundamental impact on hydrographs. The chalk rocks of Southern England are highly permeable, reducing river channel flow levels. Over clay soils, or regions of impermeable rock a drainage basin's storm hydrograph will have a flashy response to rainfall as surface quick flows are dominant.

Drainage density

This is a measurement of the relationship between drainage basin area and the total length of river channel in the basin. It is expressed as kilometres per square kilometre (km/km^2). The higher the value, the more efficiently the basin will be drained and the flashier the hydrograph. (For summary, see ***Resource 8.***)

2. Temporary features

Previous (Antecedent) conditions

This is the state of the basin just before the storm event. For example, has there been recent rainfall in the area that may have saturated the soil or alternatively has a period of drought left the soil and groundwater stores depleted and able to take in and store a large proportion of the precipitation?

Vegetation cover

A basin covered in vegetation will have a high evapotranspiration output, so reducing the quantity of precipitation transferred by the river system. The type of vegetation, grass or trees, and seasonal variation will have a strong impact on the shape of hydrographs. In summer, deciduous trees intercept 60% more water than in winter and the rainfall that infiltrates the soil is more likely to be taken up by tree roots for growth than to reach the river channel directly. A single mature oak tree uses up to 230 litres of water each day in the summer growing season.

3. Land use

Vegetation change

Human activity has altered the nature of the land surface across the British Isles, not only the obvious development of urban landscapes, but even the vegetation cover of the remote rural landscape. In the mountains of central Wales two rivers, the Severn and the Wye, have their sources in adjacent basins on the slopes of Plynlimon. In many respects the two basins are very similar in size, shape and underlying geology. They are also subject to very similar rainfall conditions. However, since the 1950s one has been extensively used to grow coniferous forest on a commercial basis. The other basin has a vegetation cover of upland heath and moor used for hill sheep farming. The outcome of this variation in land use is seen in the respective hydrographs for the two basins (***Resource 9***).

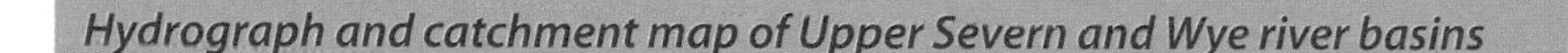

Hydrograph and catchment map of Upper Severn and Wye river basins ***Resource 9***

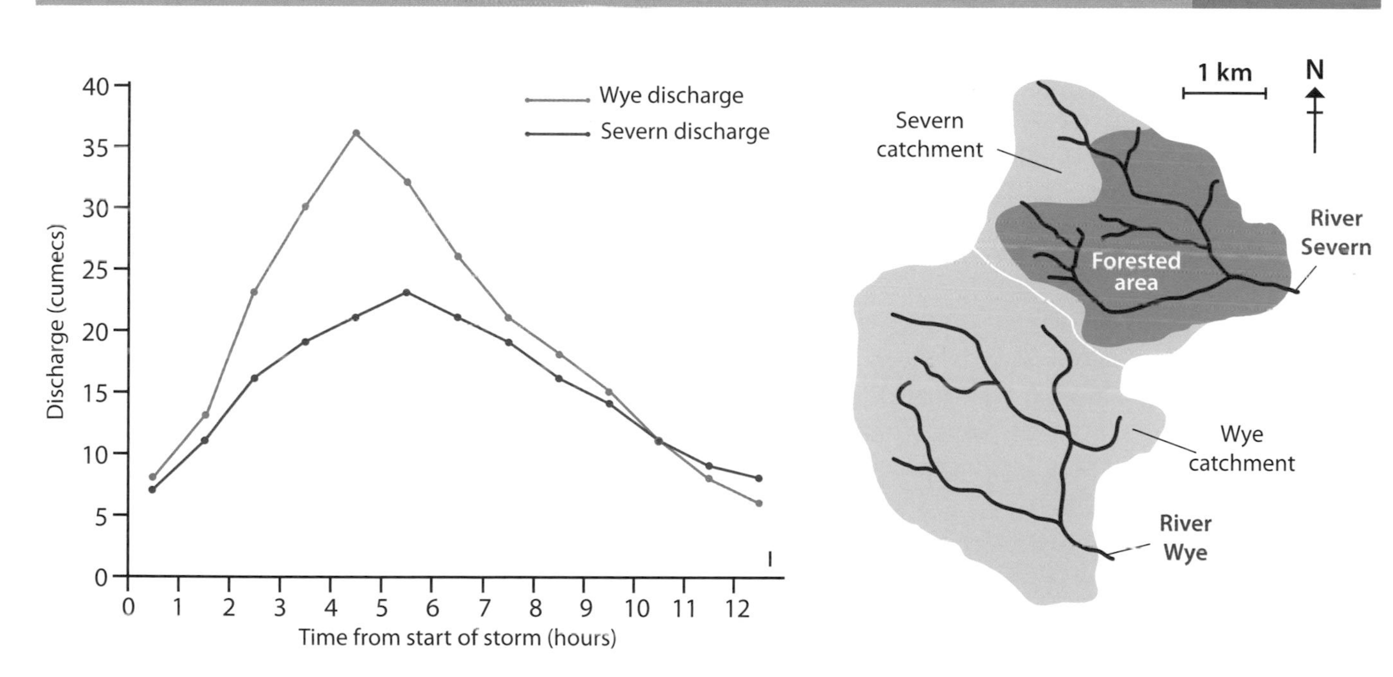

With reference to **Resource 9**: ***Exercise***

(i) Compare the response of the Severn and Wye river basins to the same rainstorm. (Figures from the graphs and the correct hydrograph terms, **Resource 6**, should be used.)

(ii) Explain the difference between the two hydrographs.

Urbanisation

While the planting of trees – afforestation – modifies basin response to a less flashy hydrograph (***Resource 9***), the replacement of natural surfaces by artificial ones has the opposite effect. Two characteristics of urbanisation impact basin response: firstly, the replacement of vegetated soils with less permeable surfaces such as tiled roofs, tarmac roads and concrete paving. Even today, when strict planning has prevented the unplanned expansion of towns and cities in the British Isles, the replacement of permeable with impermeable continues. In-fill housing and the demands for off-road parking both contribute to the increased proportion of impermeable land cover. Secondly, urban environments are provided with highly engineered drainage systems. Buildings have guttering and downpipes, roads are cambered and have kerbside drains, while beneath the ground a complete system of pipes direct water efficiently into natural rivers and streams. Even these natural routes have been channelized and modified for rapid drainage.

River processes and landforms

Rivers at work

Drainage basins not only involve the movement of water and sediment, but there are also transfers of energy through the system. Water on land has potential energy, which becomes kinetic or movement energy when it starts to flow down slopes. In river channels this energy is used for three types of work: friction, erosion and transport. Energy is lost by **friction** between the water and the bed and banks of the channel; this is most obvious where the channel has large boulders as often seen in mountain streams. **Erosion** is the wearing away of the bed and bank of the channel and **transport** is the carrying away of this eroded material. Material carried by a river is called the **load**.

Erosion

Most erosion by rivers takes place during high flow or flood periods when the volume and velocity is higher and the river energy is at a maximum. This is why some rivers seem unchanged over many years because flood conditions may be rare. Fluvial erosion is commonly divided into four distinct processes, though all may well operate simultaneously during periods of high flow.

1. **Abrasion** (corrasion) occurs when rock fragments carried by the river wear down the bed and banks of the channel. Large boulders scrape the rock bed while sand and gravel can smooth the surfaces much like the action of sandpaper. In high flows the impact can be dramatic, and the swirling of pebbles in hollows of a solid rock bed can create potholes.
2. **Hydraulic action** is the erosive power of the water itself. It is most effective on loose (unconsolidated) bank materials and can often undermine the bank on the outer side of meanders. It can also weaken solid rock by forcing air into cracks, especially at waterfalls and rapids.
3. **Solution** (corrosion) occurs when rock or minerals are dissolved by the water. This is particularly effective on carbonate-based rocks such as limestone and chalk. Unlike the previous two processes, this form of erosion has more to do with the chemical nature of the water than its volume or velocity.
4. **Attrition** is the wearing down of the river load itself as particles strike each other and the bed and banks. The particles will then reduce in size and become more rounded as they travel

downstream. Attrition therefore, is not a process that causes change in the shape of the river channel.

These processes may result in:

- vertical erosion in mountain streams as abrasion and hydraulic action lower the channel bed giving a steep sided valley; or
- lateral erosion caused by meandering rivers creating a wider shallow valley.

Transport

Rivers gain their load either by the erosion processes or from material falling or washed in from the valley sides. The sediment load carried by a river can be described in three parts, representing the **method of transport**, **solution load**, **suspended load** and **bedload**. **Resource 10** illustrates each of these, and the four processes involved in their movement downstream. The movement of bedload downstream is described either as a rolling motion, termed **traction** or a bouncing, skipping action, termed **saltation**.

River transport processes **Resource 10**

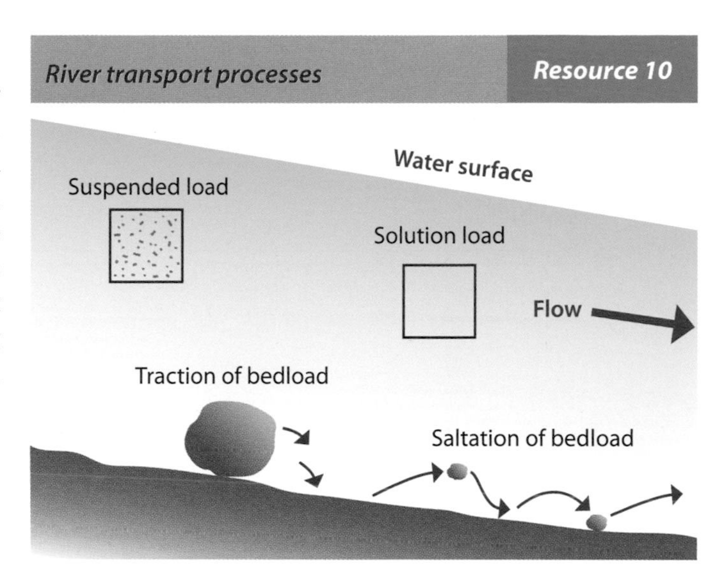

Deposition

Rivers need energy to overcome friction and to erode and transport their load but when a river loses energy the load can no longer be carried and so deposition occurs. River velocity is a good indication of river energy so where and when rivers slow down, deposition is likely. The most obvious place for a river to slow is where it reaches its mouth and flows into a lake or the sea. Here sediment falls to the bed starting with the heaviest, largest particles such as gravel and sand while finer sediment – silts and clays – may be carried further. More detail of this process is given in the description of delta formation later in this chapter. But rivers also slow down along their course. A common site for deposition of river load is on the inner bank of meanders, known as **point bar deposits**. As a river channel swings around, the water flowing on the inner bend has less distance to travel and slows down. As a result, the processes of erosion and deposition can occur at the same time on opposite river banks only a few metres apart. Any material deposited by a river is termed **alluvial**. When a river slows down in its course it may start to deposit sediment on its bed and small islands called **eyots** may form. The river channel can be divided into several smaller channels in a process called **braiding** (***Resource 11***).

Braiding of the River Krishna, Andhra Pradesh, India **Resource 11**

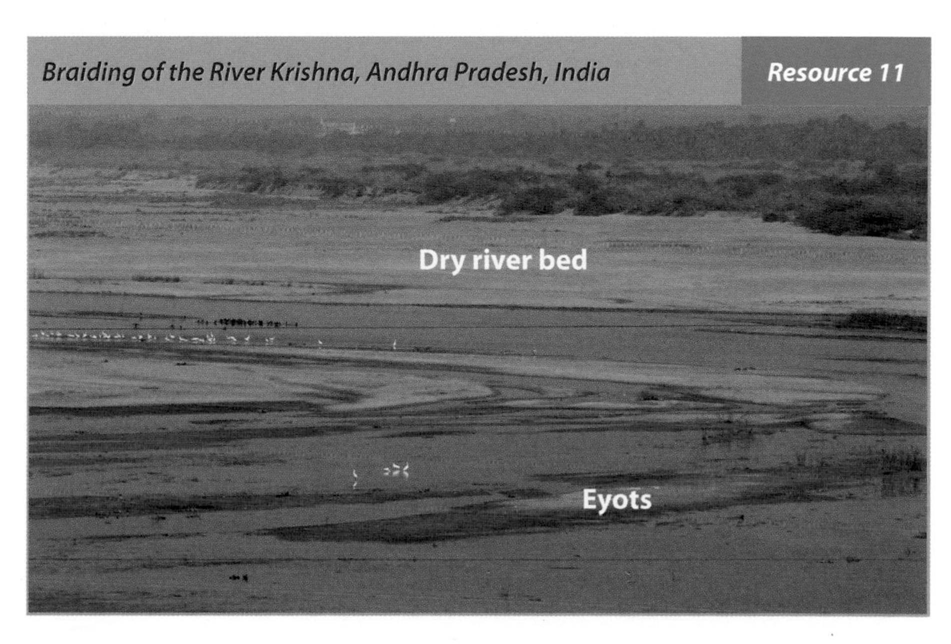

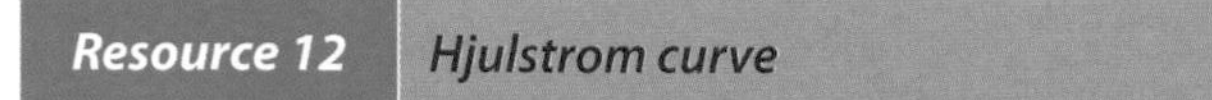

Resource 12 *Hjulstrom curve*

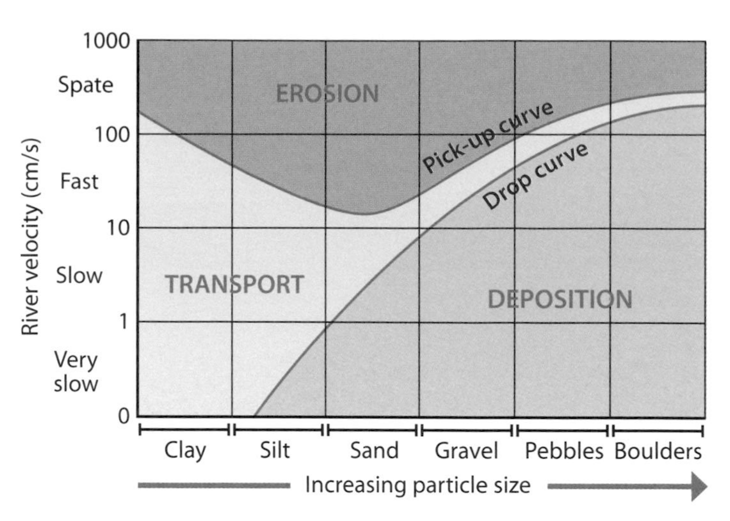

Hjulstrom curve

Simple logic suggests two concepts:

A) The larger the particle, the more energy needed to erode or transport it; and

B) Erosion will need more energy than transport, ie once something is lifted it will require less energy to keep it moving.

Resource 12 illustrates these two concepts, where the vertical axis represents energy as river velocity and the horizontal axis is sediment size.

When real rivers and streams are studied, the actual relationship does not hold true for the smaller sediment. Hjulstrom's curve shows that small clay and silt particles require much more energy than expected to lift them up, although the energy needed to transport them is, as expected, low. What is it about these small particles that causes this?

Exercise

With reference to the Hjulstrom curve (**Resource 12**), explain why the critical velocity for lifting (eroding) any size of sediment is always greater than the velocity to transport it?

There appear to be two factors that cause the apparent anomaly of the Hjulstrom curve.

- Firstly, clay or silt particles have a natural cohesion, and tend to bond together so requiring more energy to lift (erode) them. Once separated, the individual particles are very easily carried and clay will remain suspended in water even at very low velocities.
- Secondly, clay and silt particles can be tightly packed. Imagine a river bed covered in gravel; its surface will be rough and varied and the water flowing over it will be disrupted and turbulent. If the same bed was covered by a flatter layer of clay particles, the flow would be smooth with little friction. The clay is less likely to be disturbed by the water flow.

River landforms

Waterfalls

Perhaps the most dramatic and well known river landform is the waterfall. The name succinctly describes this feature as it is when the water in a channel falls vertically down a rock face. There are several possible reasons for the initial formation of a waterfall, but their features and their active processes are similar. From the majestic Niagara River falls to those of Glenariff, County Antrim the same essential features can be seen. A common creative feature for waterfalls is a band of resistant rock (cap rock) across which the river runs. As the water falls vertically, it rapidly erodes the weaker rock and forms a deep plunge pool. Abrasion and hydraulic action combine to enlarge this pool and undercut the layer of hard rock above. Eventually, part of the cap rock collapses into the pool, providing more material for the abrasion process, and the position of the waterfall moves upstream. This gradual progression of the waterfall is termed 'retreat' and this creates a steep-sided gorge, clearly recording the former position of the fall.

Sketch cross-section and map of Niagara Falls

Resource 13

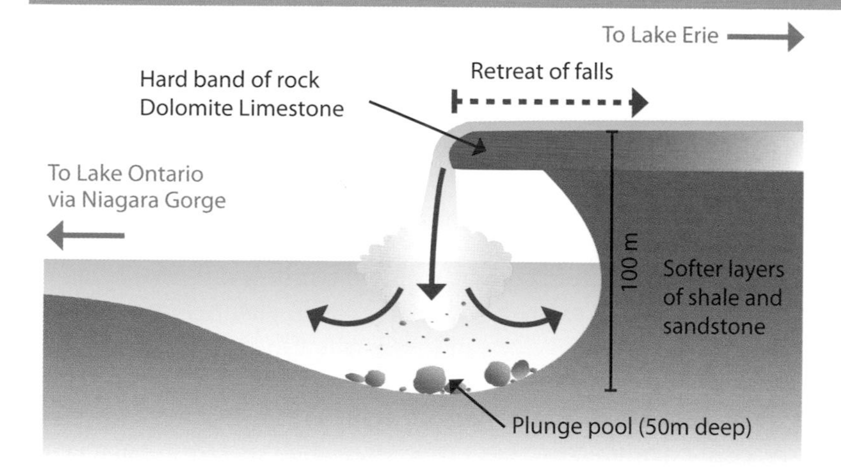

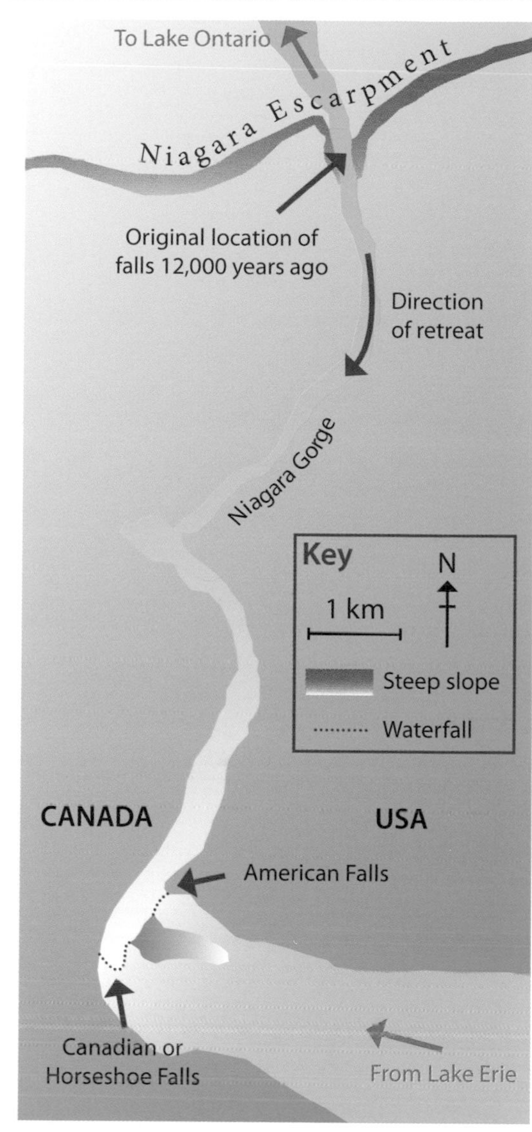

Niagara Falls: The cap rock, or fallmaker as it is termed in North America, is a hard band of limestone. At the end of the last ice age, water from Lake Erie flowed northwards into Lake Ontario. At this time a 100 m high waterfall existed, which has, over the last 12,000 years, retreated south to become the two 50 m high Horseshoe and American waterfalls. This wandering retreat is marked by the 11 km long Niagara Gorge, featured in a scene in Superman 2. In the future the waterfall will continue to retreat and reduce in height until it reaches Lake Erie by which time it will have degenerated into a **cascade** – a series of small steps and rapids – a steep river section of white water flow.

Detail of a fall, pool and gorge at Hardrow Force, North Yorkshire

Resource 14

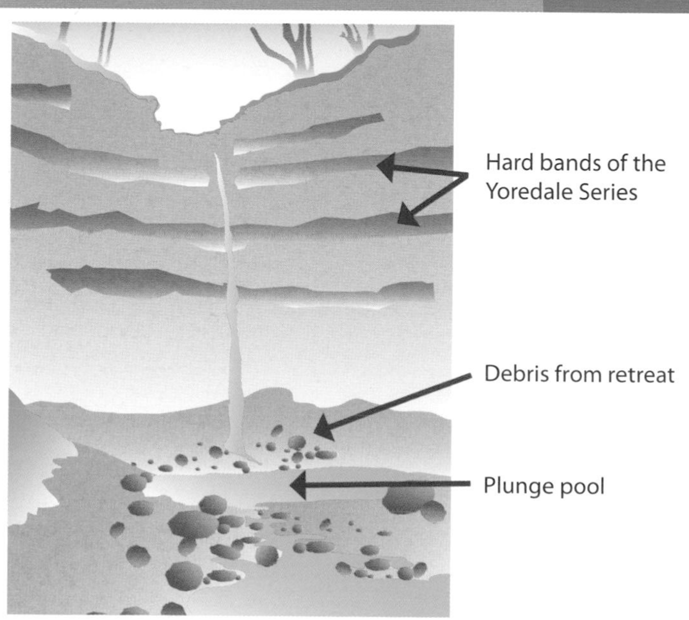

Resource 15 *Diagram of a meander*

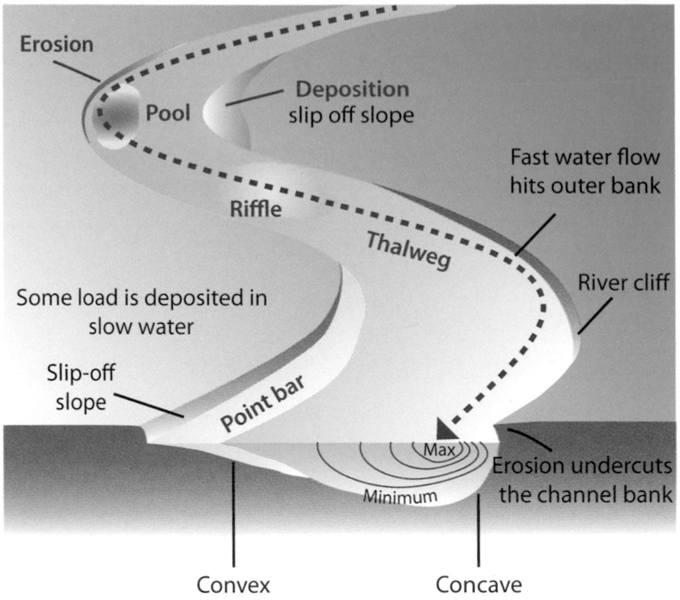

Resource 16 *Possible sequence in the evolution of a meandering channel*

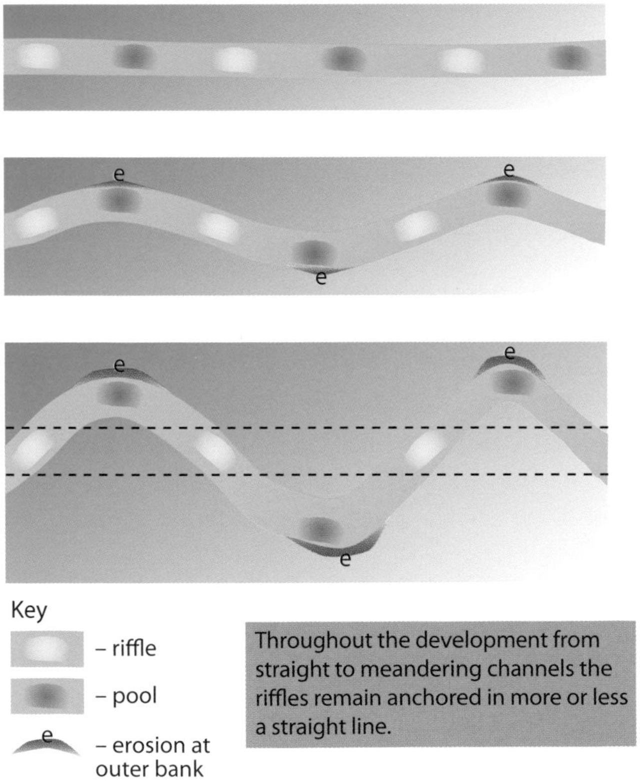

Key
- riffle
- pool
- erosion at outer bank

Throughout the development from straight to meandering channels the riffles remain anchored in more or less a straight line.

Meanders

If waterfalls are common on rivers then meanders are universal. Taking their name from a particularly sinuous, twisted river in Turkey, a meander is a river bend. Despite years of study and research there is still no agreed explanation for the formation of meanders in river channels. Many patterns have been noted, for example meanders are more common in channels where the bedrock is neither too hard nor too fragile, the gradient downstream is gentle and the load carried is not excessive. Straight river channels are rare in nature but even in these, when the line of fastest flow (called the **thalweg**) is plotted, it forms a sinuous course. Meanders are not the result of obstacles in the river path but rather they seem to allow a river to balance its energy with its load. In the early twentieth century, river engineers shortened the length of the lower Mississippi to improve flood control and river transport. To do this, they cut through the river's enormous meander loops to create short-cuts, some saving 5 km on the old channel. Within 15 years the Mississippi, through natural erosion and deposition processes, had restored most of its original length. If understanding why meanders occur is complex, it is much more straightforward to picture why, once a river bends, the meander will grow and develop.

The processes of meander formation

One pattern closely associated with meanders is the sequence of **riffles** and **pools** along a channel. Riffles are sections of channel between two meanders where the water is shallow and flows through coarse bed sediment. Pools are found in the channel bed near the outer bank of meanders (***Resource 15***); they are areas of deep, smooth water flow. Riffles appear to deflect the maximum velocity of a river towards one bank and so cause that bank to be undercut by erosion processes. On the opposite bank water moves more slowly and material is deposited. In this way the river channel does not get wider but it moves sideways (laterally). **Resource 16** shows how the riffles and pools created by the line of fastest flow in straight line channels remain as the river meanders widen and grow by lateral erosion processes. Meanders in cross-section have an asymmetrical appearance with a steep, often concave, outer bank or river cliff and a gentle inner bank called a slip-off slope on which point bar deposits are found (***Resources 15*** and ***17***).

In reality, the thalweg is only one line of flow in the river; other flows form a helical or downstream corkscrew pattern.

The asymmetrical channel cross-section of a meander **Resource 17a**

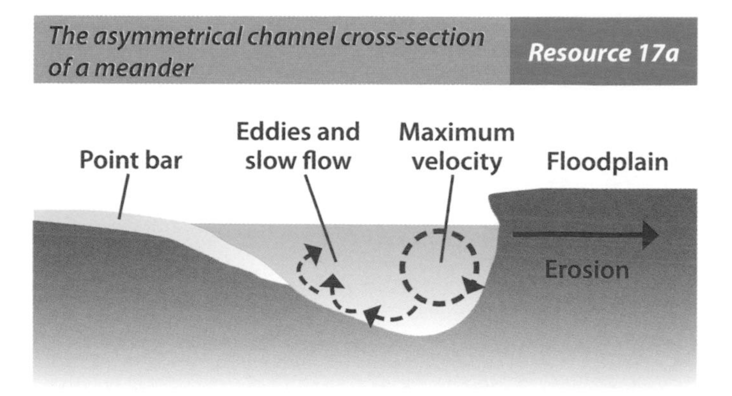

Meander features on the River Ure, Upper Wensleydale **Resource 17b**

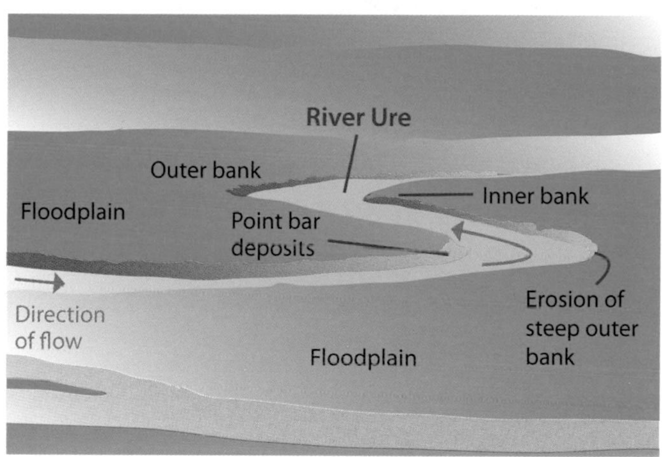

Meandering meanders

The width of a river's meanders is related to its load, slope and discharge, and lateral growth is limited. However, meanders are not fixed but tend to move down the valley, just like a wave moving along a piece of string. The reason for this goes back to the pattern of outer bank erosion. The maximum erosion point is normally downstream of the mid-point of the bend due to centrifugal forces, and so the meander channel creeps downstream. The sketch map of the Mississippi River channel (***Resource 18***) shows how, in over 80 years, a meander has grown laterally by over 1 km and migrated downstream by over 2 km. This illustrates one good reason for not using natural streams as political boundaries!

Migration of meanders downstream **Resource 18**

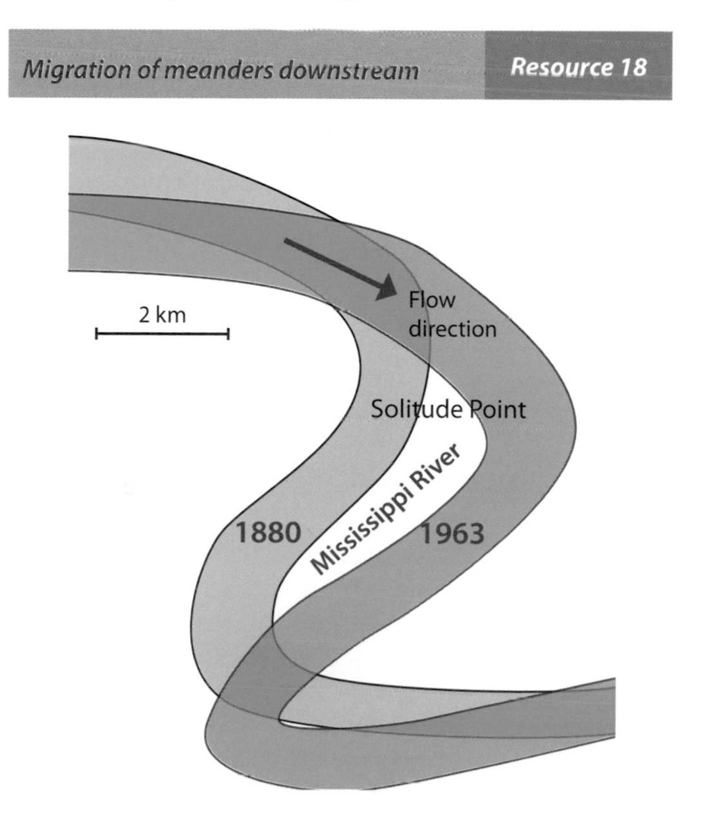

Oxbow lakes

Also known as **mort lakes** or **billabongs** (Australia), **oxbow lakes** are the remnant of a former meander that has been naturally by-passed by the river channel. The usual sequence involves a meander becoming increasingly sinuous and its 'neck' narrowing to a short distance. Then, often in flood, the river cuts through the swan's neck and forms a new channel. For a time water may flow across the new route and through the

Resource 19 *Ox-bow lake on the River Ure floodplain*

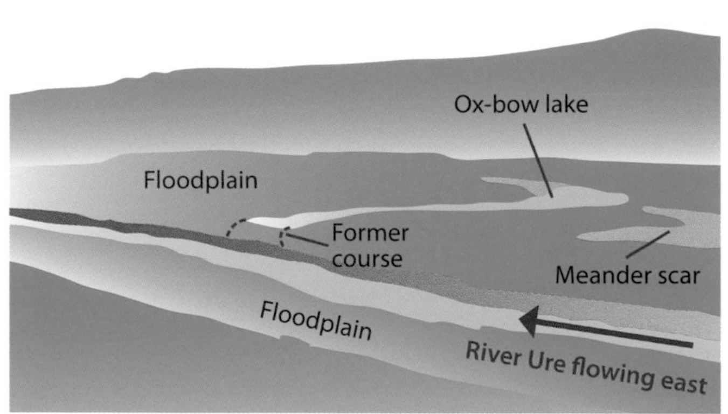

old meander but gradually deposition will block off the meander leaving it disconnected from the channel, creating an ox-bow lake (***Resource 19***). The name comes from the similarity in shape to the U-shaped piece of wood of the same name fitted under the neck of a harnessed ox. Over time plant invasion and sediment washed into ox-bows causes them to dry out. From an aerial view, their shape often remains visible as a **meander scroll** or **scar.**

Resource 20 *Sediment deposition on the floodplain*

Overbank flooding
The shifting channel
Levee
Raised bed
Point bar deposits
Yazoo
Floodplain
Sediment store

Floodplains and levees

Floodplains are most common where rivers have left the mountain stage of their course. As the name suggests, floodplains are flat areas adjacent to the river channel. They are normally made of alluvial material deposited by the river itself. As noted earlier, meanders wander laterally and downstream so over time they erode their way across a broad belt of land. This area will in part be covered by point-bar deposits from the inner bank of the meanders but also with sediment (clay, sand and silt) deposited during flood periods. In times of high rainfall or snowmelt, rivers may exceed their channel capacity, forcing some water out onto the floodplain. This is known as **flooding** or **inundation**. Naturally, as soon as it leaves the channel the water will slow down, lose energy and some river load will be deposited. As the Hjulstrom curve shows, the largest sand-sized particles will settle first, while further away the silt-sized particles will be deposited and finally the finest clays may be carried far from the channel, settling as the water infiltrates into the soil or evaporates. Such floodplains often have a convex upward profile.

One distinctive feature of these is the natural bank of coarse deposits alongside the channel itself. These natural raised river banks were first named by French settlers along the Mississippi river as **levees**. Often these deposition features have been reinforced and raised to provide flood protection. It was one of these on the Mississippi that tragically broke in 2005 during Hurricane Katrina, causing the flooding of New Orleans sited on the floodplain. Levees can cause difficulty for tributary streams as they fail to break through them to join the main channel. Often such streams will flow parallel to the channel for miles before finding a break in the levee. These are technically called **deferred tributaries** or, more memorably, a **yazoo** after a Mississippi tributary of that name.

Deltas

The ultimate output from the river drainage basin system is when the channel reaches its mouth and its energy, discharge and load meet the sea or lake. Here may be found the final river feature – the delta. Deltas are areas of deposited alluvial material often extending the coastline out into the sea. They can be found on many rivers including the world's largest, such as the Ganges-Brahmaputra, Amazon, Colorado and Nile, but are rare in the British Isles. This is because of the large tidal range and high energy environments of our coasts, which prevent deposition. The shape of deltas is largely the result of how the river load is deposited. Along with the normal processes of deposition, due to falling river velocity, the salt nature of sea water causes an electrical charge to draw fine clay particles together, allowing them to settle. This process is termed **flocculation**. A delta can be pictured as an extension of the floodplain out into the sea. One feature common in deltas is that the river itself breaks up into many channels across the delta, causing braiding (see page 15). These separate channels are called **distributaries** and they may give a river numerous 'mouths'. Based on their appearance, the two most common delta forms are the **Arcuate** and **Bird's Foot** deltas.

Common delta forms **Resource 21**

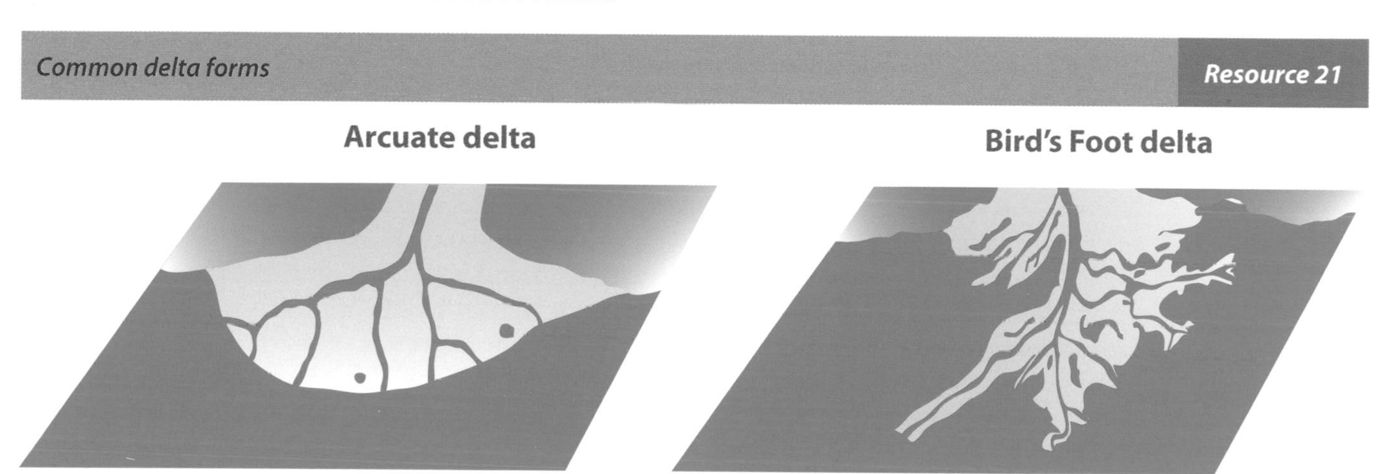

Arcuate form

The delta of the River Nile gives all deltas their name as it is shaped like the Greek letter delta. This is the **arcuate** form with a convex curved outer edge. The Nile delta is readily identified on satellite images as it forms an area of well-watered agricultural land in this desert country. Nile delta sediments are several kilometres thick, deposited over thousands of years of annual flooding, but the control of the river by the High Aswan Dam in the last 50 years has meant that less sediment is carried by the Nile to the Mediterranean. Today, the sea is invading Egypt's delta lands along the north coast, threatening this invaluable farmland.

Bird's Foot form

The Mississippi river drains almost 50% of the USA and carries a huge sediment load into the shallow, low-energy waters of the Gulf of Mexico. Here the river carries out into the Gulf depositing sediment along its channel sides. Additional distributary channels also extend, and so the bird's foot shape is created.

1B HUMAN INTERACTION WITH THE FLUVIAL ENVIRONMENT

Flooding

Case Study: The Ganges Delta, Bangladesh

While Egypt owes much to the delta of the River Nile, another tropical nation is even more dominated by a delta. The historic land of Bengal, at the northern end of the Bay of Bengal in Southern Asia, now the modern nation of Bangladesh, is 80% delta land. Most of this nation is less than one metre above sea-level and it is criss-crossed by numberless streams and rivers with rapidly changing channels. Here, two great rivers first meet each other and then the sea. Together the Ganges and Brahmaputra rivers *(see **Resource 22**)* erode and transport the growing mountains of the Himalayas grain by grain to the sea. It is reasonable to say that Bangladesh is a deltaic nation. Despite the high tidal range along the coast of the Bay of Bengal, the colossal load transported by these rivers has created and extended the shoreline. Each year these rivers carry and deposit about 1,820 million tonnes of alluvial sediment, which is more than twice the annual load of the Amazon, Nile and Mississippi rivers added together. Around 10% of Bangladesh lies under water, but every summer this increases to 25% as the land is flooded by monsoon rains and swollen rivers. These are called the good or **barsha** floods but in some years, with the bad or **bonna** floods, over 40% of the land can be inundated by water.

Resource 22 *The drainage basin of the rivers Ganges and Brahmaputra*

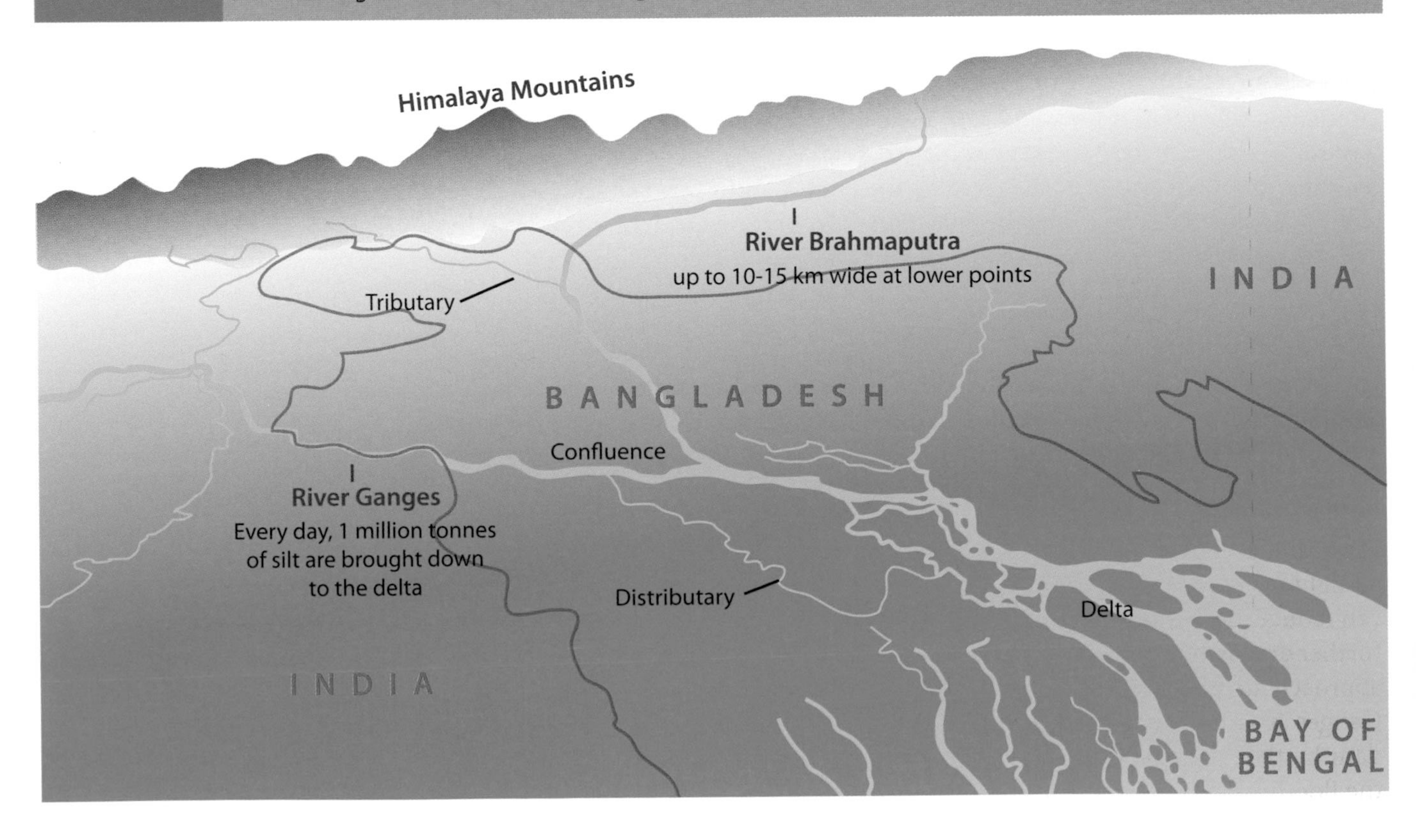

What causes flooding in Bangladesh?

Most rivers flood, but some flood more regularly than others. The reasons for flooding in Bangladesh illustrate the common causes around the world.

Physical/natural factors

Spring snowmelt

The Ganges and Brahmaputra rivers both rise in the western Himalaya Mountains; in fact their source tributaries start only a few kilometres apart. The Ganges cuts its valley across the north of India on the south side of the mountains while the Brahmaputra flows east along a parallel course to the north of the mountain chain. High mountains often cause wet climates and in these 6,000 m peaks much of the precipitation falls as snow. Each April and May snowmelt, caused by the rising spring temperatures, releases an immense input of water and silt down the 2,500 km-long river channels until they meet in Bangladesh. Here, they overtop their river banks providing water and fertile alluvial sediment for the rice and wheat lands of 'Bengal the Golden'.

Monsoon rain

Both intense and prolonged rain can cause rivers to flood. Intense, heavy rain can exceed the rate at which the soil can infiltrate water and so excess water flows quickly into rivers as overland flow. Prolonged rain may cause pore spaces in the soil to fill with water until no more can infiltrate, and again overland flow leads to rapid drainage into river channels.

The wet South East Monsoon of Bangladesh, between June and September, tends to be both intense and prolonged. Even if the rivers do not break their banks, the sheer volume of rain on this almost flat landscape causes rainfall floods ***(Resource 23)***. The monsoon climate is an extreme example of the wet/dry tropical climate found in regions that lie between the world's hot deserts and the equatorial rainforests. Between November and April (winter) Bangladesh receives about 300–400 mm of rain compared to 1250–1850 mm in the six months of summer. For comparison, Northern Ireland's total annual rainfall varies between 900 and 1,350 mm spread fairly evenly over the year.

Flood types in Bangladesh **Resource 23**

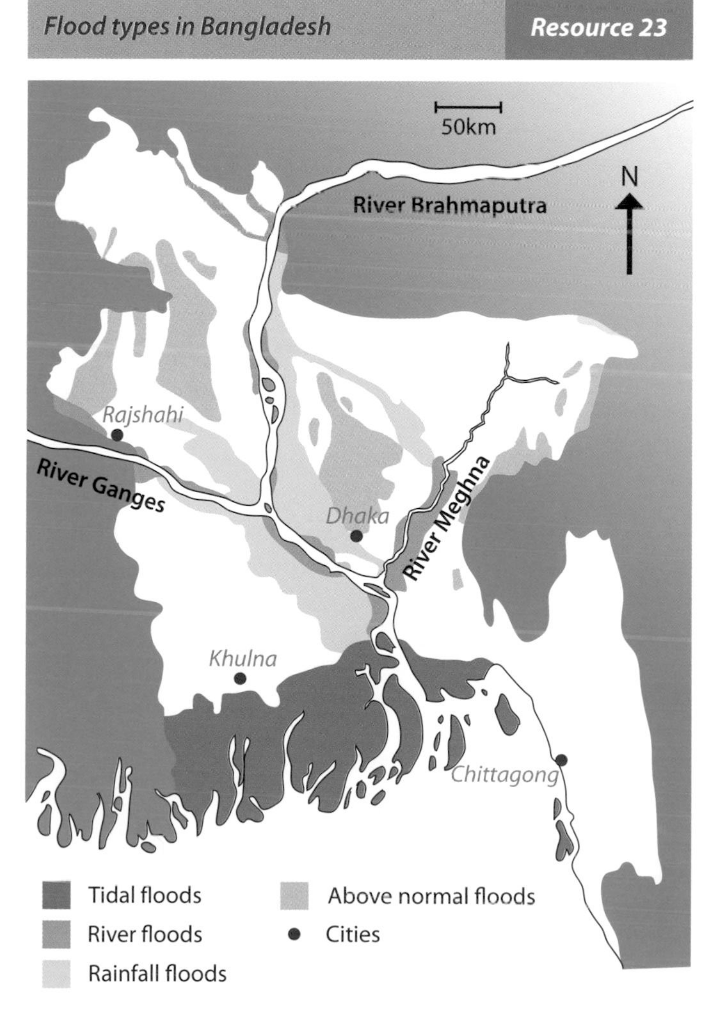

Human factors

Human activity often alters the transfers within a drainage basin in a manner that creates or increases the flood risk. Sometimes the problem is actually the result of attempts to manage the flooding issue!

Management techniques

Even though altering river courses – by making them straight or giving them artificial concrete channels – can reduce local flooding, it may only move the problem further downstream. Dams and reservoirs are a common approach to water storage and flood control. On the Ganges it is claimed that the building of the Farraka dam by the Indian Government has actually increased the flood risk as the river flows on into Bangladesh. The reasoning is that the dam holds back water flow in the dry season but releases huge surges of flow during the wet season that the downstream river channel cannot accommodate (***Resource 24***). Within Bangladesh the raising of river banks (levees) to prevent river flooding has, in places, actually extended the flood season. These banks, which prevent rivers overflowing, also prevent rainwater draining into the river, so, during the wet monsoon, areas may remain under stagnant flood water for several months.

Resource 24

Increased Occurrences of the Worst Floods

Bangladesh has become more flood-prone than it was in the pre-dam era. From time to time floods have hit with extraordinary ferocity across the country in the post-dam era. Dams are used as flood outlets during the wet season when the upstream country cannot withhold the rising flood water. Floods cause irreparable damage to crops, livestock, and above all, humans. Bangladesh is never given a warning of potential floods by the neighbouring country, forcing it to face the flood without preparation. The flood of 2000 inundated areas in Rajshahi, Nawabganj, Kustia, Satkhira, and Jessore. In Rajshahi, dead bodies were seen floating in the Ganges.

Based on an article by Shah Mohammed Saifuddin

Deforestation in the upper drainage basin

Over the last 30 years as the population of Nepal has grown, one outcome has been the large-scale clearance on the foothills and slopes of the Himalayan Mountains. The annual rate of land clearance in Nepal is stated at 1.7% per year. It has been argued that this loss of vegetation cover has increased both the rate and quantity of water discharge and sediment feeding the tributaries of both the Ganges and Brahmaputra basins. The water swells the rivers and the sediment load reduces the channel capacity. While some people have accused the mountain nations of causing floods to increase, it should be noted that the deforestation rate in Bangladesh itself is currently 3.3% per annum. Recent research shows that the only significant statistical link between deforestation and severe Bangladesh floods relates to forest clearance in the Meghalaya Hills in northern Bangladesh itself.

Urbanisation and population pressure

By 2008 Bangladesh was home to over 146 million people, making it one of the world's, most densely populated nations (1,045 people per km^2). In common with most LEDCs, it has been subject to urbanisation, and cities such as its capital, Dhaka, have grown very rapidly. Urbanisation, as we noted before, replaces permeable surfaces with impermeable, and natural drainage processes with efficient artificial drainage networks. Both of these increase the speed and volume of water moving to rivers, thereby increasing the flood risk. With growing rural populations and their demands for water, numerous additional wells have been sunk. In some places these have led to a local lowering of the water table that in turn has caused land subsidence on floodplains, making the risk of deep water floods even greater.

Climate change

If global warming does raise global sea level in the coming decades, even by 30–50 cm, the low-lying nature of Bangladesh's coast and river banks means it will be even more vulnerable to flooding.

Beneficial and detrimental effects on people, property and land

In our local context it is difficult to imagine that there are benefits to flooding. The months of May–July 2007 were the wettest in England and Wales since records began in 1766. The cost of the resultant widespread flooding, in terms of damage to property and loss of agricultural property, was estimated at £5 billion. By contrast, in Bangladesh the annual inundation of around 15% of its land area is critical to its economy. These expected (barsha) floods are prepared for using a sophisticated agricultural system involving different crops and varieties of rice developed to grow in different water depths. If no annual flood arrived it would be a disaster for the region.

The beneficial effects of barsha floods

Agriculture

In common with many rural regions subject to flooding, the event adds silt, nutrients and moisture to the soil, maintaining and improving its fertility. This rich alluvium replaces the nutrients removed by the previous year's crop harvest. In Bangladesh the climate allows for year round plant growth. In the dry season (November to May), known locally as the Rabi, wheat or dry rice is cultivated. For example, Aus is planted because it is a drought-tolerant rice strain. During the wet monsoon season (June to October), the Kharif varieties of wet rice, such as Aman are sown. In places where floods are shallow (less than 180 cm) two wet monsoon rice harvests are possible. The introduction of HYV (high yielding varieties) rice allows Bangladesh to feed its large and rapidly growing population, and even export crops in these good flood years.

Annual cropping patterns in relation to seasonal flooding in some regions of Bangladesh ***Resource 25***

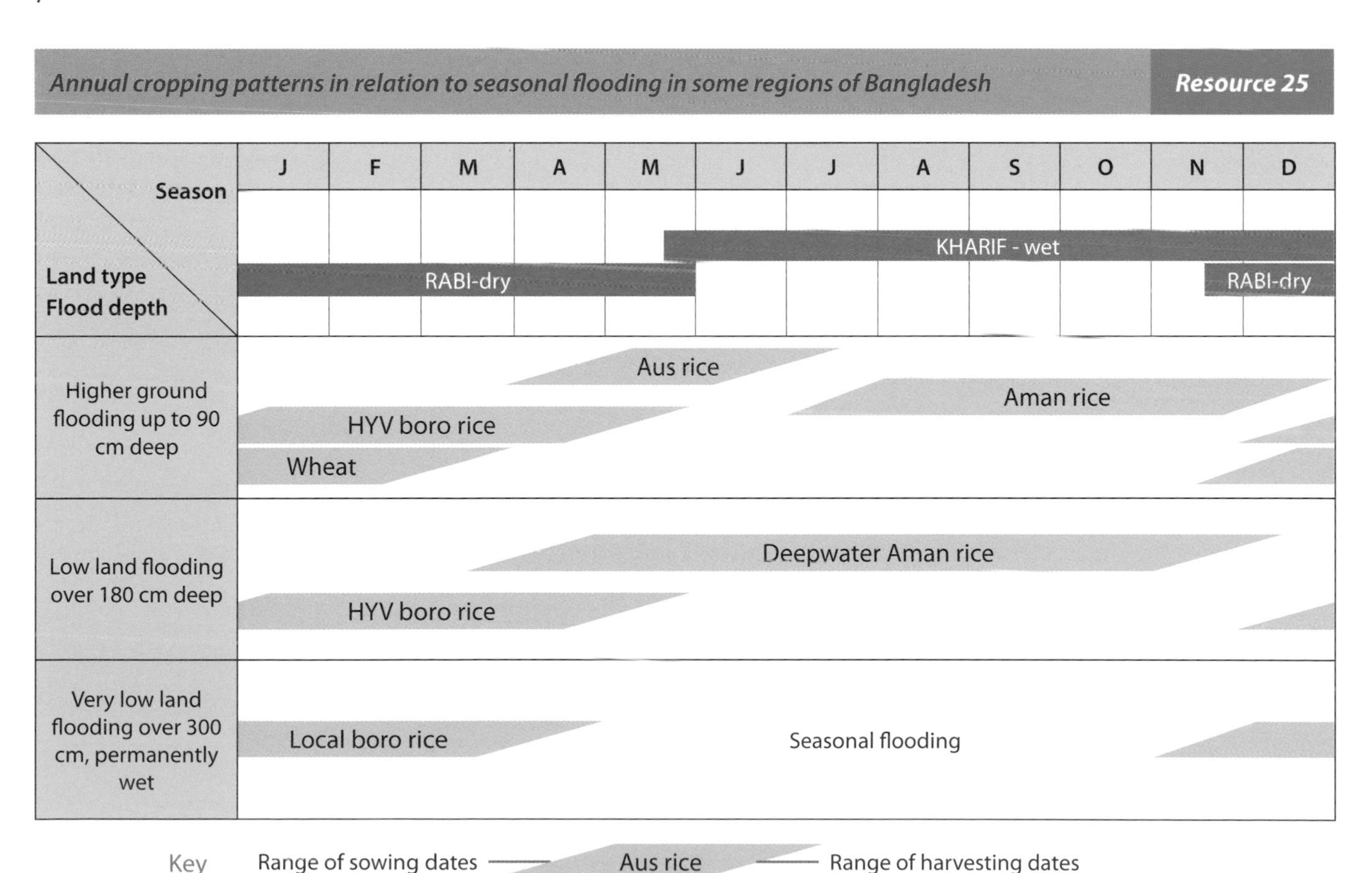

Fishing

Some 260 different fish species live in the fresh water rivers of Bangladesh. Three of these, including the hilsa fish, are a critical source of protein in the diet of the country's largely subsistence population. Flooding extends the aquatic ecosystem, creating breeding ponds and pools across the inundated floodplain. These additional water areas teem with life and represent considerable economic opportunity. Near the coast the prawn-fishing industry provides seasonal employment for many, including school children who collect small hatchlings from ponds in order to stock local fish farms. The industry provides important variety for the local wheat and rice-based diet, and an export commodity to other Asian nations.

Groundwater

Ironic as it might seem, a shortage of fresh water is a potential problem in the regularly flooded lands of Bangladesh. Firstly, in the dry season, agriculture may only be possible with irrigation: the artificial additional of water to the land. Secondly, everyday domestic needs such as water consumption, washing and cooking, place a large demand on supply. Fortunately, beneath much of the country there are layers of rock that can hold water; these are known as aquifers. In Bangladesh, wells can be readily sunk into the aquifer that lies between 6 and 12 m below the surface. These critical water sources are recharged each year by the floodwaters. As the floodwater spreads over the floodplain and delta, it first infiltrates the soil and then percolates down into the groundwater store. This is particularly important near the coast where salt water could seep into these aquifers polluting the supply. Such saline water can be toxic to both the crops grown and the people themselves.

The detrimental effects of bonna floods

Outlined below are the major problems created by abnormally severe flooding in Bangladesh.

A) People:

- Deaths
- Homelessness and refugees
- Social disruption
- Illness spread by lack of food or contaminated water, such as diarrhoea and typhoid

B) Property:

- Homes destroyed in cities and remote rural areas
- Businesses and buildings destroyed
- Food crops washed away and livestock drowned
- Transport links, including road and railway links, washed away or covered

C) Land:

- Shifting river channels
- Islands, known as chars, eroded or removed

The next section describes the 1998 floods in Bangladesh. This, or resources for another severe flood year, could be used to add specific detail to the outline of detrimental effects provided. Those floods in Bangladesh caused by tropical cyclone storm surges, such as that in 1991, are not relevant to this study.

The Ganges/Brahmaputra floods in Bangladesh 1998

The flooding between July and October of 1998 was one of the worst flood events of the twentieth century. Not only did the heavy monsoon rains raise flood levels to record heights, covering up to 65% of the nation's land, but the flood lasted for over three months; much longer than is normal. Around 300,000 homes were destroyed by flowing water, including many in the country's capital city of Dhaka. Here it was reported that over two million inhabitants were forced to flee from the filthy waters polluted by the flooded sewerage system. Nationally over 30 million people were affected. Several hundred people died from water-borne diseases, mainly the old and the very young. It was feared that epidemics of typhoid, cholera and hepatitis could follow. At least 800 deaths were caused by flooding from drowning, mudflows and house collapse.

The country's infrastructure was severely damaged with around 11,000 km of its roads washed away along with hundreds of small bridges that spanned the numerous river channels. The economic impact of the flood was estimated at over £1 billion ***(Resource 26).*** The rice crop of late summer, known as the second Kharif crop, was almost completely destroyed meaning that emergency food relief was needed for millions across the rural areas. With only three heavy lifting helicopters in the country, that meant a dependency on other countries and charities for help. The loss of the food crop was therefore both a short term relief issue and a long term economic problem as new seed had to be bought.

Bangladesh flood 1998 affected areas ***Resource 26***

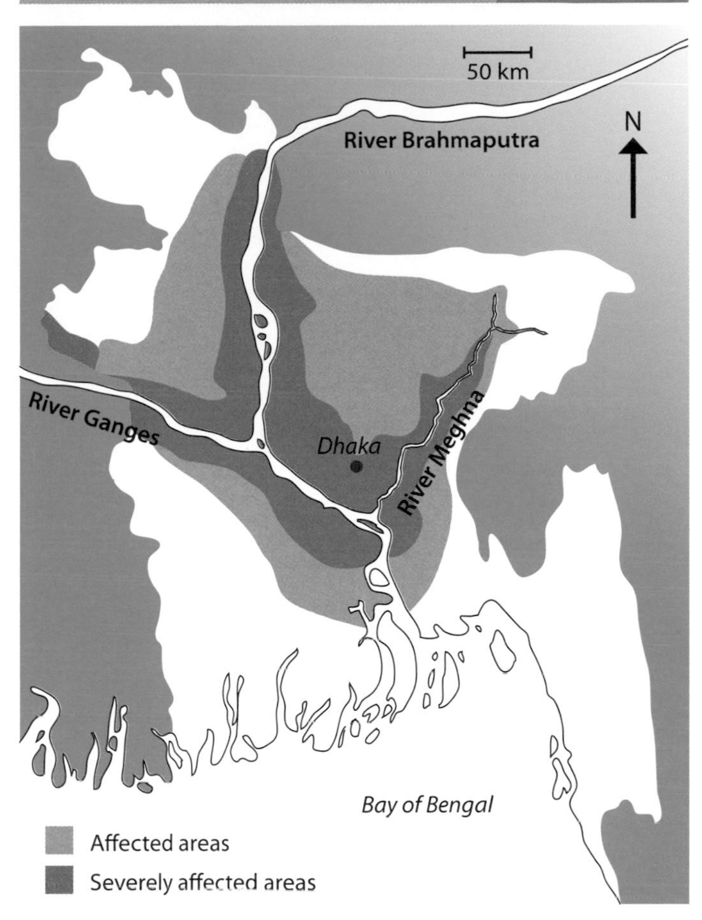

Fieldwork ***Exercise***

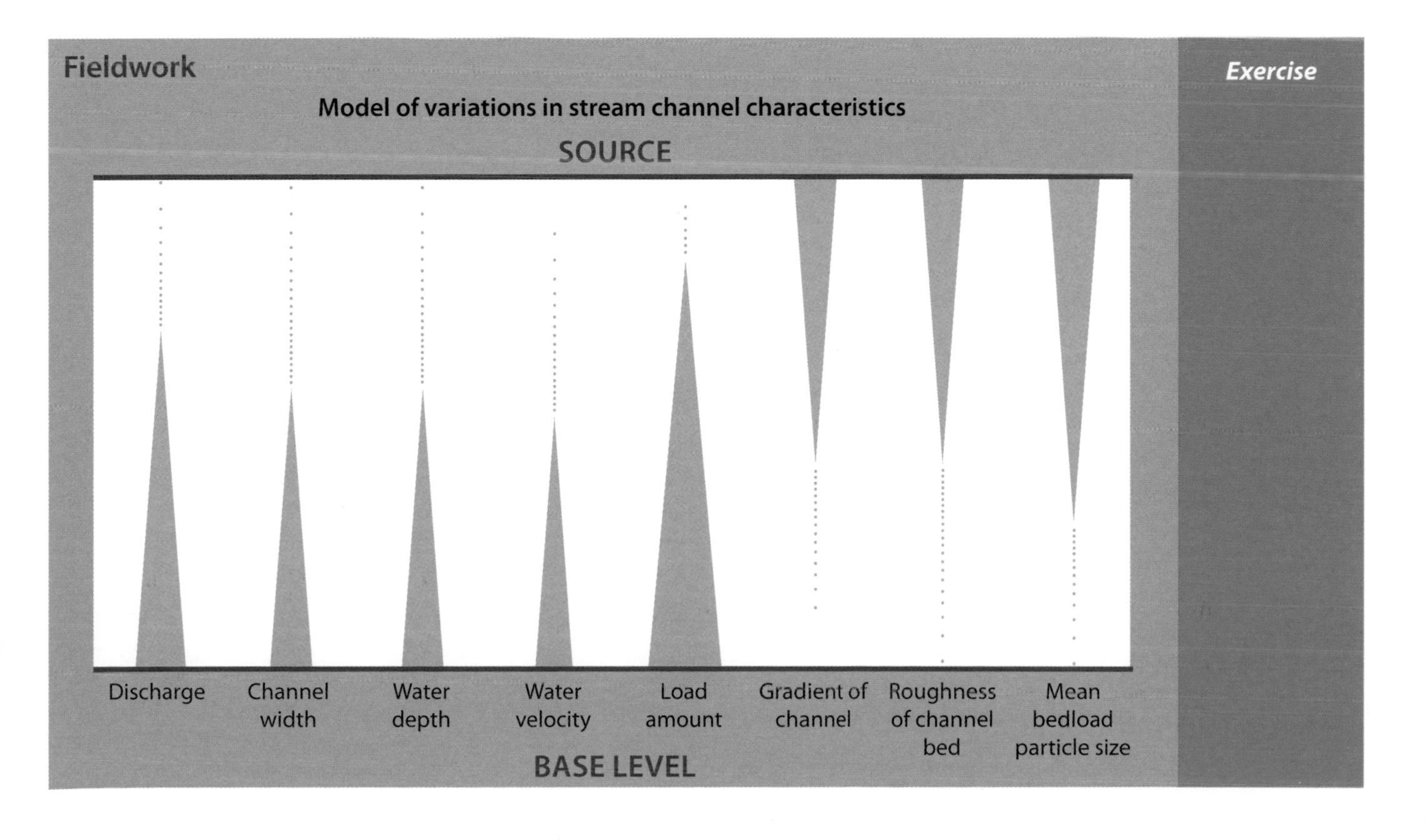

Exercise continued

Fluvial processes and landforms are a fertile area for investigation and primary data collection. River models, such as those devised by Schumm or Bradshaw, are a starting point for such first-hand fieldwork. The diagram (previous page) illustrates some of the relationships that these models identify between processes and fluvial characteristics in a drainage basin.

Because the model shows river characteristics changing with distance downstream, it is possible to examine these as correlations. Data from actual rivers can be plotted as scattergraphs and the nature and strength tested using Spearman's Rank Correlation technique (see *Skills, Techniques and Decision Making*, Reid and Roulston, 2nd edition, Colourpoint Educational 2008).

Channel and bedload characteristics on the River Shimna, County Down, 16 March 2007

Site	Distance km	Cross-section area m^2	Velocity m/s	Discharge cumecs	Bedload Mean A-axis mm
1	0.1	0.19	0.07	0.01	181.8
2	2.2	0.49	0.13	0.07	114.5
3	4.0	0.74	0.23	0.17	88.2
4	5.8	1.23	0.24	0.29	124.6
5	7.3	1.89	0.19	0.36	86.6
6	8.6	2.18	0.16	0.35	106
7	9.9	1.65	0.32	0.53	107
8	10.9	1.52	0.66	1.02	96.9
9	12.3	5.93	0.25	1.46	101.5
10	12.9	7.45	0.39	2.9	91.8

Data from Fieldwork *(River Shimna table)*

1. Using the table of data for the River Shimna:
   - Plot as a line graph the downstream variation in discharge; and
   - Construct a scattergraph of mean bedload size and site number (downstream distance).
2. Describe what each of the graphs constructed suggests about the relationship between the variables. How would you test the nature and strength of these patterns?

Questions

With reference to the diagram of the river model (previous page) answer the following:

1. The model shows that rivers normally increase their discharge from their source to their mouth. Why, based on the resource, does this appear inevitable?
2. How can discharge increase downstream even if the river has no tributary channels joining it?
3. Under what natural circumstances might river discharge not increase downstream?
4. The model suggests that the total bedload carried by rivers increases downstream while, at the same time, the average bedload particle size decreases. Write a paragraph to explain these changes describing the erosion and transport processes involved.
5. Rivers carry material other than as bedload; suggest the downstream pattern of these other elements of a river's load.

Exercise

With reference to the photograph below:

(a) Draw a sketch diagram based on the photograph and identify the following features: the inner bend; the outside bend; deposition on the slip-off slope; erosion at the river cliff and the thalweg (line of fastest flow). Note: the river is flowing towards the observer.

(b) Draw an annotated cross-section diagram of the meander. Describe and explain its shape.

INSIDE THE EXAMINER'S MIND

Study the section of the specification provided.

Elements	Elaboration	Spatial context requirement
(b) Interaction with the fluvial environment may have benefits and harmful effects for people	(i) Causes of floods and their effects on people, property and the land.	For (i) – the beneficial and detrimental effects of flooding in a large-scale drainage basin or its delta (Regional/National Scale study).

Attempt to write two examination style case study questions that an examiner could set based on this section. Could the examiner use resources to examine this section?

If so:

— What sort of resources might be used?

— What questions might be asked, based on these resources?

2A THE ECOSYSTEM AS AN OPEN SYSTEM

The living world is sometimes called the Biosphere and it is confined to a narrow zone on the surface of the earth and within the oceans. Geographically, life can be divided into different regions most often described by its vegetation, such as the tropical rain**forest** or the temperate **grass**lands. These vast zones are known as Biomes. At a smaller scale, anything from a small pond to a forest, and the way in which plants and animals interact with their surroundings is termed an **ecosystem**. Vegetation, animals, people and bacteria form the **biotic**, or living elements of an ecosystem while the physical environment in which plants and animals exist is the non-living, **abiotic** habitat. Water, light heat energy, oxygen, carbon dioxide, rocks and soil nutrients are all abiotic features. **Resource 27** illustrates the interlinked nature of the four key elements within any ecosystem namely: vegetation/plants (flora), animals (fauna), soil and climate. Soil is the mixture of weathered rock fragments (inorganic) and decomposed organic matter.

Resource 27 *The interaction between biotic and abiotic components and each other*

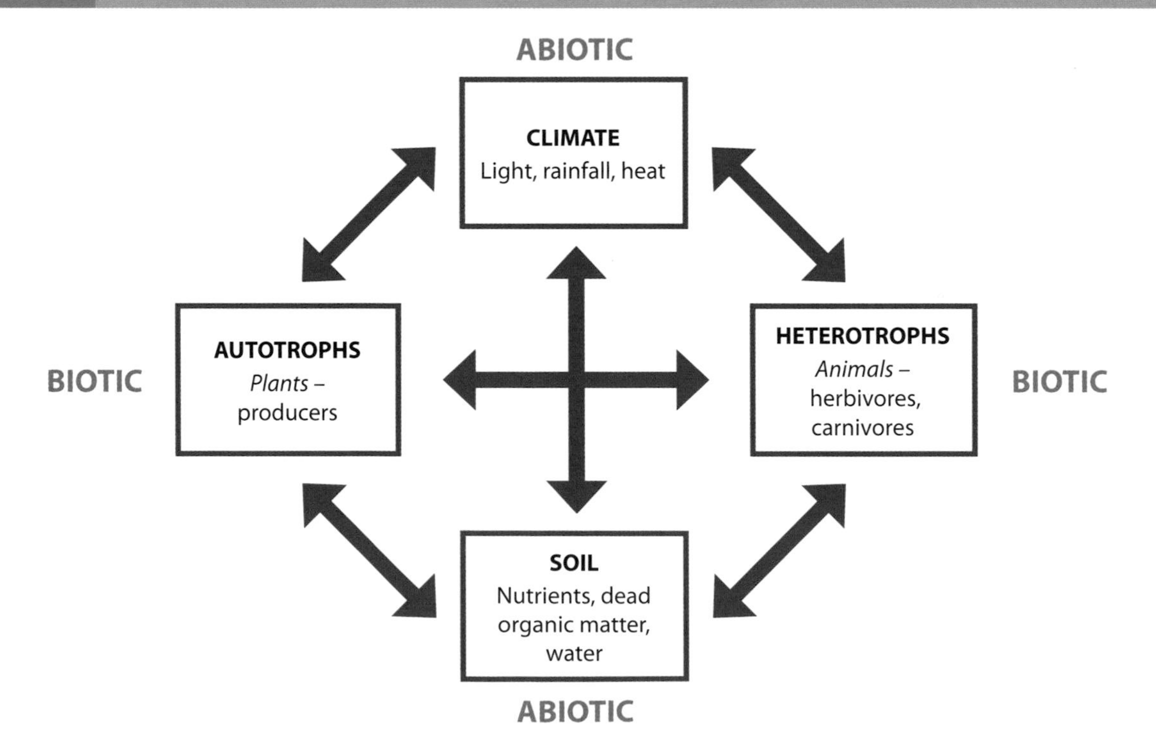

In common with a drainage basin, an ecosystem can be viewed as an open system. The stores are the animals, plants and soil, and the transfers involving energy and materials are in the form of nutrients: plants and animal foods.

Ecosystems as open systems with examples of inputs and outputs ***Resource 28***

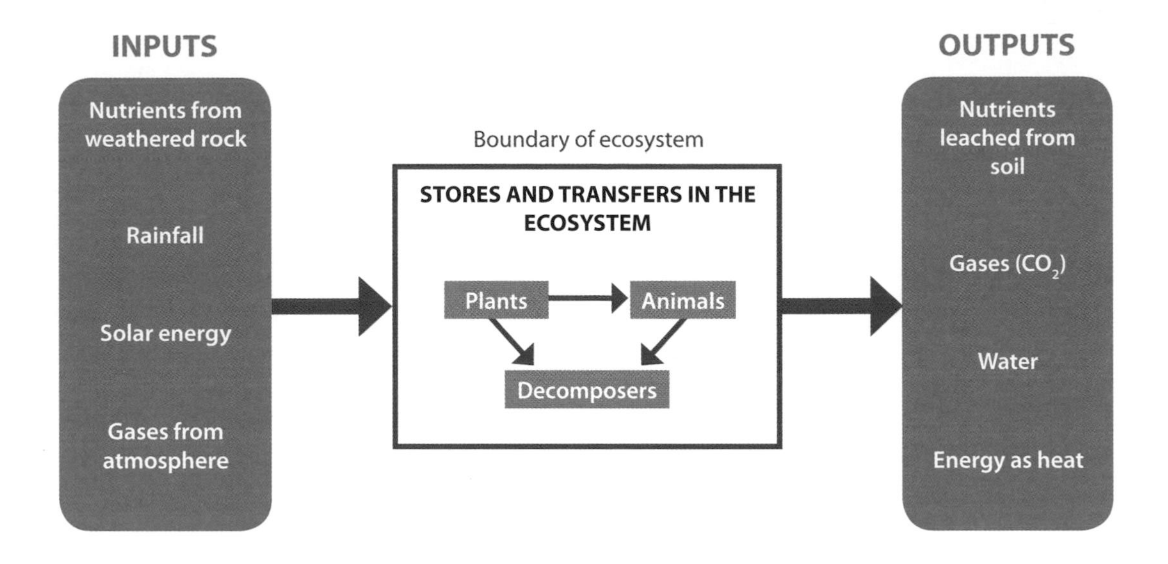

The key process in ecosystems and all life on earth is **photosynthesis**. This is the conversion of light energy into food energy, such as carbohydrates, using water and carbon dioxide. Only plants are capable of this task and are therefore known as the **producers** of an ecosystem.

Ecosystems: energy flows and trophic structure

Energy flows

The sun is the primary source of energy for all living things. As solar energy is held only briefly in the biosphere before returning to space, ecosystems need a continual supply. Heat energy from the sun cannot be retained by plants and animals, though it warms their abiotic environment. Light energy from the sun is used by green plants or **autotrophs,** in the vital process of photosynthesis. Light, chlorophyll, heat, water and carbon dioxide are needed for this process on which all life on earth depends. Photosynthesis creates chemical food energy, stored in carbohydrate compounds, which is then available for the wider ecosystem. An example:

Carbon dioxide + water + solar energy ⟶ glucose + oxygen (released)

Trophic structure and material transfers

A feeding chain exists when chemical plant energy is passed through an ecosystem, that is by animals grazing, hunting or scavenging. In most ecosystems there are four stages and each energy store is called a **trophic** (energy) **level.** The **first trophic level** is occupied by the ***autotrophs*** ('self-feeders', ie producers/plants). All other levels are occupied by **consumers** or ***heterotrophs*** ('other feeders'), namely the animals: either **herbivores** (plant eaters), **carnivores** (meat eaters) or **omnivores** (plant and meat eaters).

Herbivores form the **second trophic level**, carnivores the **third** and higher carnivores or omnivores (including humans) in the **fourth trophic level**. Ecosystems also have a fifth group of organisms that operate at each trophic level. These are the ***detrivores*** or decomposers of dead organic material. Dead organic matter, or DOM, is the remains and waste product of plants and animals. The decomposer organisms include earthworms, springtails, slugs, mites, fungi and microscopic bacteria.

An ecosystem's **trophic structure** is often illustrated as a pyramid (***Resource 29***). Each box of the pyramid represents one trophic level and the relative size of the box can indicate the total

A key concept to remember is that, while Material and Energy both flow *through* an ecosystem's trophic structure, only Material can be recycled.

amount of energy stored in that level. **Biomass** refers to the amount of living organic material (plants and animals) and is a measure of the energy stored. The huge decline in biomass between each stage is because consumption between each level is never 100 per cent efficient. Animals do not eat all parts of a plant and digestion processes do not convert all material. Thus, energy and matter is lost at each transfer between trophic levels. Also, within each trophic level, energy and material is used up by life processes or **respiration.** This includes heat energy from activity, and material lost as waste products, eg shed skin, leaf fall and faeces.

Resource 29 *The Trophic Pyramid – energy flow through an ecosystem*

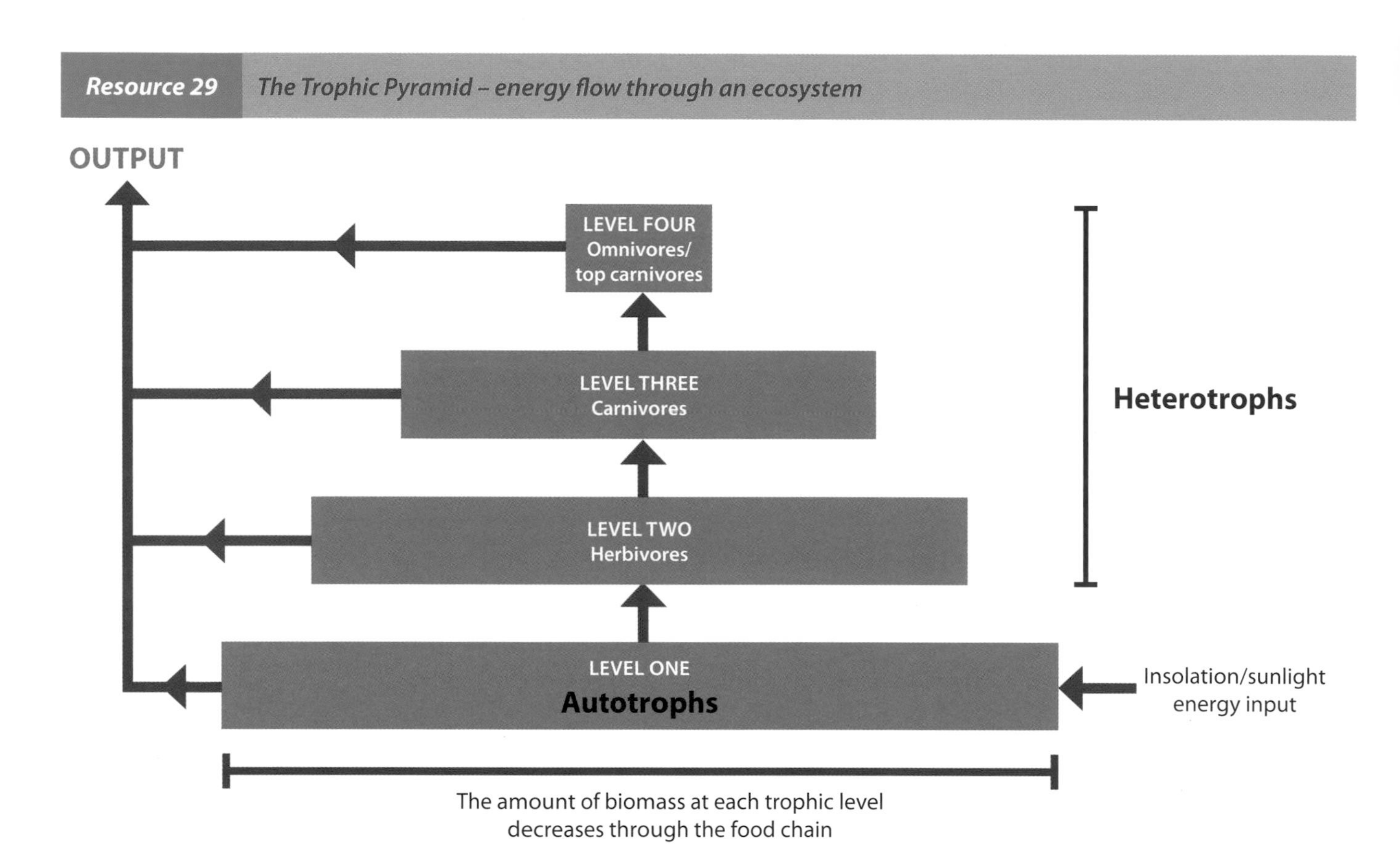

Exercise

Resource 30 shows a freshwater pond.

Using the details of its biotic and abiotic components attempt the following exercise.

(a) What are the boundaries of the system?

(b) On a copy of **Resource 29** add examples for each trophic level from the freshwater pond ecosystem.

(c) Complete the diagram by adding some inputs and outputs appropriate to the pond ecosystem.

A freshwater pond: a simple ecosystem **Resource 30**

Nutrient cycling

Nutrients are the chemical compounds that all living organisms require to survive and grow. The specific nutrients and quantities needed vary but the most significant are the macro-nutrients such as carbon, nitrogen, potassium and calcium. The sources of nutrients are the atmosphere (gas and dust), the underlying geology and the soil. The transfer of nutrients is commonly by water, including rainfall and soil moisture. Once inside a natural ecosystem, nutrients are normally retained by a highly efficient re-cycling system. Taken up from the soil in water through the roots of plants, the nutrients pass onto herbivores then to carnivores through the trophic structure. In and between each trophic level, these nutrients may be released to the decomposers that break down dead organic material and make the nutrients available again in the soil. Studies in tropical rainforests have shown that in the recycling of nitrogen, levels of 99 per cent efficiency can be achieved. Inevitably, some nutrients are lost (output) from an ecosystem such as the washing away of fertile soil or the migration of animals to another area.

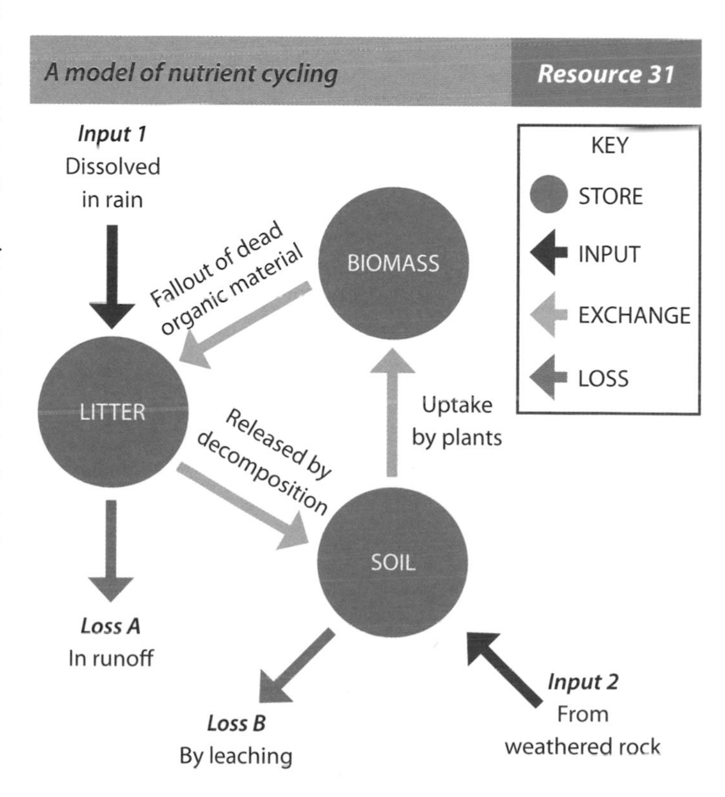

A model of nutrient cycling **Resource 31**

Within an ecosystem, nutrients are found in one of three natural stores: biomass, litter and soil. The first two stores have already been described, the third – litter – is the accumulated dead organic material normally found on the soil surface. Between these three stores, nutrients are transferred in a single (unidirectional) cycle: soil to biomass to litter to soil. The nature of these transfers is shown in **Resource 31**. Also illustrated in this diagram are the original sources of nutrients, inputs from the atmosphere and underlying rocks and potential outputs by water across the surface or through soil. Leaching is the process by which water passing down through soil carries nutrients beyond the reach of plant roots or out of the soil altogether. This model diagram of nutrient cycling can be used to show differences in nutrient stores and flows between different ecosystems. The circular stores can be drawn to show the proportion of nutrients they hold in a particular ecosystem and the width of transfer arrows to indicate the rate of transfer. Two contrasting examples are shown in **Resource 32**.

Resource 32 *Nutrient cycling in two contrasting ecosystems*

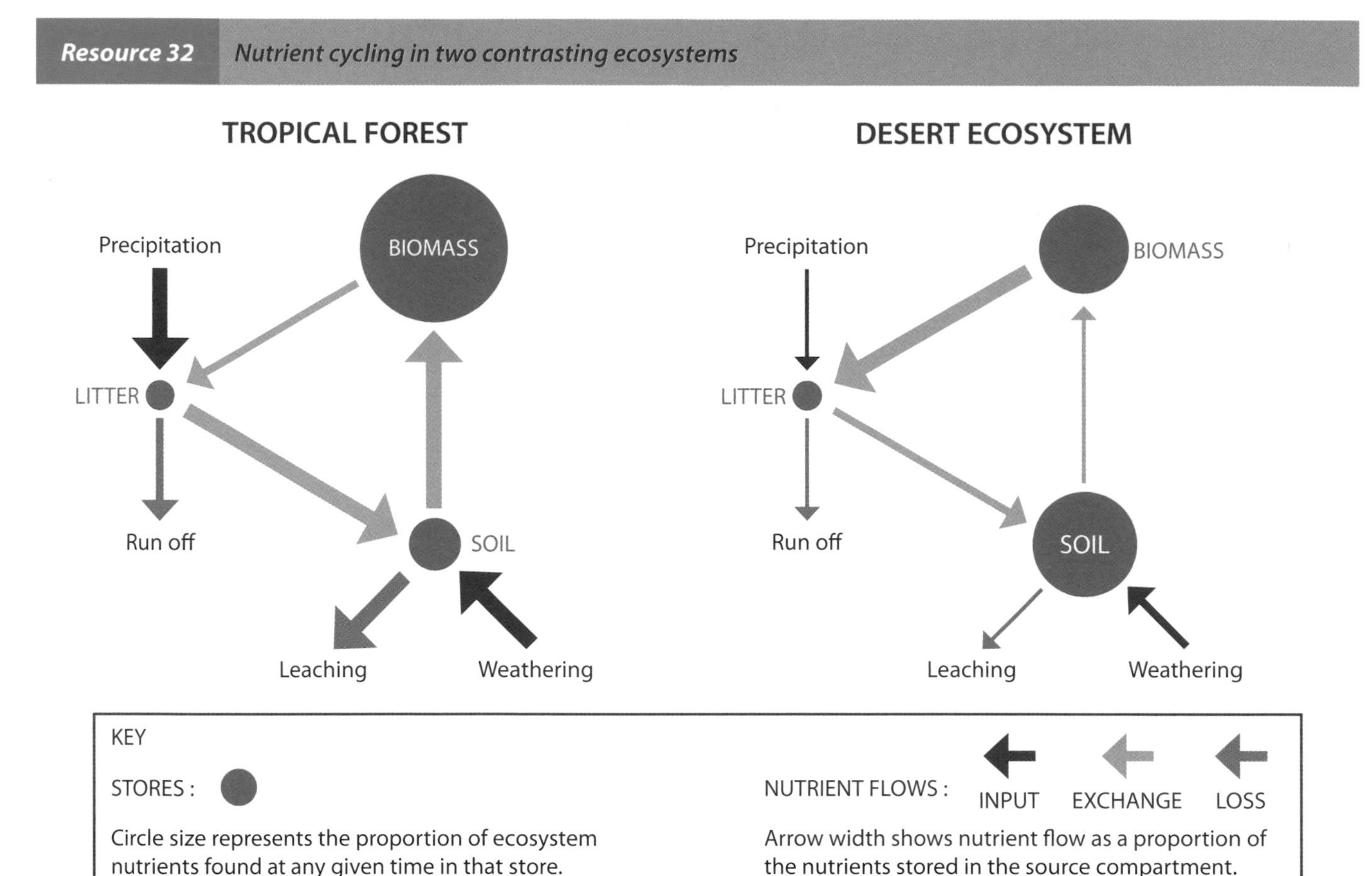

CASE STUDY: A deciduous woodland ecosystem

Across Ireland and Great Britain human development over the last 5500 years has meant that original natural ecosystems rarely survive. Most of these islands would, under natural conditions, be covered in wild wood dominated by a variety of tree and plant species. The exact assemblage of the vegetation would depend on local conditions including climate, relief and soils. In isolated pockets across the landscape, small remnants of this natural woodland have been identified as ancient or probably ancient woods (continuously wooded since at least 1600). Examples in the north of Ireland include Banagher Glen, County Londonderry, Breen Wood and Straidkilly, County Antrim, Redburn and Rostrevor forests, County Down and at Crom and Marble Arch in County Fermanagh.

Breen Wood is the case study used in this section but the principles and illustrations could be modified and developed for any of the other examples noted.

An exemplar case study: An Oak Woodland – Breen Wood, County Antrim

If left to nature much of Ireland would be dominated by oak woodland, especially the sessile but also the pedunculate species of oak. Today, the few ancient remnants of such woodland cover less than 0.1% of Northern Ireland. At Breen Wood, narrow valleys, acidic soil and a harsh environment have helped to reduce human impact on the natural vegetation. Its name, meaning 'Fairy Palace' in Irish Gaelic, may suggest another reason for it being largely untouched. The wood is one of the finest remaining areas of ancient sessile oak woodland in the North.

The abiotic environment

Breen Wood has been designated as an SAC – Special Area of Conservation – over a 36 hectare site, of which two thirds is wooded. It lies on the north facing slope of a river, about six miles from the town of Ballycastle.

Lying between 130 and 190 m (400–600 feet) above sea level, Breen Wood has moderately high annual rainfall of around 1600 mm and the mean monthly temperature varies from 4°C in January to 15°C in July. Most days of the year are cloudy and frosts are common during winter months. Trees require large amounts of water but the cold soil conditions of winter prevents tree roots from taking up the water needed to support growth. Deciduous trees have adapted to this problem by shedding their leaves in autumn to reduce transpiration – the loss of water – from their leaves. A second factor limiting vegetation growth in the area is the poor quality of the soil. The underlying parent rock is basalt, a dark fine grained igneous rock formed by massive volcanic outpourings around 50 million years ago. This basalt is low in nutrients and soil developed on it tends to be thin, rocky and acidic, with a pH around 4.5. In addition, the rainfall rapidly leaches nutrients from the soil maintaining its low nutrient status. Deciduous woods are often associated with brown earth soils but in upland oak woods, such as Breen, the leaching produces a podsol. Podsols often have a thin surface humus layer below which has a 'washed out' appearance.

As a result of these environmental factors, the range of plants and animals in Breen Wood is limited and the trees tend to grow slowly and have a stunted appearance. Two-hundred-year-old oak trees on the higher slopes of Breen Wood have attained only half the height they would have in lower lying areas.

Breen Wood **Resource 33**

Sketch map location of Breen Wood Nature Reserve **Resource 34**

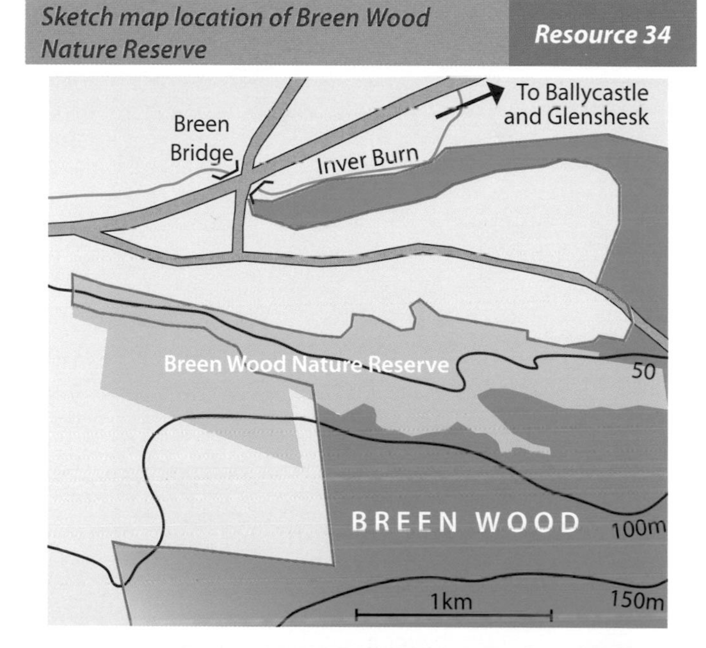

Soil profile in Breen Wood **Resource 35**

The biotic environment

Autotrophs

As mentioned, the dominant tree species in Breen Wood's vegetation cover are the oaks but also common is another deciduous tree, the downy birch. Below the tree canopy is a shrub layer which is mainly composed of rowan, hazel and holly trees, while beneath and around these is the field layer of bilberry, bramble and numerous species of fern. On the soil surface, great wood-rush, wood-sorrel, mosses and ferns form a blanket cover. The appearance of the deciduous woodland changes greatly with the seasons; in early spring the bare trees allow sunlight to reach the forest floor and for a few weeks bluebells and anemones produce a sea of blue and white flowers. Fungi are common on fallen tree trunks, while mosses, lichens and ferns often cover the branches and trunks of dead and living trees.

Resource 36 *Tree species in Breen Wood (oak, downy birch, holly and rowan)*

Resource 37 *Ground cover in Breen Wood (wood sorrel, ferns and mosses)*

Resource 38 *Birch Bracket Fungi in Breen Wood*

Heterotrophs

Once more, the harsh environment is a limiting factor on the animal life of these woods. Many insect species are found in and around the vegetation, including at least fourteen species of butterfly such as the Orange Tip and Speckled Wood. The native red squirrel is the herbivore most closely associated with the oak woods though the introduced larger grey squirrel species threatens its future.

Other heterotrophs are the woodland birds, including the Goldcrest, and Great, Blue and Coal tits that forage the forest for their diet of seeds and insects. Goldcrests are the smallest bird and feed on insects in the tree canopy. These small birds can become the prey of larger carnivores passing through, such as buzzards and sparrow hawks. On the ground the largest mammals are the badger, the stoat and the fox. All three are mainly nocturnal animals. However, stoats are carnivores – eating rodents, birds and eggs – while foxes and badgers are omnivores. Foxes prey on small mammals such as squirrels and birds, but also eat grass and fruit. Badgers consume earthworms, small birds, fruit, nuts and the root bulbs of plants.

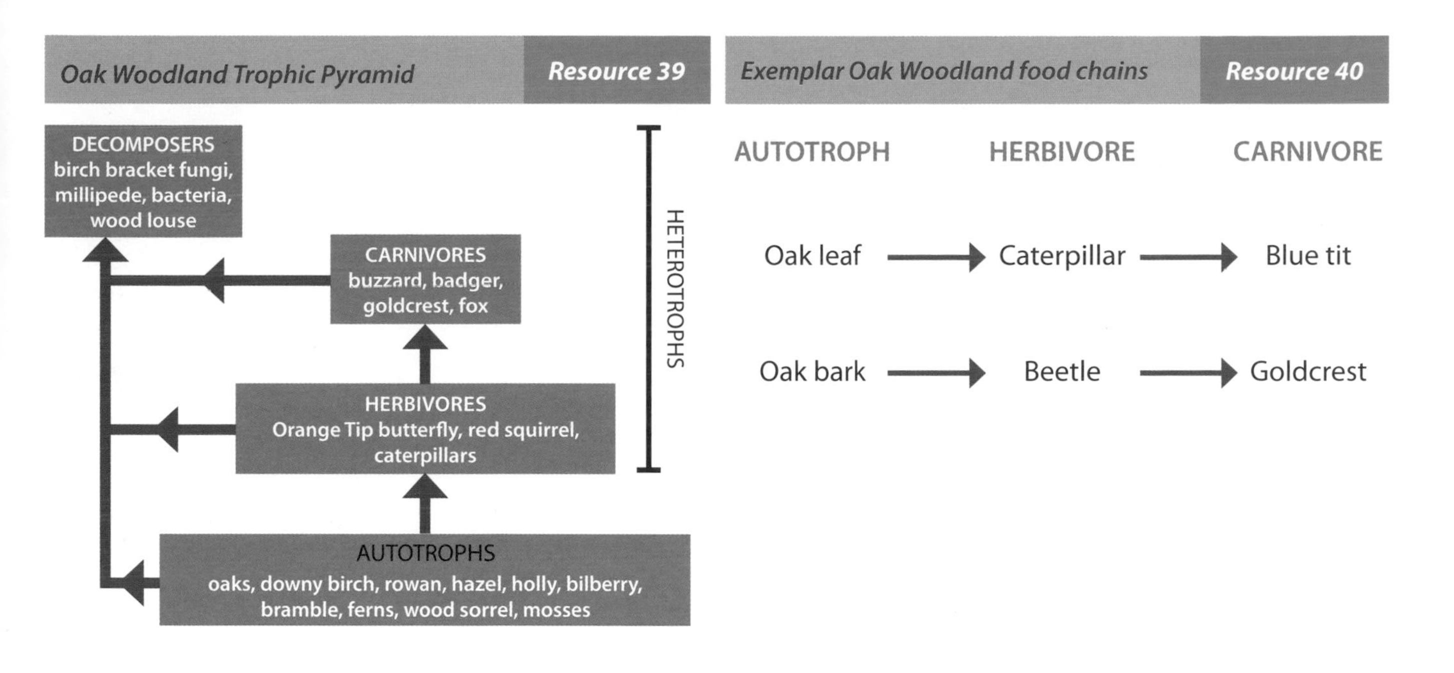

Create other food chains appropriate to the oak woodland ecosystem in Breen Wood.	***Exercise***

Nutrient cycling

The harsh abiotic habitat of cool temperatures and acidic soils makes survival difficult for many species of soil organism. Few earthworms are found and consequently litter, such as the autumn leaf fall, may take years to fully decompose and return their nutrients to the soil. The nutrient input from the underlying rock is low, leaving the rainfall as the main source of new chemical nutrients. As noted, the release of nutrients from the litter to soil stores is slow but with autumn leaf fall a large flow from the biomass to the litter store does take place.

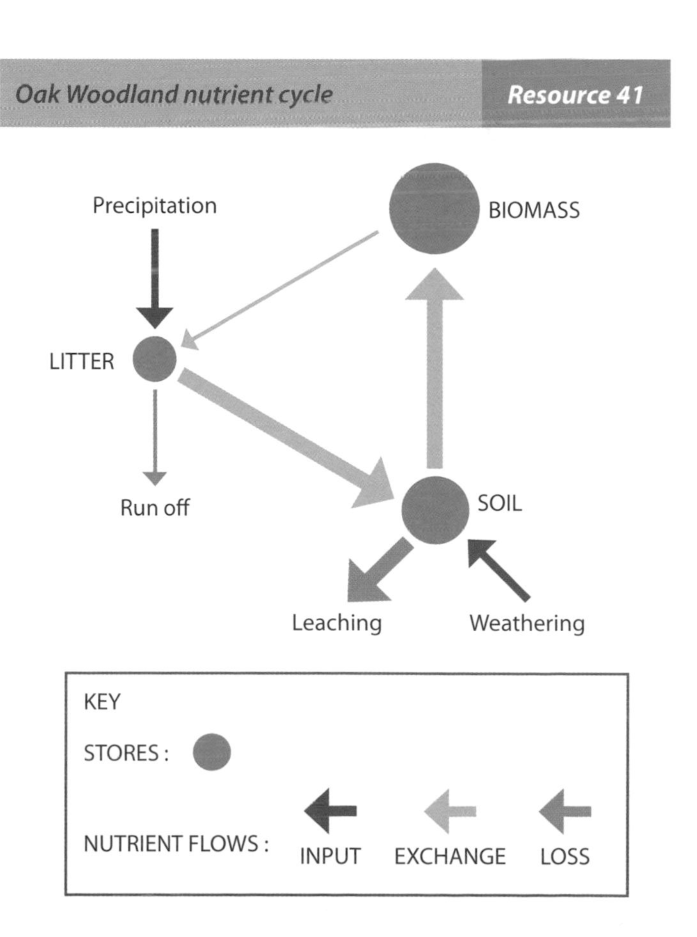

2B PLANT SUCCESSION

As any gardener will know when presented with a newly-available area of rock or soil, plants will quickly invade. This process is called **colonisation** and the first plants to arrive are termed **pioneers**. This is the first step in a sequence of changes that will eventually produce a balanced community of plants called the **climatic climax vegetation**. This is the process of **succession**; the whole sequence is called a **sere** and each step in the process is a **seral stage**.

Resource 42 *The simple Primary Succession concept and terminology*

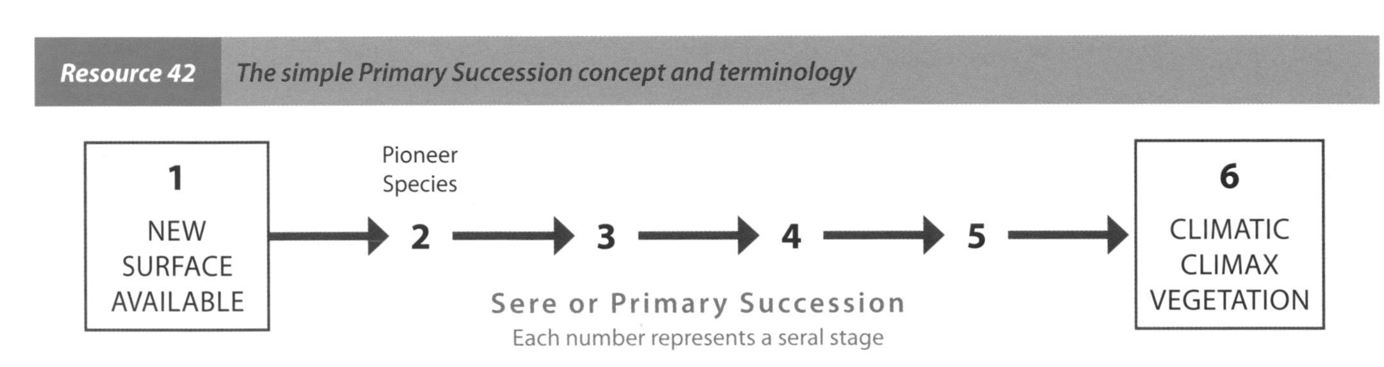

As the name suggests, a climatic climax community is the group of plant species that are best adjusted to the climatic factors that exist. As noted earlier, most of Ireland would be forested if left to nature, so even an abandoned garden would undergo succession to reach deciduous woodland in a century or so.

The succession process helps explain why weeds can invade a newly-available site but shrubs or trees cannot be established until later in the sequence.

Resource 43 *Plant succession in an abandoned field in a West European climate*

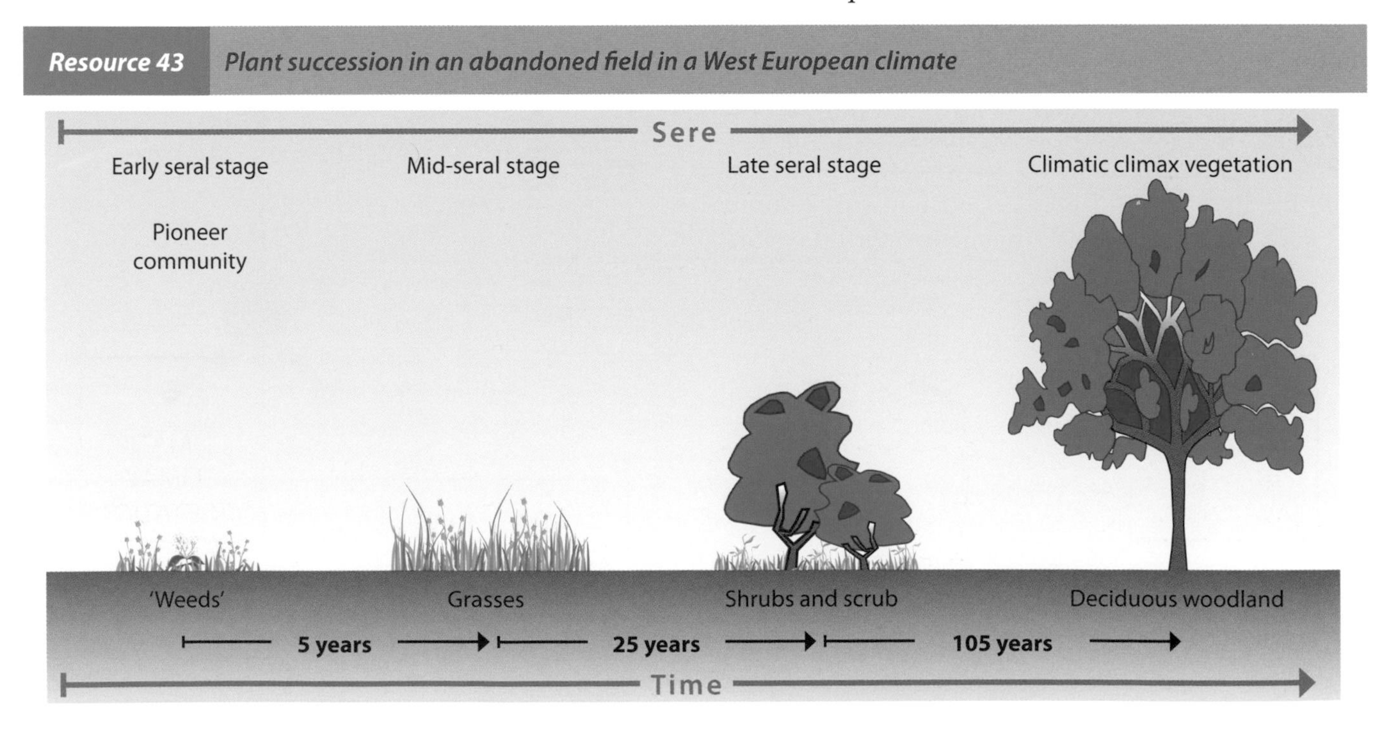

The plants that most easily adapt to the unique environmental conditions of the site will colonise a location. Each plant species in a community has a specific range of tolerance of the abiotic conditions. Climate and soil factors are the most important influence over the successful establishment of plant communities. These include sunlight, moisture, soil fertility and stability.

1. Not only is the *amount* of sunlight available important but also its *duration* and *quality*. The duration of sunlight affects the flowering of plants. The intensity of light affects photosynthesis and the growth rate.
2. The *availability of water* is important for the survival of most life forms but plants require water for a number of life processes – germination, growth and reproduction.
3. The presence and depth of the soil, along with the range of nutrients available, will influence which plants thrive.

In their turn, the pioneer community of plants and each successive seral stage alters the environment in such a way to permit new species to occupy the environment. These alterations of the environment involve changes in the microclimate and soil conditions of the site. Plants provide shade and protection for others, and after they die their decomposition improves the soil's structure and its supply of nutrients.

An example: Following a landslide, the surface of boulders are exposed to the atmosphere. On this bare rock surface there is no soil, so most plants cannot be established. Lichens and simple mosses can survive on rocks by obtaining nutrients from the rock itself and rainwater. These plants help to weather and break-down the rock surface allowing other species to be established. Plant roots help to bind the loose rock fragments as a thin soil, and plant stems and leaves shade the surface from the bright sun and shelter it from the wind. When these plants die they add organic material that holds nutrients and water in the soil. Over time more species are added, but the pioneer species of lichens and mosses die out as the taller plants shade out their sunlight. And so succession continues.

A climatic climax community is the result of a long period of plant succession. Such communities usually exhibit a good deal of species diversity and thus are relatively stable systems. If a succession continues uninterrupted by natural events or human activity from pioneer to climax, it is termed Primary succession. On the other hand, if it is disturbed this renews a succession sequence as Secondary succession. Natural disturbance may involve fires, landslides, flows of lava or volcanic ash fall, severe floods or storms. In 1987 a single storm brought down over one million trees in South East England. Human disturbance related to tropical deforestation has caused secondary succession of plant communities in the tropical rainforest.

Accidental and managed interruptions to succession **Resource 44**

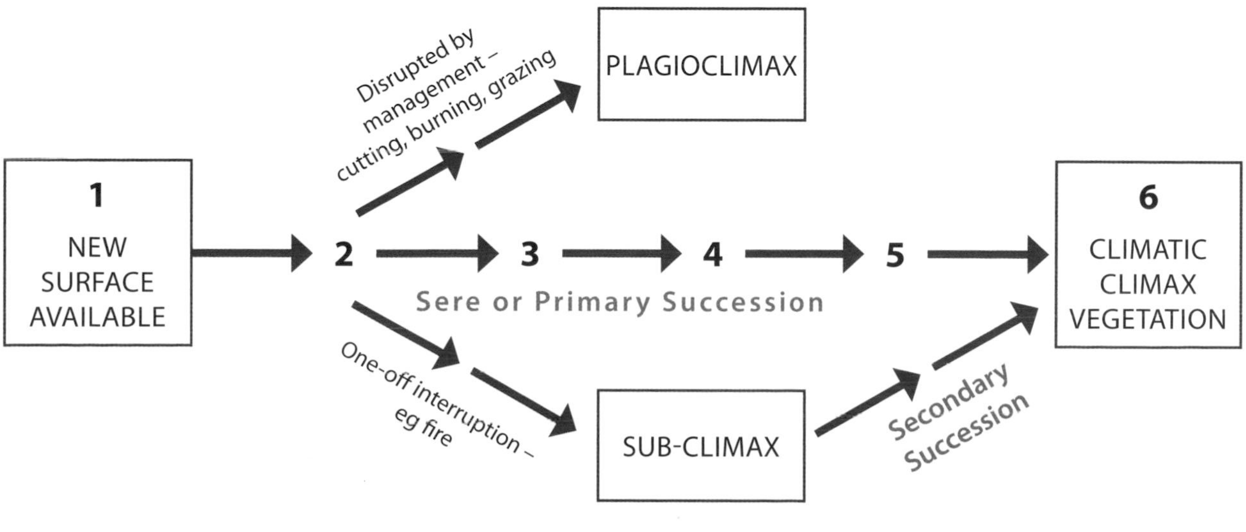

Plagioclimax

Plagioclimax is a term used to describe the vegetation community that develops when a climatic climax or succession is changed by human activity. The term implies that it is not just a disruption to the natural vegetation but rather a management system that prevents any natural change and 'arrests' the vegetation system.

Most landscapes in the British Isles today are plagioclimaxes, including school playing pitches, farm fields and planted forests.

Clearance, burning and grazing are the three most common human activities which create plagioclimax communities.

Much of upland Britain is covered by moorland often dominated by a few species such as heather and rough grasses. In nature most of these areas would undergo succession to woodland, but the use of the land for animal grazing (sheep and cattle) and burning to improve pasture, prevents the growth of tree saplings. In some upland areas the carefully managed burning of heather moorland (the muirburn) is designed to maintain the population of red grouse. The moor landscape then appears as a patchwork of different colours with heather at different ages. In shooting season, game bird hunters pay large sums to shoot grouse over these moors.

Resource 45 *Muirburn on The Fleak, Askrigg Common, North Yorkshire*

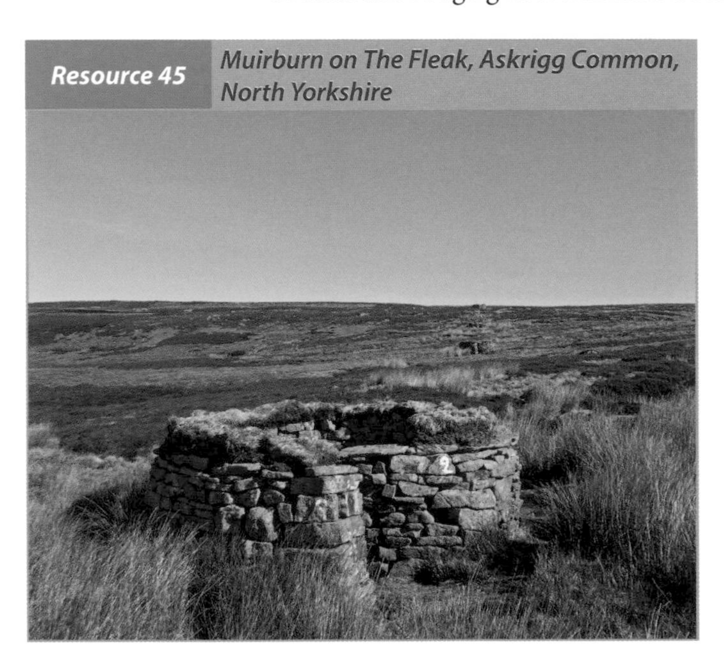

Resource 45 shows the patchwork pattern of vegetation produced by muirburn management. In the foreground is a shooting butt, one of a line, stretching to the horizon.

There are four main types of primary succession: two based on land and two on water. It is suggested that under natural succession most of the island of Ireland would reach a climatic climax vegetation community of oak woodland.

Resource 46 *Primary successions in a West European climate*

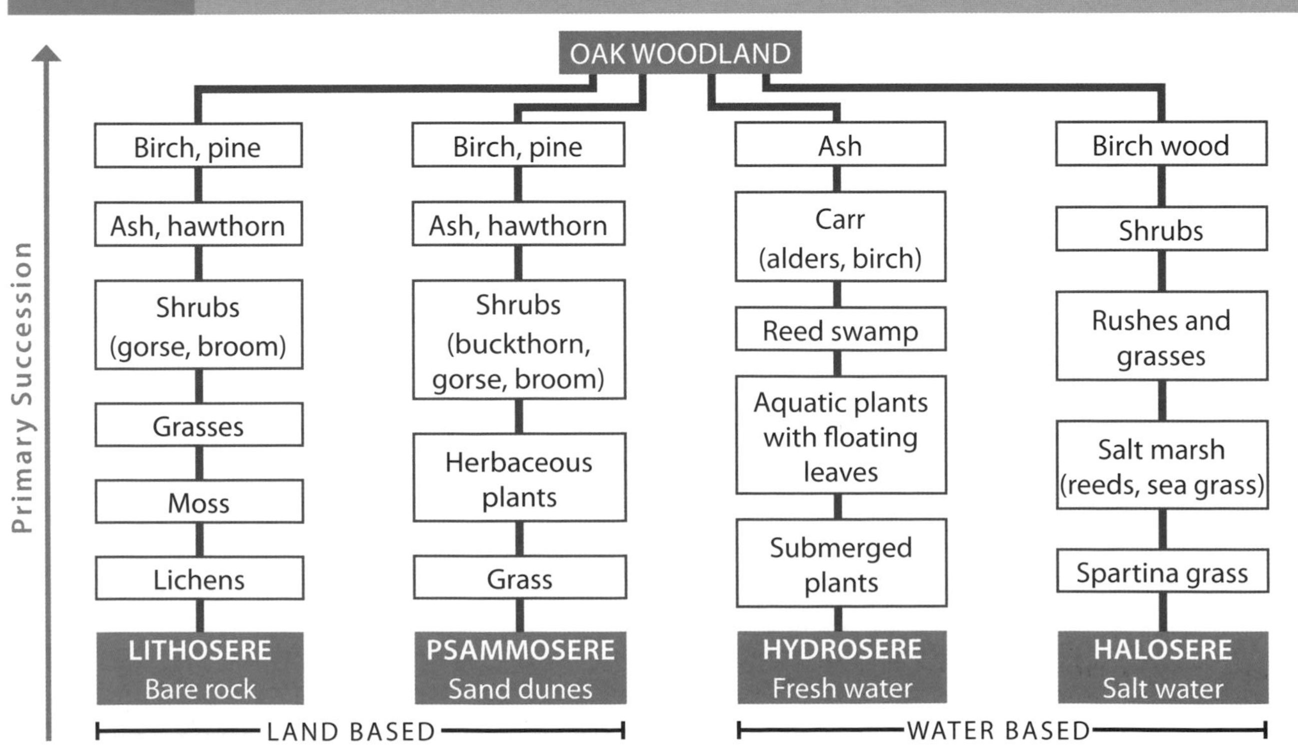

CASE STUDIES

The north of Ireland provides a remarkably broad range of environments in which succession processes may be studied. Some are particularly useful in that many seral stages can be studied at the same location. In places along the coast, sand dunes are developing behind wide sandy beaches, and elsewhere, freshwater lakes and ponds are gradually being filled in by sediment and plant debris.

Each of these dynamic landscapes provides an opportunity to witness the colonisation progress from the invasion of the pioneer plants to the climax vegetation.

Case study of hydrosere succession: Hollymount, County Down

Pond succession, County Down **Resource 47**

Around the edges of fresh water ponds or lakes, seral stages of succession can be seen. Such water bodies are gradually being filled in by sediment, washed from surrounding slopes or brought in by streams, making the shores shallower and more fertile. In the deeper water are found submerged and floating plants, duckweed and water fern along with plants rooted in lake bed sediments such as water-lilies. In the shallows, plants such as rushes, reeds and sedges dominate. These are hydrophytes – plants that can absorb oxygen through their roots even in waterlogged soils. Such marsh plants grow out into the lake – their roots trapping silt – and when they die, their remains gradually build up on the floor of the lake. When the sediment rises to the water level, the environment is described as a fen or

Model of hydrosere succession **Resource 48**

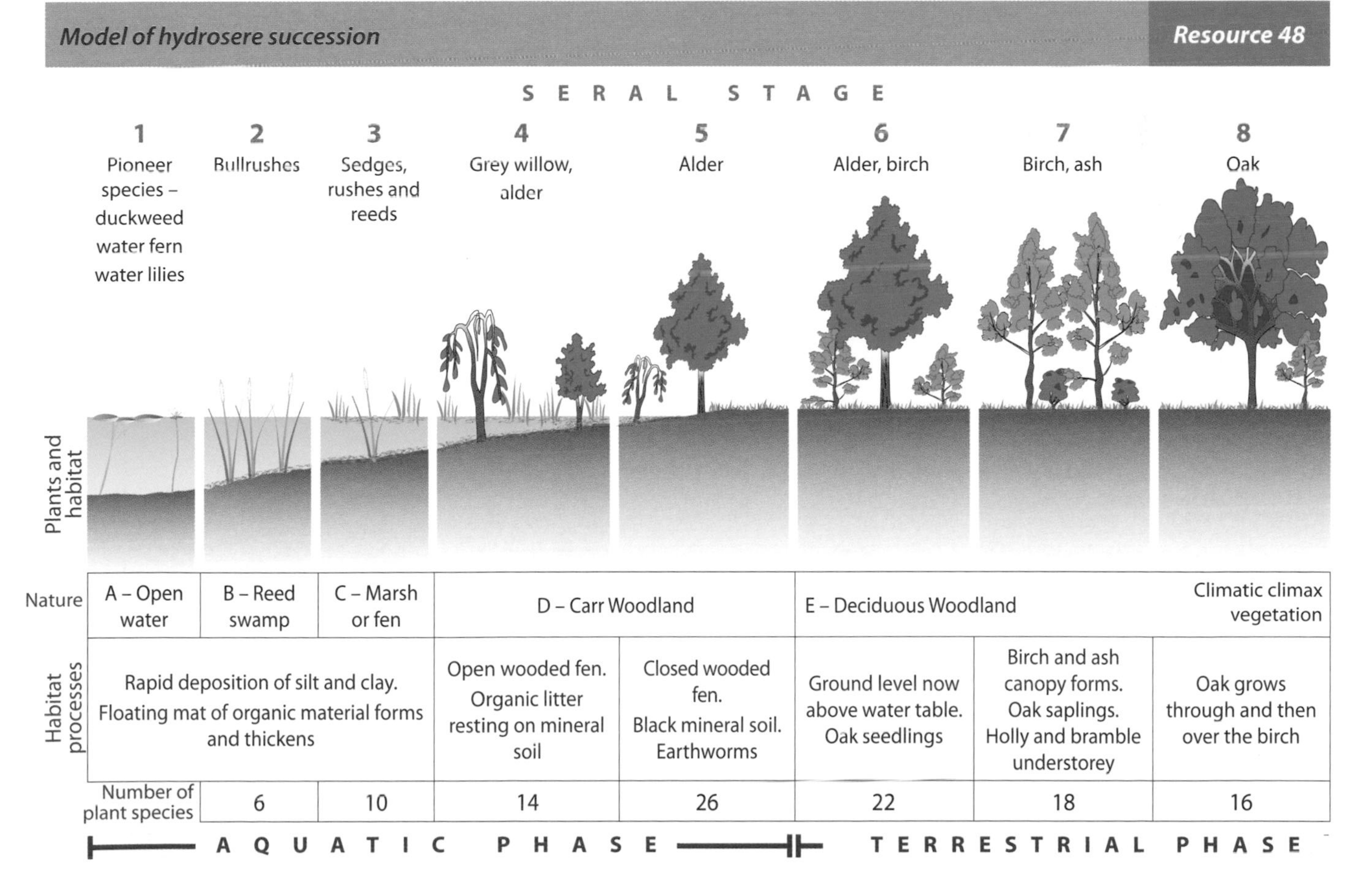

	1	2	3	4	5	6	7	8
Nature	A – Open water	B – Reed swamp	C – Marsh or fen	D – Carr Woodland		E – Deciduous Woodland		Climatic climax vegetation
Habitat processes	Rapid deposition of silt and clay. Floating mat of organic material forms and thickens			Open wooded fen. Organic litter resting on mineral soil	Closed wooded fen. Black mineral soil. Earthworms	Ground level now above water table. Oak seedlings	Birch and ash canopy forms. Oak saplings. Holly and bramble understorey	Oak grows through and then over the birch
Number of plant species		6	10	14	26	22	18	16

AQUATIC PHASE — TERRESTRIAL PHASE

carr, and water-tolerant tree species such as alder and willow take root. The succession moves from the aquatic (water) based section of its sequence to a terrestrial (land) based one. As the lake shrinks, the drier soil conditions allow other trees to become established, including ash. Eventually, through a sequence of seral stages a mixed oak woodland climax vegetation cover can be attained.

Resource 49 *Hollymount location sketch map*

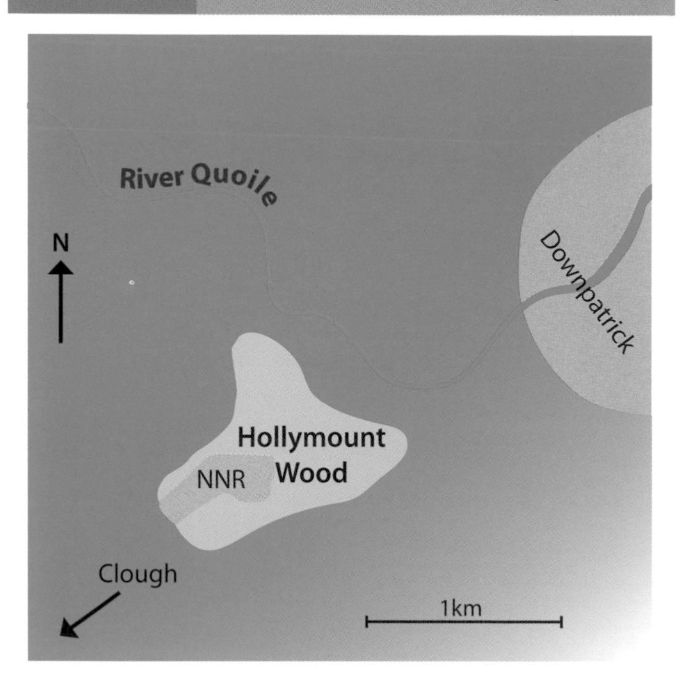

Around Hollymount NNR (National Nature Reserve), a few kilometres west of Downpatrick, a hydrosere succession can be observed. Here lies one of Ireland's best examples of an alder carr woodland environment with its damp vegetation community. In common with much of County Down the area is covered by low, elongated, oval shaped hills of glacial origin named drumlins. These produce a chaotic drainage pattern including inter-drumlin hollows in which rainwater gathers. In the Hollymount area there are inter-drumlin lakes, partially filled-in ponds, marshes or fens and dry drumlin slopes; in fact, all the seral stages associated with a fresh water hydrosere occur in this location.

Near the lake shore, water-lilies and duckweed with their large and tiny leaves, respectively, are found merging with the marsh plants – soft rush, great reedmace and great pond sedge. In the wet woodlands the ground flora includes reed canary-grass and in particular tussock sedge. Above these is a dense canopy formed by alder and grey willow trees. Although wet and prone to floods, the soil beneath the trees has a neutral pH and is moderately rich in base nutrients unlike the acidic waterlogged peat soils also found locally. Beyond the inter-drumlin hollow and on the more freely drained drumlin slopes, the succession continues with alder and willow carr replaced by oak and ash trees, with an understorey of holly and bramble.

Resource 50 *Seral stages at Hollymount*

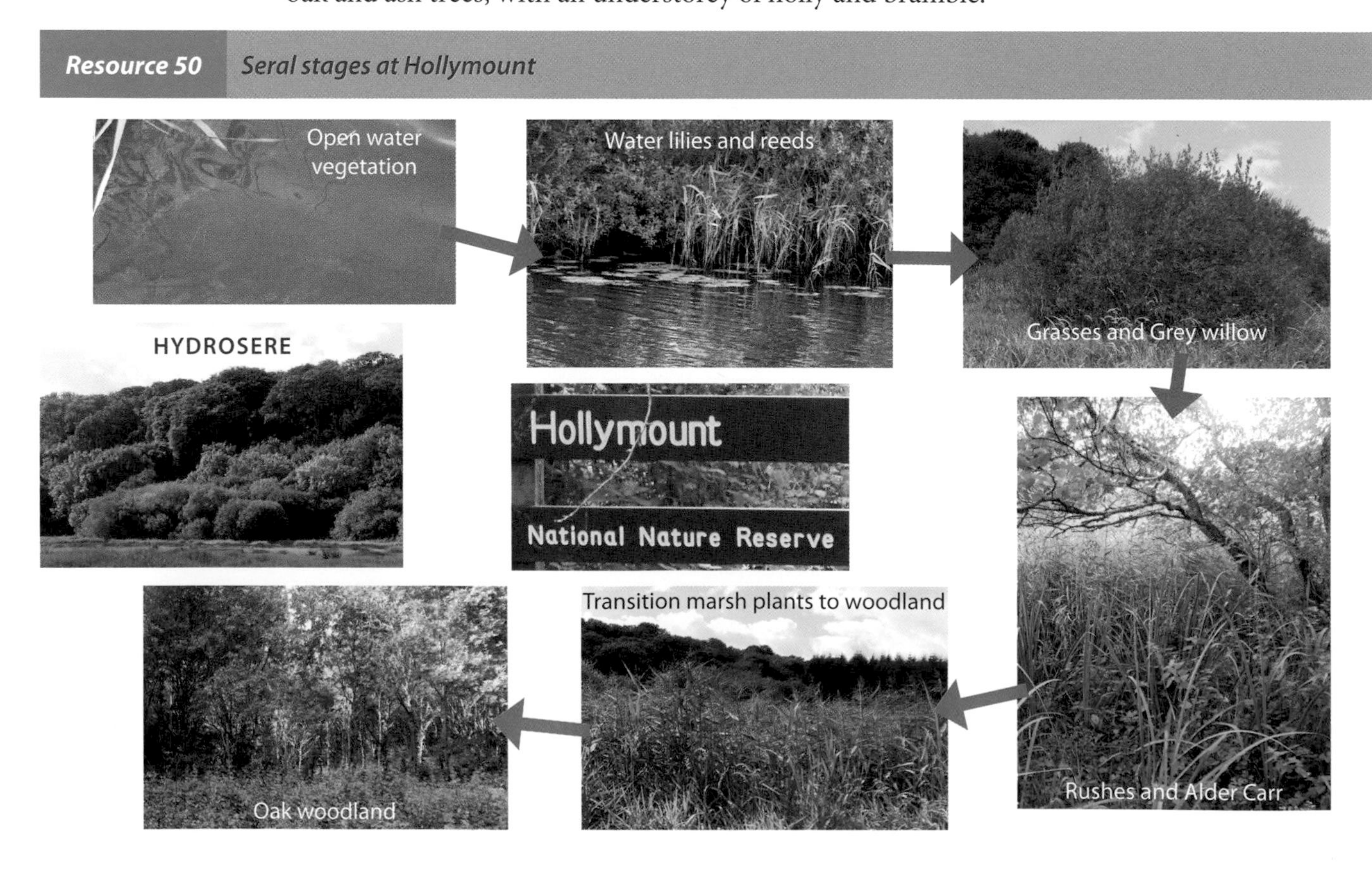

Case study: A web based enquiry of a Psammosere – a sand dune succession

Dunes succession, County Donegal **Resource 51**

Around 20% of the Irish coastline is composed of beaches backed by areas of sand dune. In the north of the island alone 2000 hectares of dunes are found. Most are covered by vegetation and a sequence of stages in their physical and ecological succession can be identified.

Sand dune formation

Dunes are formed by the deposition of sand on the landward side of wide beaches. Waves and especially onshore winds carry material above the tide line where it may be trapped behind small obstacles forming small embryo dunes. These small dunes, often less than a metre in height, can provide a home for a group of specialist pioneer plants including sand couch grass and lyme-grass. These species can tolerate the low nutrient, salty (saline) and windswept conditions of this area. As the roots systems of these species help to stabilise the dunes, more sand can be trapped and the dunes grow. One plant species above all others holds the key to the growth of sand dunes here: marram grass. This remarkable plant not only survives burial by sand but thrives on it, growing rapidly through sand forming a complex rhizome root network, binding the dune in place. As the cross-section diagram suggests (***Resource 52***), over time a series of ridges can be created parallel to the shoreline. These ridges are often described by their appearance as follows: nearest the sea are the embryo dunes, followed by the fore dunes, then yellow dunes and grey dunes. Between these ridges lie low sheltered areas termed dune slacks. In these, the conditions may be wet or dry but are always more sheltered and provide a habitat for plants, insects and animals less suited to the dune ridges. In Ireland such environments have developed over the last 6,000 years, and today some dune systems stretch over a kilometre from the sea.

A walk inland from a beach through dunes is essentially a journey through time starting with the initial colonisation of the embryo dunes to the final stage in the sere – the climax vegetation community.

Cross-section of a typical sand dune system **Resource 52**

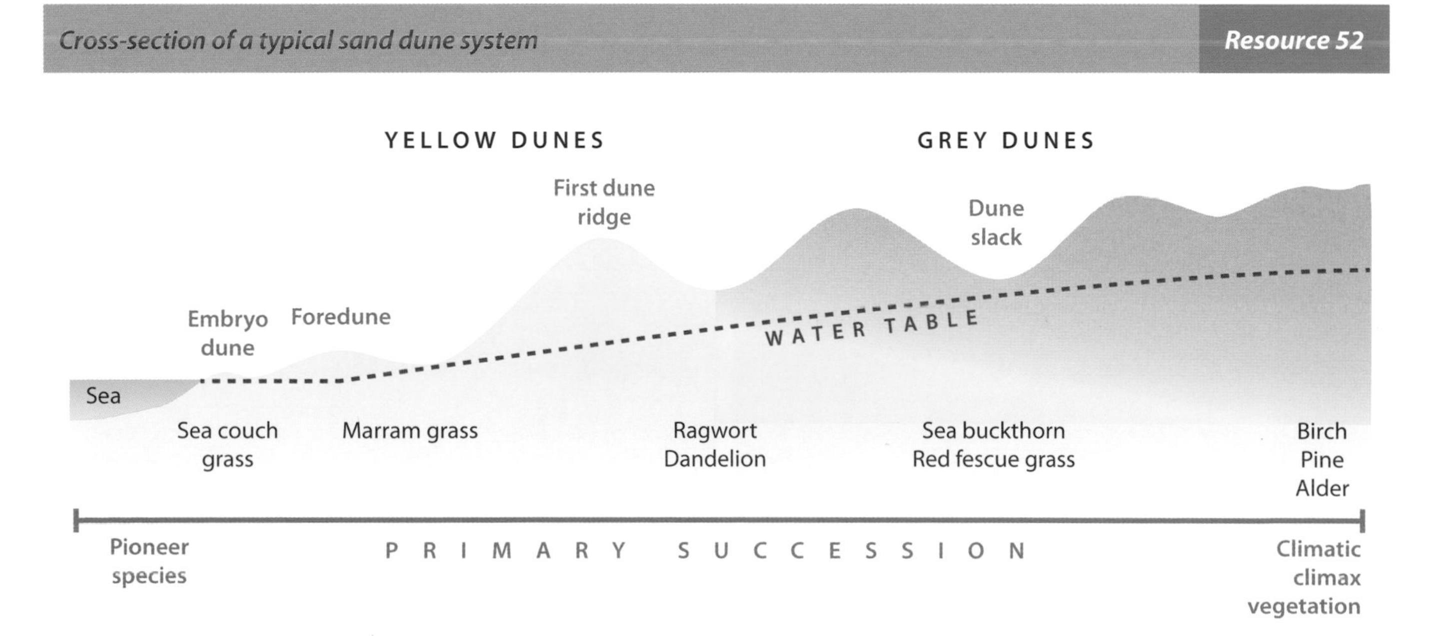

Resource 53 *Overview of Murlough Dunes*

The internal processes of succession are illustrated by the table (***Resource 54***) of changing conditions across the dunes. Taken at five points increasingly distant from the sea, patterns emerge concerning the changing micro-environmental conditions. Each point may be regarded as a seral stage.

Resource 54 *Changes in ecosystem characteristics across a sand dune system*

Characteristic	Embryo dune	Foredune	First ridge	Second ridge	Third Ridge
Distance from sea	5m	25m	80m	140m	280m
Height above sea	1m	3m	8m	10m	15m
Key plant species	Sea couch grass	Marram grass	Marram grass Ragwort	Red fescue, Sea buckthorn	Tree species pine, birch, oak
% Ground cover	5	20	70	86	100
Plant height (cm)	4	56	92	125	Over 500
Soil pH	8.0	7.5	7.0	6.0	5.0
Number of species	4	6	11	34	26
% organic matter	0.5	1.1	2.5	3.3	4.2

Resource 55a *Goats on Murlough's grey dunes*

Resource 55b *Burnt dune vegetation*

Expected changes in characteristics moving inland:

- The proportion of bare sand decreases as more species invade and conditions improve;
- The proportion of the soil that is organic in origin from plants and animals increases;
- The pH, measuring soil acidity, decreases from the salty alkaline environment of the shore as decomposing dead plant material adds organic acids;
- The ability of the soil to retain water improves as organic material is added to the free draining sand;
- The range of plant, insect and animal species found increases although the pioneer species themselves are replaced; and
- The height and stability of both the dunes and the vegetation increases.

Exercise

Discussion:

To what extent does the table of data reflect the expected changes?

Useful Psammosere website

The following site describes the sequence and development of sand dunes and their ecosystems.

http://www.countrysideinfo.co.uk/successn/primary.htm

Useful sources

A local study: Magilligan, County Londonderry

This second site is a virtual fieldtrip of the Magilligan Dune system in County Londonderry.

http://www.geographyinaction.co.uk/Magilligan/Mag_intro.html

Food for thought: How might global climate change impact on climatic climax communities?

2C HUMAN INTERACTION WITH ECOSYSTEMS

All ecosystems across Ireland and Great Britain have been impacted by thousands of years of human activity, either indirectly or by direct management. As noted, when this happens plant succession is interrupted, leading to secondary succession, or arrested succession producing a plagioclimax. The sand dunes systems of the north of Ireland rarely show a climax vegetation, not least because management is often attempting to retain the health of the dune environment by preventing the establishment of trees. The National Trust at Murlough Nature Reserve in Dundrum use controlled animal grazing and burning to prevent succession (***Resource 55a*** and ***55b***). Undoubtedly, the main reason for human interference with natural ecosystems is agricultural production. For one global ecosystem (biome) in particular this has been significant in modern times – the mid-latitude grasslands.

The mid-latitude grassland ecosystem

A - Distribution and the abiotic environment of temperate grasslands

The temperate grasslands occur in two main types:

1. In the centre of large continents	2. To the east of continents
North America – the **Prairies** Eurasia – the **Steppes**	Argentina – the **Pampas** South Africa – the **Veldt/Veld** New Zealand – **Tussock Grasses**

The temperate interior (far from sea) climate tends to be dry (400–800 mm annually), warm or hot in summer, and cool/cold in winter.

Resource 56 *World map of the distribution of temperate grasslands*

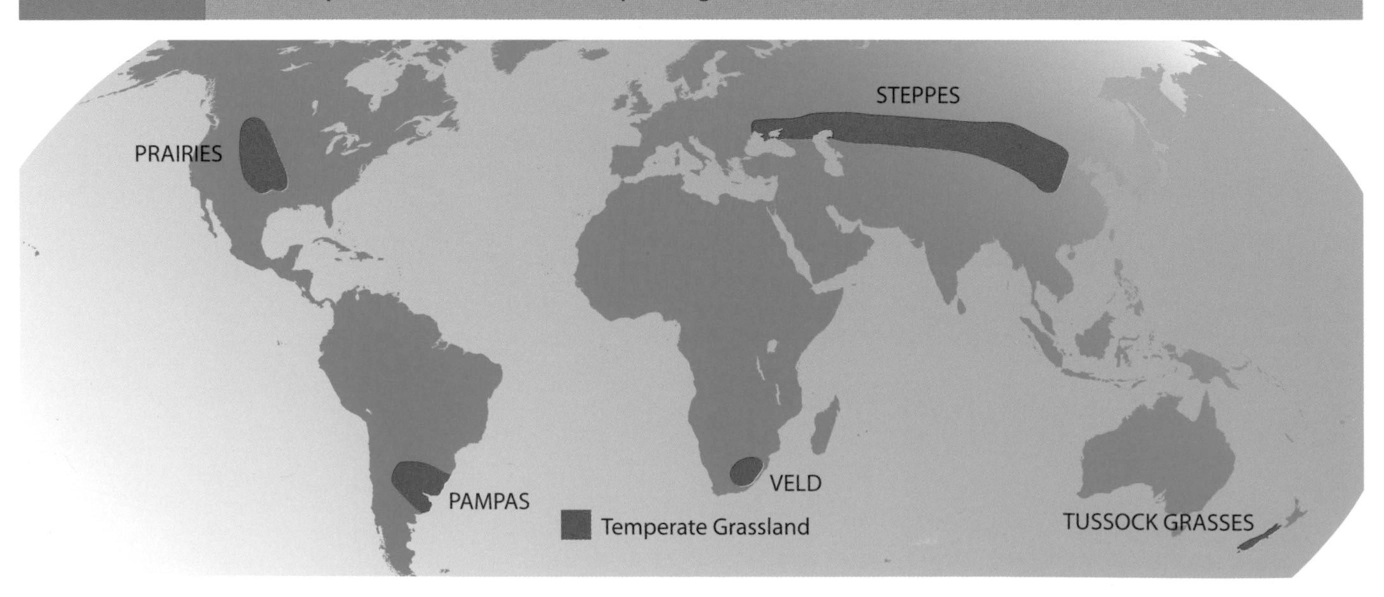

The soil of these areas is called a Mollisol or chernozem meaning 'black earth'. The climatic climax vegetation is grass and this decomposes to produce neutral or slightly alkaline soils with rich mull humus. This is quickly mixed into the upper horizons of the soil by earthworms and other soil fauna, giving a deep, dark, nutrient-rich A-horizon. Along with the crumb/granular structure of this A-horizon, this makes the mollisol/chernozem extremely fertile.

In spring the winter snows melt, causing mild leaching through the soil, but in summer the low rainfall totals and high temperatures draw water upwards in the soil by Capillary Action. These seasonal vertical movements of water help create the deep and dominant A-horizon. Capillary action also deposits nodules of calcium carbonate in the B-(sub-soil) and C-(parent rock) horizons. Commonly, the parent material is alkaline in nature, containing rocks such as limestone or a wind deposited material called loess.

Under natural conditions soil erosion, either by the intensive rainfall of summer thunderstorms or strong winds, is rare as the deep roots of the annual grasses are very effective at binding the soil into a deep sod.

The natural climax vegetation is grass, some species of which can be several metres tall; the mix depends on the local climate. Trees do not survive except alongside rivers and streams. This is due partly to the low total rainfall and also the fact that natural fires and the grazing animals kill the young tree saplings.

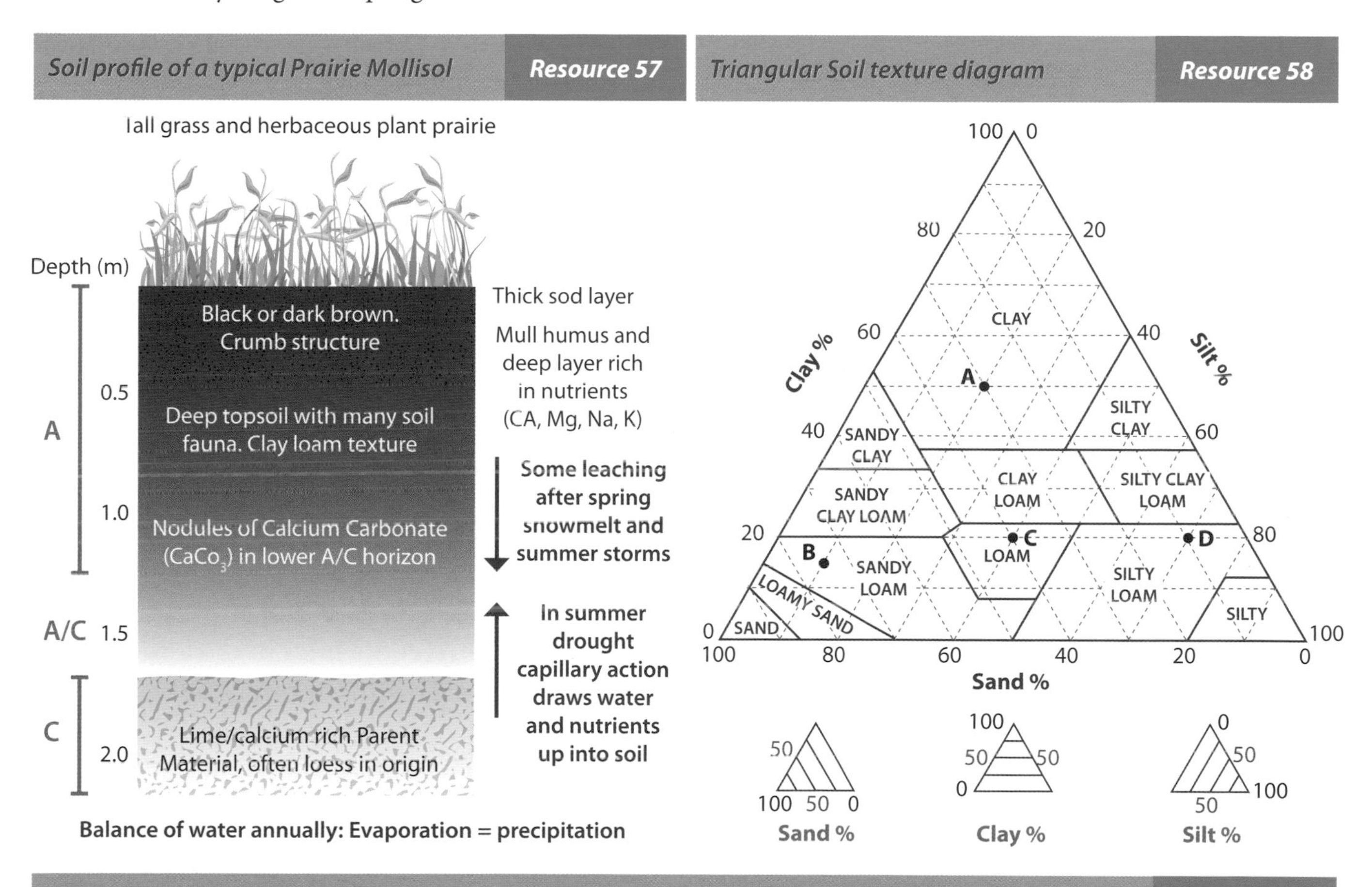

(i) Copy and complete the table for the four soils marked A–D on **Resource 58**.

Exercise

Nature \ Soil	A	B	C	D
% Sand			40	
% Silt				
% Clay	50	15		
Type	Clay			Silty Loam

(ii) What type of soil has the texture 50% sand, 20% silt, 30% clay?

Resource 59 *Table of Mollisol characteristics*

Soil Character	Description	Mollisol character	Reason for Mollisol characteristic
COLOUR	Soil colour reflects chemical composition. Red indicates iron oxide whereas pale grey or ashen soils indicate nutrient loss by leaching.	Very dark even **black** through much of its depth. Chernozem means 'Black earth' in Russian.	The tall grasses that die back in autumn, including their many roots, provide a large influx of organic matter. Soil is the main nutrient store in this ecosystem.
DEPTH	This reflects the age of the soils and the rates of weathering (inorganic material) and decay (organic material).	Most have a depth between **1 and 2 m** (compared to the 30–90 cm of local garden soils)	Mollisols are **young** (post Ice Age) soils but weathering of the parent material is fairly rapid, especially in summer and much organic material is added.
TEXTURE	The mix of grain sizes in a soil: large (Sand); medium (Silt); and small (Clay). A loam is a well balanced mix of these (***Resource 58***).	Mollisol texture varies but most have a **clay-loam** mix.	Texture is partly controlled by the underlying parent material. Wind-blown fine sands, termed **loess**, are commonly found under mollisols.
PROFILE	Soils are described as having a three layer (horizon) vertical structure: the **top soil** or **A-horizon**; the **sub-soil** or **B-horizon** and the **C-horizon** or parent material.	Mollisols have a deep **A-horizon**, often lack a **B-horizon** and the **C-horizon** is often a base rich rock, eg limestone or loess. **Calcium nodules** are found deep in the A-horizon.	A **lack of leaching** – due to limited rainfall and high summer temperatures – allows a deep build up of organic material for the A-horizon. This is also due to rapid decay of organic material by active soil organisms. Upward or capillary movement of water in late summer causes the calcium carbonate nodules to be deposited.
ACIDITY	This is a measure of the concentration of hydrogen ions in the soil – in general the more ions the less fertile the soil. Low pH (4) values indicate acidity and high values (8) represent alkaline soils.	Mollisols are generally **neutral** or **slightly alkaline (pH 6–7)** down through their profiles.	The relatively high pH values are maintained by the **abundant** supply of **organic nutrients** from parent material, the recycling of the tall grasses and the limited degree or lack of leaching.
STRUCTURE	Not to be confused with texture, this is a description of how the soil grains group or clump together. Variations are **Blocky** (square), **Platy** (flat) or **Crumb** (breadcrumb). Crumb is best for retaining water for healthy growth while also promoting free drainage.	Mollisols commonly have a **crumb** structure. When rubbed between the hands the soils form loose circular clumps with open pore spaces.	The mixed texture of sands, silts and clays are combined with the **humus** (decayed organic material) and organic acids. The lack of leaching allows the nutrients and organic 'glues' to stay in the soil.

Resource 60a *Map of North America prairies*

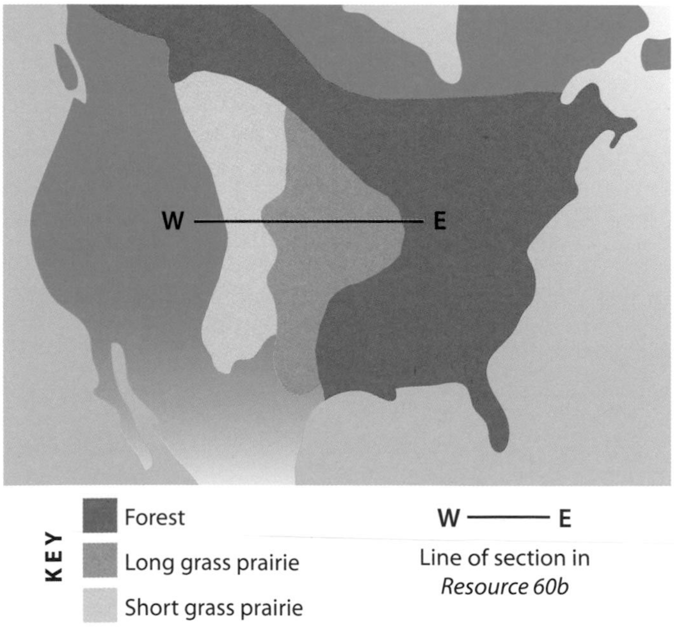

W —— E

KEY
Forest
Long grass prairie
Short grass prairie

W —— E Line of section in *Resource 60b*

B - The impact of human activity and attempts to manage mid-latitude grassland

CASE STUDY: The Great Plains of the USA – the Prairie grasslands

In the USA, annual rainfall totals generally fall from the east to the west coast. As a result the forests of the Eastern USA are replaced in the centre with grasslands. In the east the grasses are tall – switch and big blue stem grow to 1.5–2 m. Further west are the short grass prairies with blue grama and buffalo grass. Originally, the prairie was home to huge numbers of a few species of large herbivores. Bison and pronghorn antelope lived in herds for protection from their predators, wolves and coyotes. Likewise, smaller herbivores such as jack rabbits, prairie dogs, meadowlarks and prairie chickens were abundant and became the prey of carnivorous hawks and eagles.

A cross-section through the Prairies and High Plains — **Resource 60b**

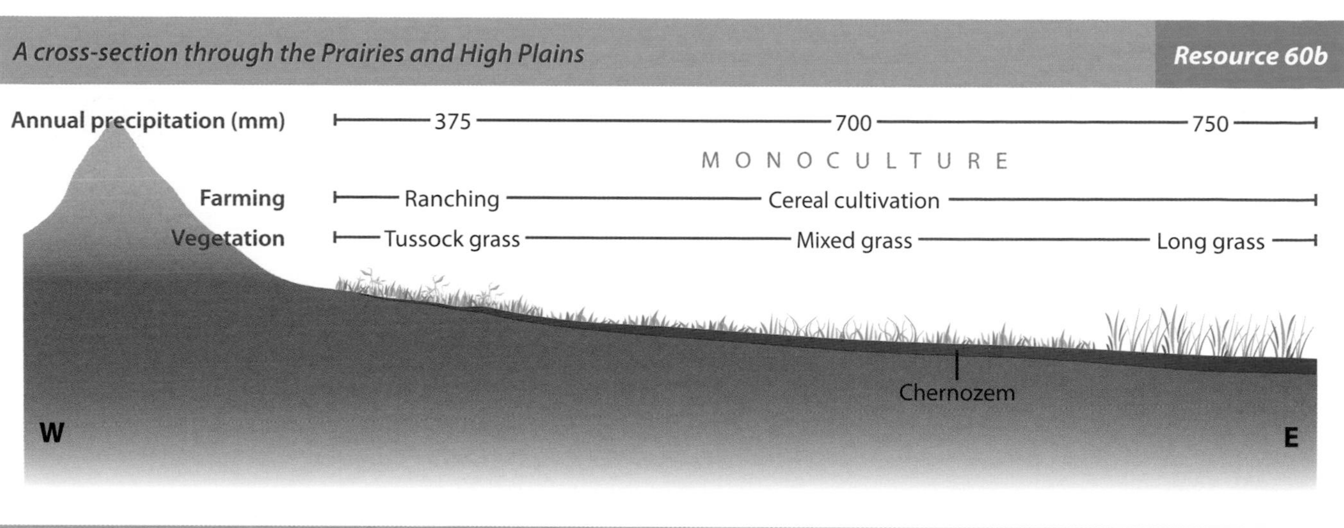

Copy and complete the diagram using appropriate examples from the text. — **Exercise**

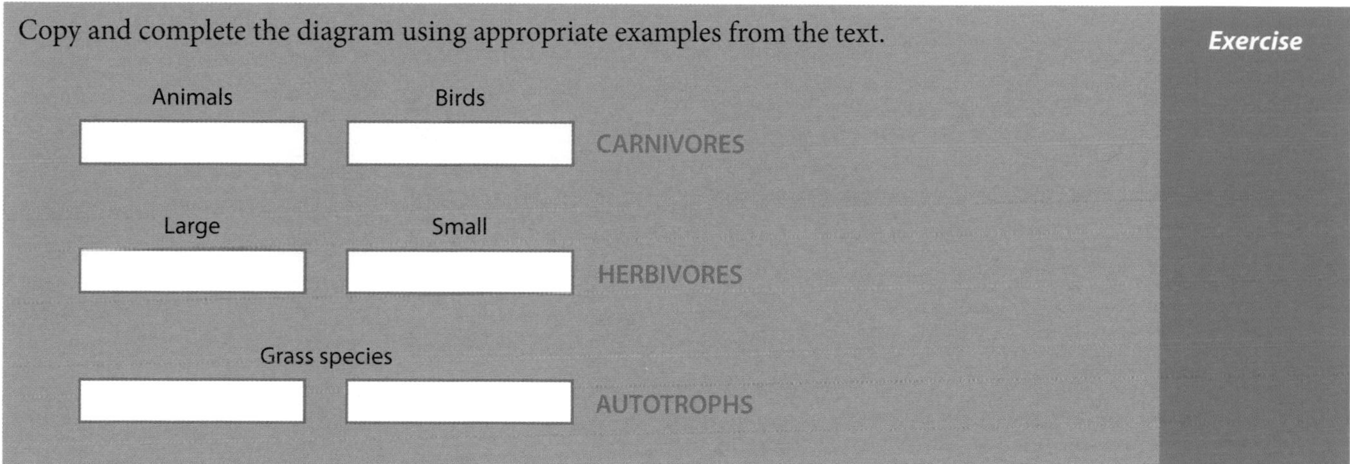

People and the Prairies

The Native American population had existed in harmony with this grassland environment for many hundreds of years before the arrival of Europeans. They had developed a close interdependent relationship with the dominant herbivore species especially the bison herds, and maintained a sustainable hunting and farming economy.

- The first approach by Europeans to the grasslands in the mid 19th century (1850–70s) was to remove the native herbivore species (bison and antelope) in a frenzy of destruction and replace them with their own domesticated herbivores – cattle. This was the infamous cowboy era of the Wild West.
- By the 1890s the use of the steel plough saw cattle ranches and cowboys forced further west on to the drier high plains, as cereal farming on an extensive scale took hold. Wheat and maize (American corn) replaced the natural and planted grasses of the prairie. The sod cover was broken and a monoculture system was developed. The sound of the first iron ploughs cutting through the dense mat of roots in the topsoil was described by Native Americans as the screaming of Mother Earth.

Monoculture = an agricultural system where the land is used repeatedly for the same crop.

- Some cattle ranching remains on the drier western high plains towards the Rocky Mountains. But it has been the use of the remaining prairie for extensive cereal production, aided in dry areas by irrigation, that has created a series of disastrous physical and economic events such as the Dust Bowl eras of the 1890s and especially the 1930s.
- Today, it is argued by some that the agricultural use of most prairies is no longer viable and that they should be restored to their former state.

Resource 61 *Summary of Prairie history of ecosystem change*

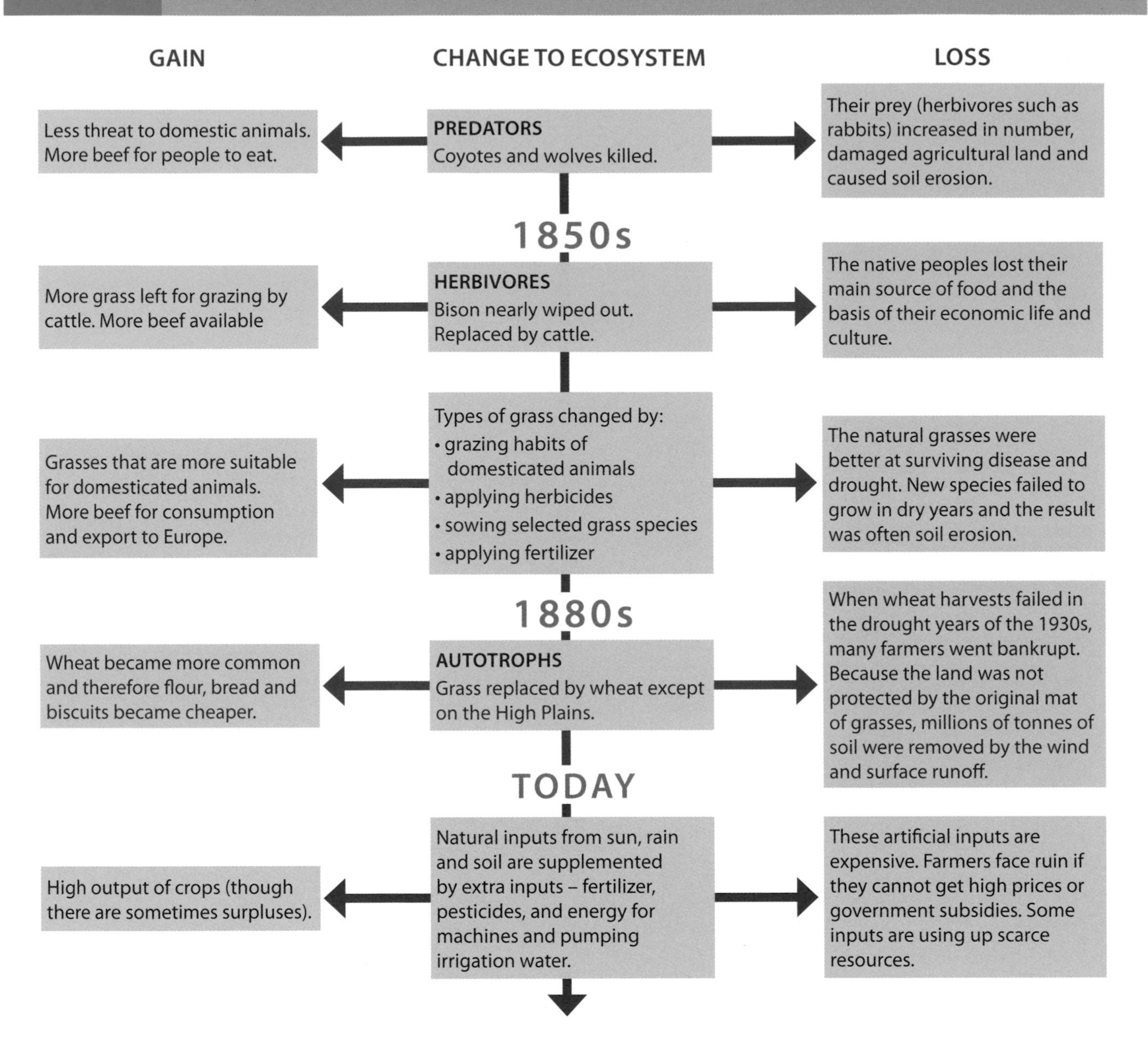

Monoculture and soil erosion

The US Environmental Protection Agency (EPA) believes that overgrazing by livestock (cattle) in the western prairie is the source of 28% of the soil erosion problem. In the east, soil erosion is blamed on the use of long term monoculture.

Monoculture involves the continuous and repeated use of soil for the annual production of a single crop. In the North American prairies the natural fertility of the mollisol ensured farmers many years of good harvest yields until the nutrient supply began to fail. This, combined with drier years, caused depletion of the soil moisture content, the wilting of crops and even failure of harvests. The loss of humus is particularly destructive to these soils as once their structure is lost the top soil becomes dry and loose (friable). Cereal farming by its nature means that the land is left bare after harvest and so the soil surface is not physically protected from the elements of wind and rain. The roots of cereal crops are shallow and less dense than the natural grass mat so their binding action is less than the roots and rhizomes of the native prairie species. The situation is compounded by poor management practices; for example, by burning the crop stubble which remains after harvest, the soil is left exposed to wind and rain. In mollisols, leaching – except for the spring snowmelt – is not a major process, but the removal of the dense organic mat in the soil and the harvesting of cereal crops promotes more downward water movement. This allows leaching of soluble nutrients from the deep A-horizon into the subsoil beyond the reach of plant roots (***Resources 57*** and ***59***).

The Dust Bowl years of the 1930s were one extreme example of these poor farming techniques and in some places the prairies have been reduced to near desert environments or have become so deeply eroded that restoration is too expensive or impossible (the Badlands of Dakota).

The Dust Bowl – 1930s

Between 1933 and 1939, many parts of the Prairies suffered extreme soil erosion, resulting in the region being referred to as the 'Dust Bowl'. The winds carried away the fertile top soil in 'black blizzards'. A number of factors caused the problem, including:

- **A series of long hot summers**
- **Long periods of strong winds**
- **Overgrazing**
- **Rapid expansion of wheat farming**
- **The introduction of tractors**

The Dust Bowl occurred because of a fatal combination of climatic extremes and farming methods which were unsuited to the rich but delicately balanced soil environment. Soil erosion was the result of two processes: 1) Wind erosion (aeolian) and 2) Water erosion (fluvial) caused by splash, sheet wash and gully erosion. Vast amounts of several states were destroyed by this erosion and even today, despite the introduction of conservation techniques, soil erosion is a serious problem.

Changes to temperate grasslands due to arable farming monoculture — ***Resource 62***

Natural grasslands	Grasslands used for annual cereals
Mainly perennial grass species	Monoculture of annual cereals
Some annuals and bulb plants	Surface ploughed breaking the root mat
Soil surface composed of dense root mat present all year	Need for the application of inorganic fertilisers, herbicides and fertilisers
Root mat retains moisture	Reduction in soil moisture
Soil particles bound by root mat	Few roots to bind the soil particles
Limited nutrient loss by leaching	Increased loss of nutrients by leaching
Soil rich in organic material	Decrease in soil organic matter
Limited soil loss by wind erosion	Increased wind erosion
Limited surface runoff and sheet erosion	Increased surface runoff and sheet erosion

Soil erosion and harvest yield — ***Resource 63***

KEY

1. Minor loss of topsoil - yields can be maintained by adding fertilizer.
2. Moderate topsoil loss - overall yields reduced despite increasing fertilizer use.
3. Topsoil lost from large parts of fields, serious gullying (water related) or blowouts (wind related) develop. Yields decline rapidly.
4. Erosion rapidly removes topsoil and fields are no longer economic to farm.

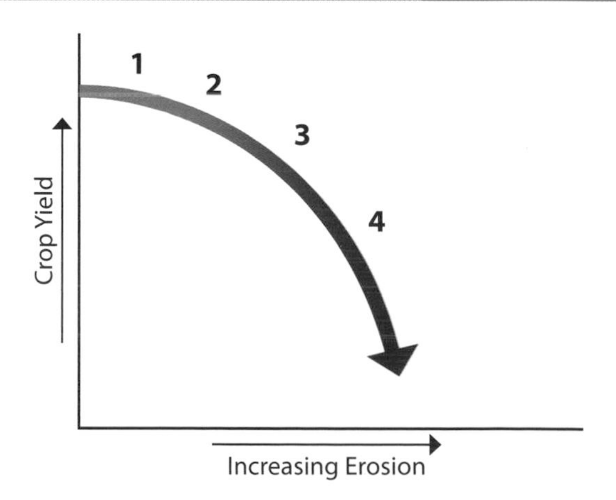

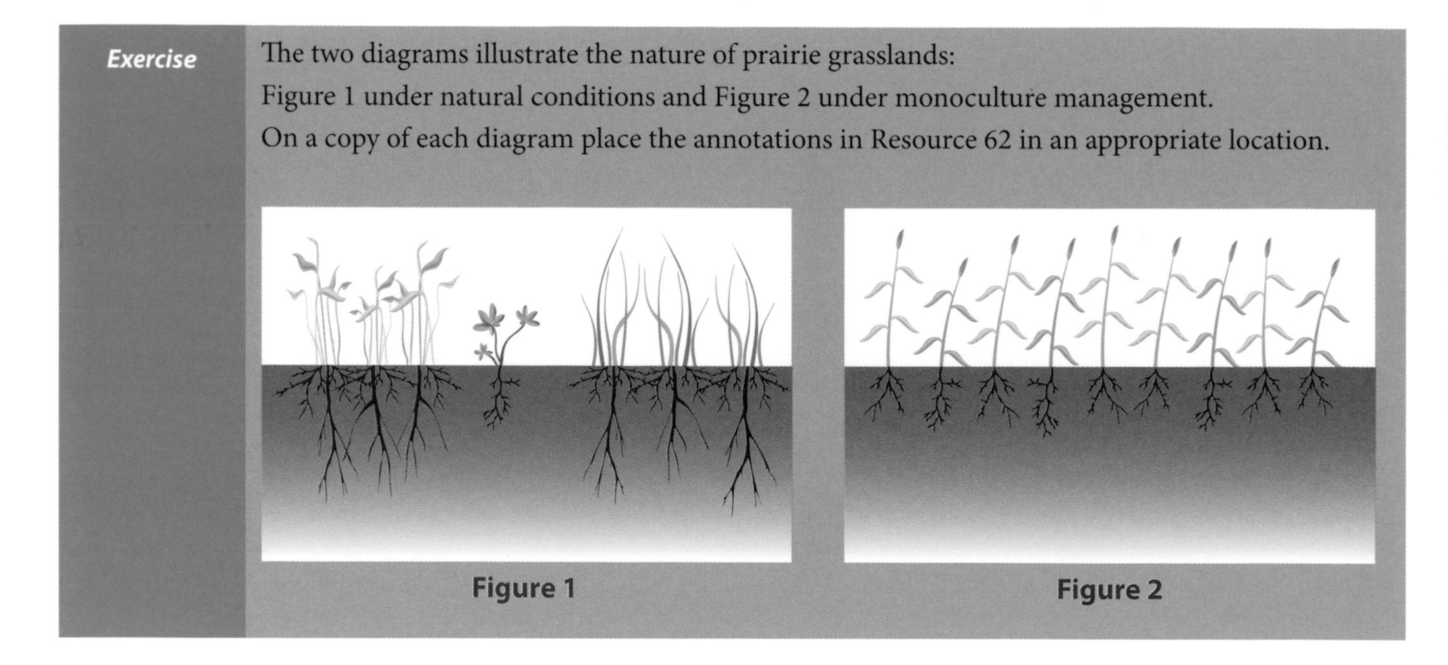

Figure 1 Figure 2

Modern monoculture management

Following the Dust Bowl years, prairie farmers were encouraged to use dry farming techniques to help protect their soil. These included shelter belts, terracing, contour farming and strip cultivation (***Resource 64***). Fortunately for the farmers, though less so for the environment, in the 1940s an enormous aquifer was discovered lying below much of the dry prairie grassland. An aquifer is a natural store of water in permeable rocks beneath the soil, which can be tapped by wells and pumping systems.

The Ogallala aquifer may be the largest of its kind in the world and its presence meant that the prairie farmers were no longer dependent on the variation of rainfall. A ready supply of water allowed the wheat and maize monoculture to resume and the 'Bread-basket of the world' was back in production. Farm management became more and more dependent on technology as these examples illustrate:

- Boom-arm sprinkler systems were introduced where a 400m-long, centrally pivoting arm moves around spraying water onto the soil below. These huge circular systems are visible from space and ten boom-arm systems use the same quantity of water as a city the size of Belfast;
- All the water used has to be pumped from beneath the ground and the costs for fuel especially for petrol and diesel have increased more rapidly than wheat prices; and
- The ongoing monoculture means that the soils have to be recharged with artificial fertilisers including nitrogen, potassium and phosphates. These are products of the chemical industry requiring energy both in production and in application to the soil.

By the 1970s, farming on the North American prairies had become the most technologically and energy-dependent system in the world. Only in a few special areas had the native species of grass, insect and animal survived and so most of the region had the appearance of a man-made artificial environment. In recent years a disturbing reality for many prairie farmers is that the Ogallala aquifer is under threat; over extraction of its water has reduced the height of the water table with even deep wells drying up.

Economically, farmers in the prairies have been struggling to survive. While costs have soared, their incomes have remained fairly static and the main social impact in the last three decades has been the amalgamation of farms and the decline and abandonment of farming settlements. Across the mid-west from Texas to Kansas farming settlements are shrinking and crop yields have started to fall. Each year, soil loss reduces the value of prairie crop production by an additional $6 million.

Some soil conservation techniques **Resource 64**

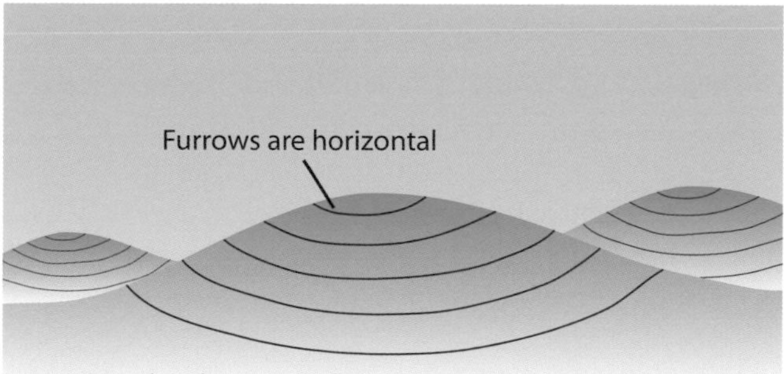

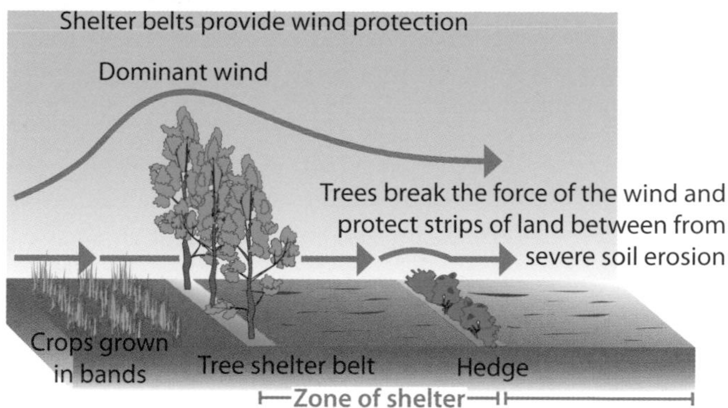

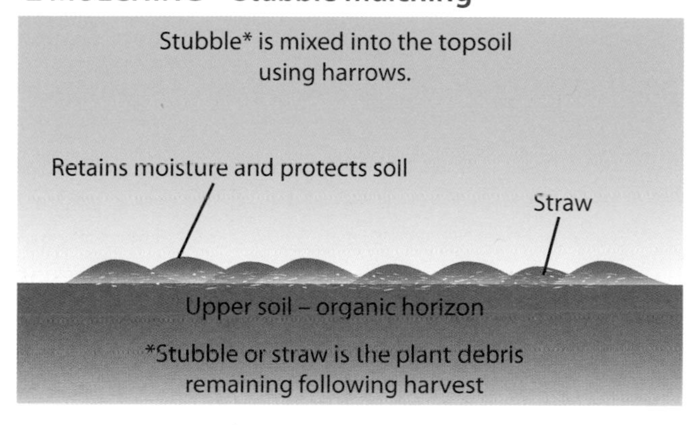

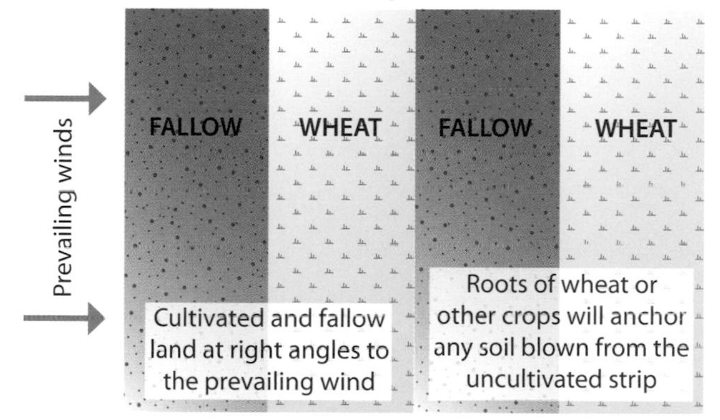

Alternatives to monoculture management

In response to the cycle of boom and bust that the prairie economy has endured, some environmentalists are calling for the abandonment of monoculture farming and the restoration of the natural ecosystem. One movement called the Buffalo Commons has encouraged the return of natural species to the region. Deer, pronghorn antelope, coyotes, prairie dogs and even cougars and bears have returned to roam their natural grassland environment. Elsewhere some farmers seek more sustainable agriculture. Buffalo ranching has developed to supply the demand for organic red meat, and some have even returned to dry farming techniques in which the lower yields are at least offset by the reduced dependency on energy and technology costs.

Management for soil conservation (Resource 64)

Controlling erosion means protecting soil from the destructive effects of wind and water. To achieve this, a number of conservation-farming methods may be used.

- **Mulching**. Mulch is any material left covering a soil. Often it is dead organic material. Placing mulch on the soil surface has a number of benefits – including protecting the soil surface, increasing water infiltration, and greatly reducing soil loss. When the plant stubble that remains after harvest is left on fallow lands in the prairies, the annual soil loss is much less than when fields are left bare.
- **Contour Farming**. In contour farming, plough furrows and crops are run across a slope, creating a ridged surface which traps soil and water. This is effective on gentle or moderate slopes.
- **Strip Cropping**. When used to control wind erosion, strip cropping alternates erosion-resistant grains or fodder crops with more easily eroded crops. These run at right angles

to the direction of the prevailing winds. The impact of the wind is lessened and erosion reduced.

- **Cover Crops**. These are planted to protect the soil when the main crops do not cover the ground. The cover crop adds organic material to improve the physical condition of the soil.
- **Shelterbelts**. The idea of a shelterbelt is simple – it acts as a windbreak, reducing erosion by restricting the wind speed. Shelterbelts are most effective when used with other soil-conservation methods. While trees have often been used, they are slow-growing and plants such as tall wheatgrass or flax have recently been planted in strips as windbreaks.

Both mulching and shelterbelts also help to retain moisture in the soil, another key to maintaining the health and productivity of the prairie grassland.

Exercise

Past AS questions on management and mid-latitude grasslands

Select any one of the following past paper AS questions on mid-latitude grasslands and attempt to write an answer. This may be done as an open book exercise or under examination conditions; in the latter case a time limit of 15 minutes could be set.

1) With reference to your case study of an area of mid-latitude grasslands, explain how human activity can have both positive and negative impacts on the ecosystem. (June 2004) (12 marks)

2) With reference to your case study of an area of mid-latitude grasslands, describe and explain how people have attempted to mange this ecosystem. (June 2005) (12 marks)

3) Describe how the characteristics of mollisols/chernozems have influenced the management of an area of mid-latitude grassland that you have studied. (January 2006) (12 marks)

INSIDE THE EXAMINER'S MIND

Study the section of the specification provided.

Elements	Elaboration	Spatial context requirement
(c) Human interaction with ecosystems.	(i) Characteristics of mollisols/ chernozems, characteristics of mid-latitude grassland; (ii) Monoculture, soil erosion and soil conservation.	(i) and (ii) Study of the impact of human activity and attempts to manage an area of mid-latitude grassland, eg N. American Prairies (regional/national scale)

Attempt to write two examination style case study questions that an examiner could set based on this section. Could the examiner use resources to examine this section?

If so:

— What sort of resources might be used?

— What questions might be asked, based on these resources?

3A ATMOSPHERIC PROCESSES

'It's bright today', 'it's getting breezy', 'it's warm at the moment', or 'it's misty over the hills'; a stranger to these islands might be puzzled by the subject of such comments they overhear. The 'it', of course, is the weather – meaning the state of the atmosphere. The mixture of gases that form a thin layer around the earth not only supplies the vital material for human life – oxygen – it also controls the temperature, air flow and water supply.

Most scientific textbooks describe the atmosphere as being composed of 78% nitrogen, 21% oxygen and the remaining 1% includes carbon dioxide, ozone and the noble gases. This is an accurate analysis of Dry air but in reality air is rarely dry and water as a gas (water vapour) can make up to 4% of air by volume. All atmospheric gases play a role in the nature of weather and climate but in fact water vapour may be considered as the most significant. Dust and ice particles are another small but vitally important component of the atmosphere.

Resource 65 The structure of atmosphere based on temperature change

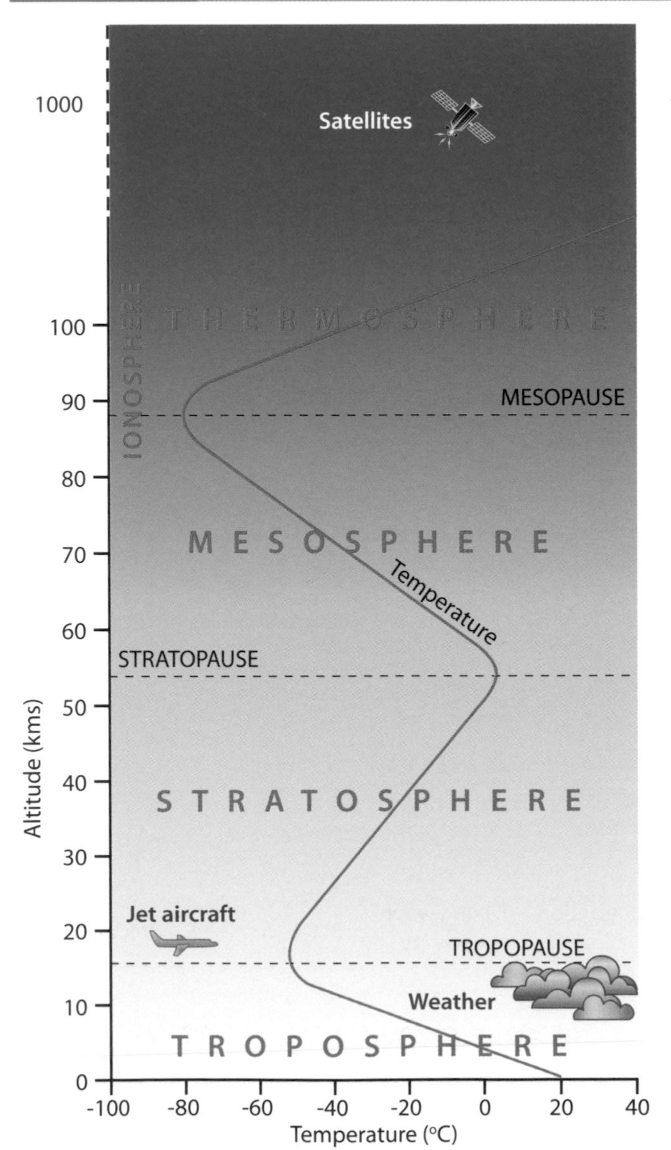

Resource 65 shows the structure of the atmosphere based on physical changes in temperature with height. Most atmospheric gas and the weather systems are confined below the Tropopause, in the lowest layer known as the Troposphere. This varies in its thickness from around 8 km at the poles to 16 km at the Equator. Most long distance air travel takes place above this level, partly to avoid the variable weather it contains.

The global energy balance

The global energy balance can be described as an open system in which the earth itself and the atmosphere form the system as illustrated in **Resource 66**.

The earth/atmosphere system receives its energy from the sun as light and heat; this radiation input is called **insolation**. Most of this energy is in the form of short-wave solar radiation and at any one time half the earth is receiving this input. As insolation enters the earth's atmosphere some of it is absorbed or reflected but much, about half, reaches the earth's surface. This energy heats the surface, which then emits long-wave energy back into the atmosphere where it is absorbed by the gases in the air. In short, the atmosphere is not heated directly by the sun but indirectly by the earth from below.

Energy exchange in the earth-atmosphere system — *Resource 66*

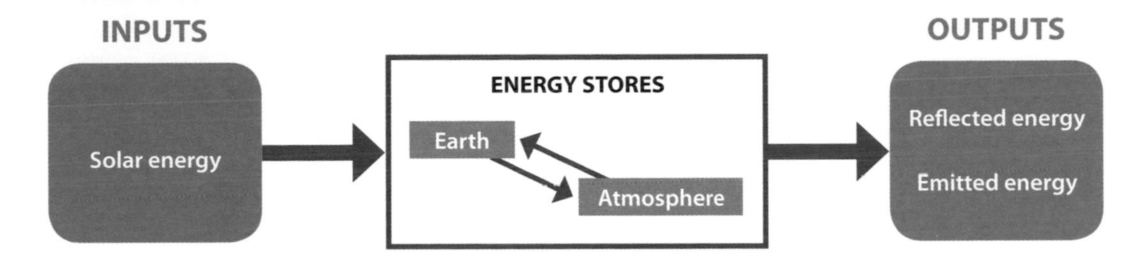

Resource 67 shows the energy pathways of insolation entering the earth-atmosphere system; the figures represent the average values if the insolation is taken as 100 units.

Cloud and dust particles in the atmosphere both absorb and reflect energy. Some gases including ozone and carbon dioxide can absorb short-wave energy. The reflective index of clouds or the earth's surface is known as the **albedo** and is expressed as a percentage of incoming radiation that is reflected. The average surface albedo value is around 6% but this varies widely from fresh snow at 85% to the oceans at less than 2%.

Overall, there must be an annual balance between the solar input and the loss of energy to the earth-atmosphere system or the world would either be heating up or cooling down. **Resource 68** illustrates the routes by which the energy absorbed by the earth or atmosphere is returned to space.

The response of the earth-atmosphere system to solar insolation — *Resource 67*

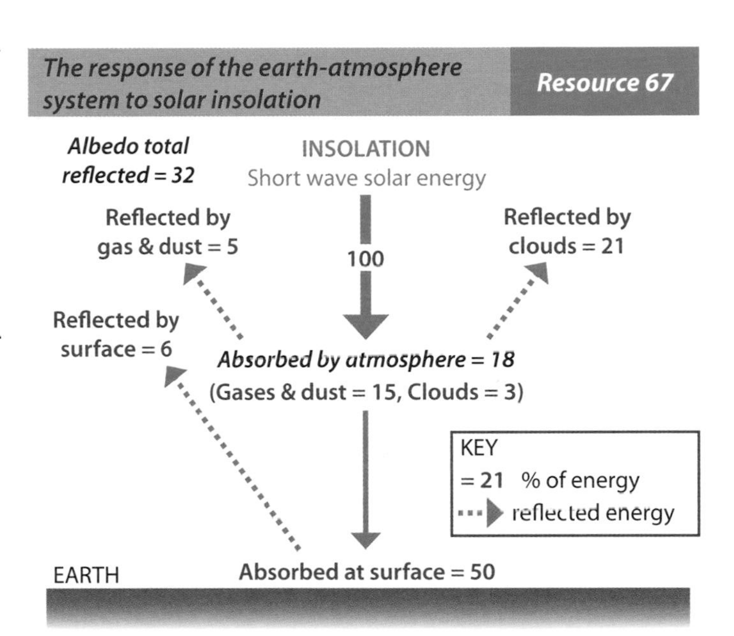

On average 68% of solar insolation is absorbed by the earth-atmosphere system. The other 32% is reflected back to space from clouds, dust and the surface. Albedo is the reflective index of the surface. Insolation is solar energy short waves of light and heat.

The loss of energy absorbed by the earth-atmosphere system — *Resource 68*

The energy absorbed is returned to space either:

1) directly by the 'radiation window' or

2) by transfer to the atmosphere by:

a) long wave radiation b) latent heat c) convection

Hot surfaces such as the sun emit radiation as short waves whereas cooler surfaces, such as earth, emit long wave radiation. Most gases cannot absorb short wave radiation but can absorb long wave. This is why the atmosphere is not heated directly by the sun but rather by the earth from below!

Latent heat is energy stored when water changes from a liquid to a gas – for example in the process of evaporation.

Convection is the transfer of energy by rising currents of warm air.

In order to maintain a balance all 68% of incoming solar energy must return to space as shown in the diagram.

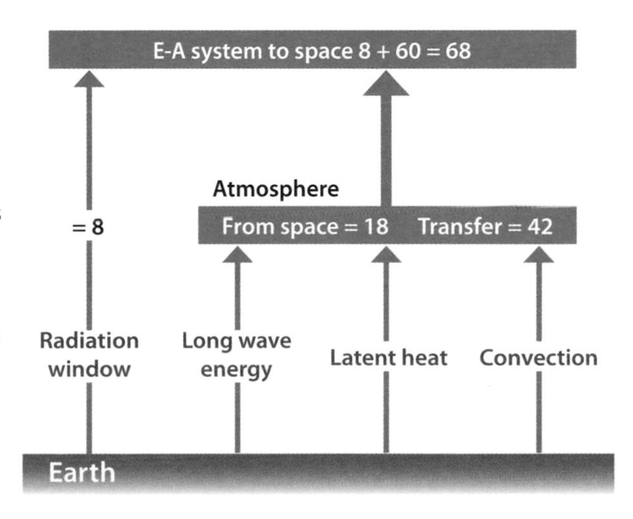

The heat budget

The heat budget diagram, **Resource 69,** shows that due to the variation in day length and seasons, some regions receive an energy surplus over a year while others have a net deficit. Anywhere between 40°S and 40°N of the Equator the earth-atmosphere system has a surplus of heat energy each year. Beyond this, from 40°N to the North Pole and from 40°S to the South Pole there is an annual deficit of heat energy. If this was the end of the story then the Polar Regions, including most of Europe, would be getting colder and colder while the tropics would get hotter. This is not happening, so somehow the surplus energy from the tropics must be transferred to the other regions to retain the balance. This heat exchange involves two transfer mechanisms – vertical and horizontal heat transfers.

Resource 69 *Variation in the energy budget of the earth-atmosphere system*

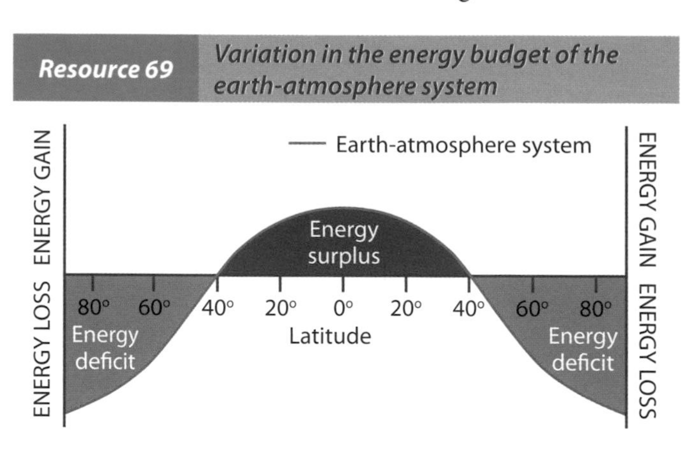

Global patterns of insolation and temperature

Latitude (distance from the equator) is the most important factor in the pattern of surface temperature across the earth. Within the tropics, day length is constantly around 12 hours and the sun travels high across the sky. At higher latitudes, day length varies over the year and the height of the sun is lower, meaning its energy is spread over a larger surface area (***Resource 70***). Other factors that influence temperature patterns include:

- Altitude – temperatures fall with height above sea level at an average rate of 0.6°C every 100 m.
- Land and sea – being transparent, the sea can absorb insolation to greater depths than the land surface. Consequently, the sea tends to warm up and cool down more slowly. Coastal areas tend to have moderate temperatures while the interior of continents often have extreme seasonal variation.
- Prevailing winds and ocean currents will cause local heating or cooling.
- Cloud cover.
- Aspect – the angle of slopes in relation to the sun.

Finally, because the overhead sun migrates, or appears to, between the Tropics of Cancer and Capricorn (23½°N and 23½°S), the energy budget varies over each year. Some of these factors can be seen on the world temperature maps (***Resource 71***).

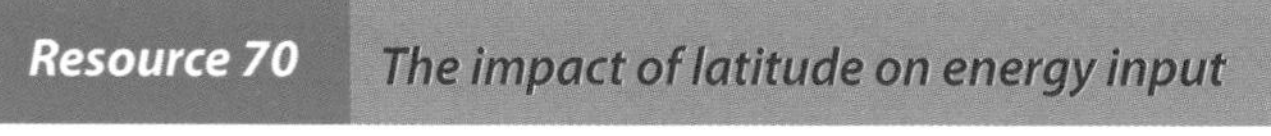
Resource 70 *The impact of latitude on energy input*

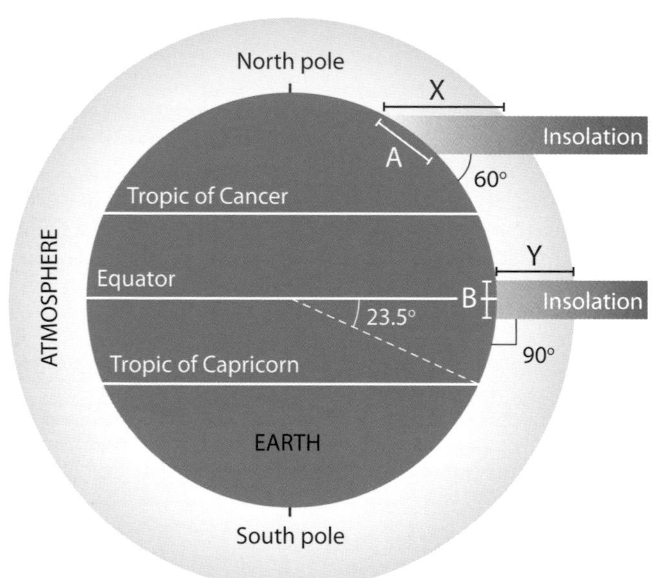

X and Y – Rays pass through different depths of atmosphere.
A and B – Same quantities of insolation heat different size areas.

Vertical heat transfer

Heat energy is transferred vertically from the earth's surface into the atmosphere, cooling the earth while warming the air. There are several mechanisms for this transfer including:

- Radiation – long wave energy
- Conduction – energy transfer by contact
- Convection flows – rising thermals of warm air

World maps of mean sea level temperature (°C) for January and July — Resource 71

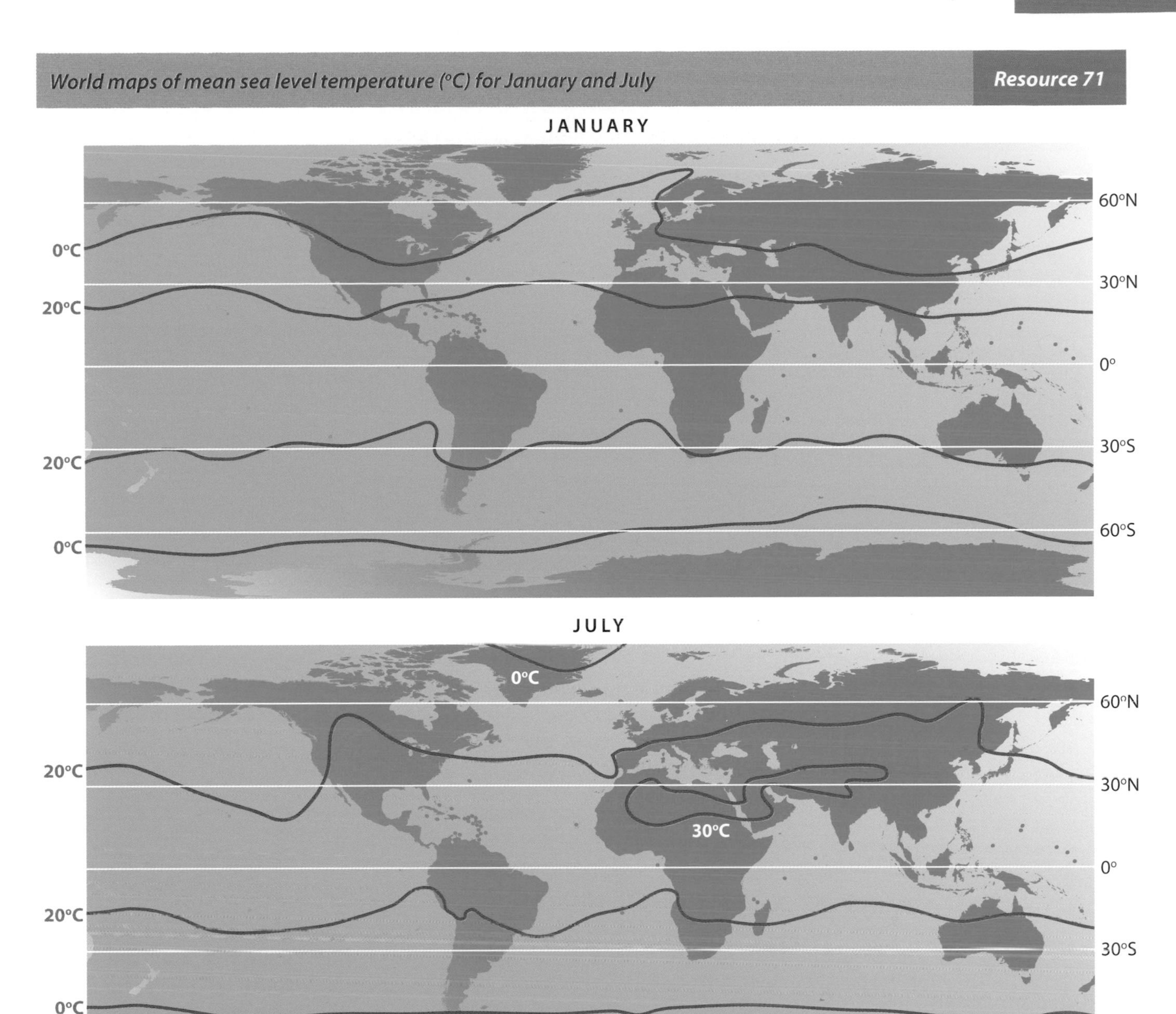

Horizontal heat transfer

The two natural routes for exchanging heat energy across lines of latitude are winds in the atmosphere and currents in the oceans. Research suggests that winds are the more important mechanism but that ocean currents still account for up to 35% of the energy exchange.

1. Ocean currents

These are complex flows that depend on heat energy and variations in salinity but the most consistent ocean currents form a circular pattern in each of the world's oceans. Ocean currents that flow away (north or south) from the Equatorial region are warm and these are found in the western parts of the world's ocean basins. Currents that move towards the equator carry cooler water towards the tropics and are therefore cold currents, usually in the eastern ocean basins (***Resource 72***). In the North Atlantic a warm current crosses over from west to east to produce the North Atlantic Drift, which warms the west coast of Europe, including Ireland (50°–55°N); meanwhile off the coast of Canada, the cold Labrador current brings icebergs and

Resource 72 *The main ocean currents*

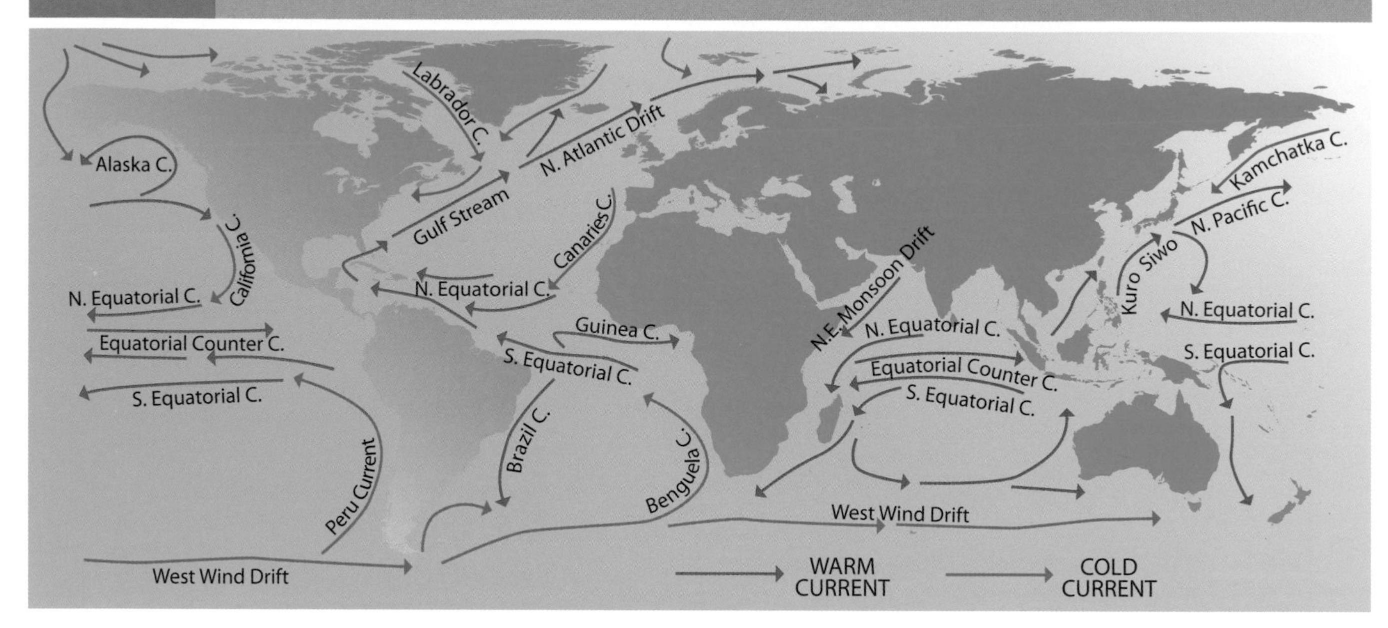

freezing conditions to Canada at latitudes much further south. The Titanic built in Belfast at 54°N was sunk by an iceberg at 42°N, more than 14 degrees closer to the equator! Perhaps the most powerful example of the influence of ocean currents is the flow of water in the central Pacific known as the El Niño. Across the world this occasional event triggers unusual and extreme weather, including storms and droughts, often with large scale impacts on society and the economy.

2. Winds

The majority of the horizontal heat exchange required to maintain the earth's balance is by airflows across latitude. Winds carry surplus energy from the tropics in two distinct ways. Firstly, in the form of heat, that is, the air which is warmed by the hot tropical land or sea surface. In Western Europe such warm tropical air often arrives from the Atlantic as the warm sector air in mid-latitude depressions. Secondly, heat is transferred as latent or stored energy in water vapour carried by the wind. Over tropical sea areas such as the Caribbean vast amounts of water evaporates into the air. This change of water from a liquid to a gas (known as a phase change) requires energy from the surrounding environment. This energy is then stored as latent heat in the water vapour and can only be released when the process is reversed, that is, when the water vapour condenses back into a liquid, forming clouds and rain. If the air holding the water vapour is moved from the tropics to beyond 40°N or S before it condenses, the stored latent energy can then be released into these energy deficit regions. **Resource 73** shows global wind patterns.

Latent Energy – cooling by evaporation

This process is the mechanism that people and many animals use to regulate their temperature in hot environments. Perspiration onto the skin surface encourages evaporation which takes energy away from the surroundings, so cooling the surface.

World maps of mean pressure and common winds for January and July **Resource 73**

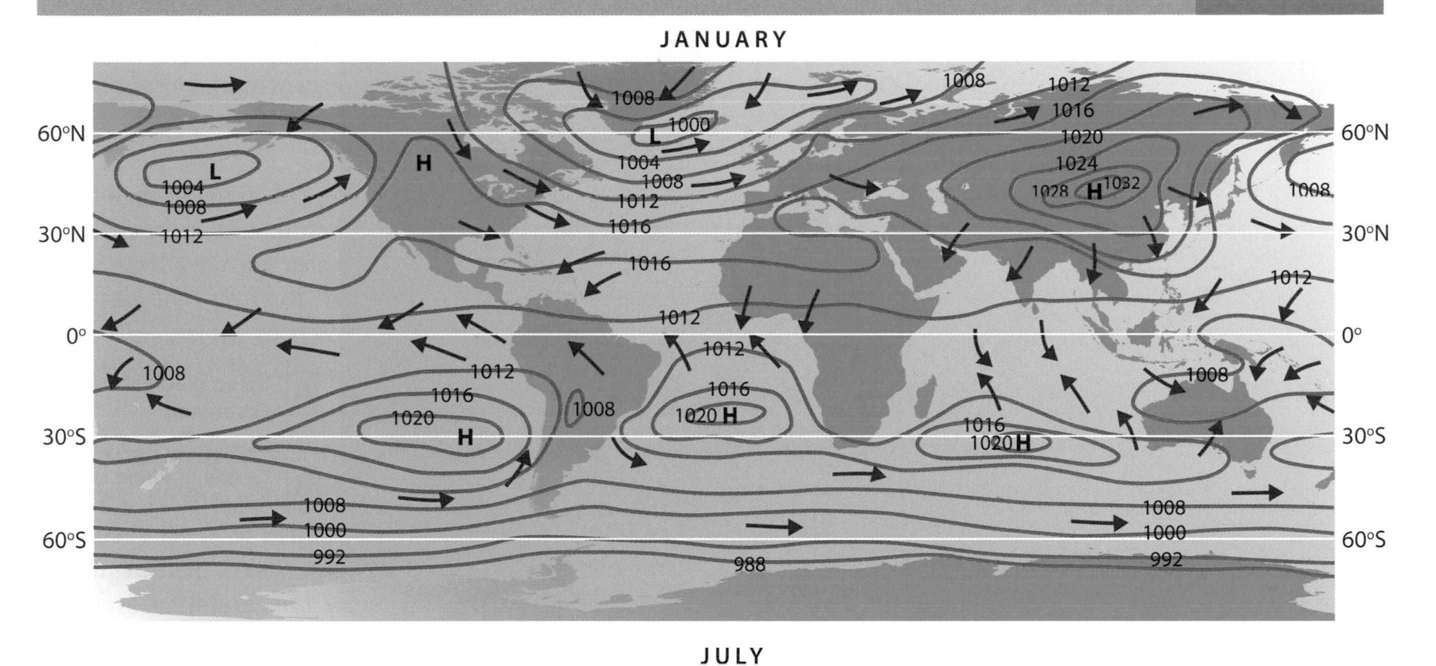

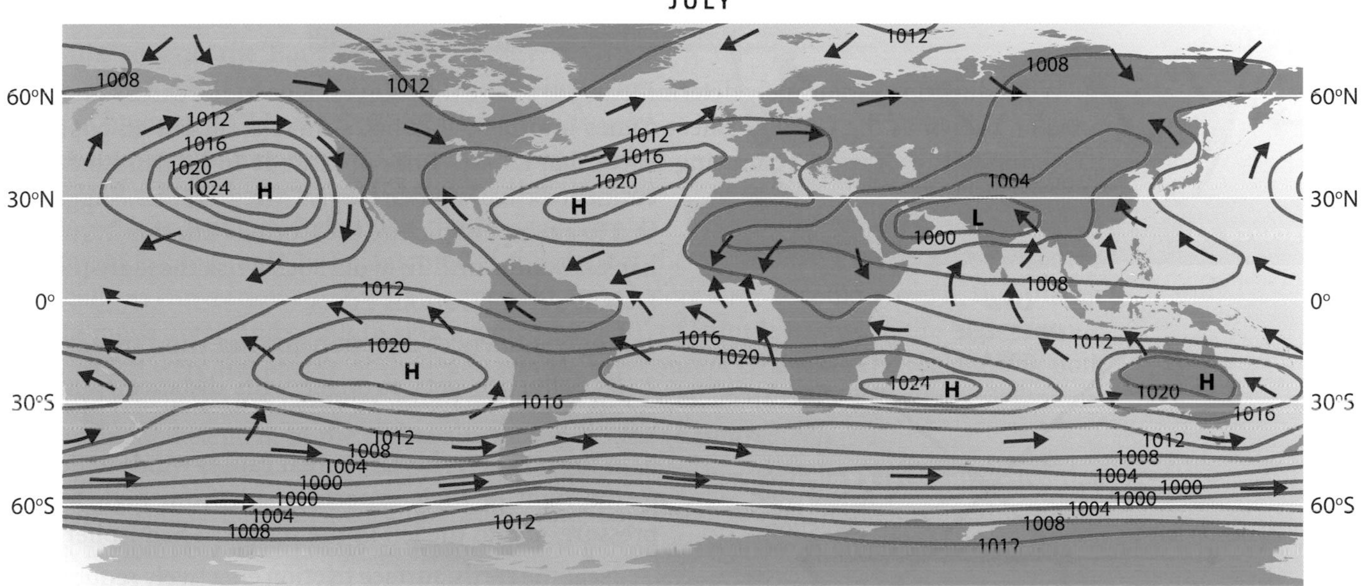

Winds speed and direction

'From high to low the wind doth blow' says an old sailing adage accurately reflecting the key role of pressure difference in causing air flow. Atmospheric pressure is simply the weight of the air at any point. At sea level the atmosphere's average weight is one bar, which is divided into one thousand units – millibars (mb). Weather maps or synoptic charts use lines of equal pressure, isobars, to show variation in pressure over the surface. The difference in pressure between two places is the pressure gradient and the greater the difference, the faster the air will flow between them. On a synoptic map this can be seen by how widely isobars are spaced (see A and B on ***Resource 74***). At sea level atmospheric pressure normally varies between 1040 and 950mb. Pressure is measured by a barometer.

The only other force that influences wind speed is friction. Friction caused by the earth's surface can slow winds speeds in the lowest 1,000 m of air; above this, it has little effect. Consequently the fastest winds in the troposphere are found above 5 km, where flows of 400 kmph are not unusual. Slowing by friction depends on the nature of the surface below; wind speeds over the sea can be high compared to those over rough land terrain.

Resource 78 *The tricellular model of global circulation*

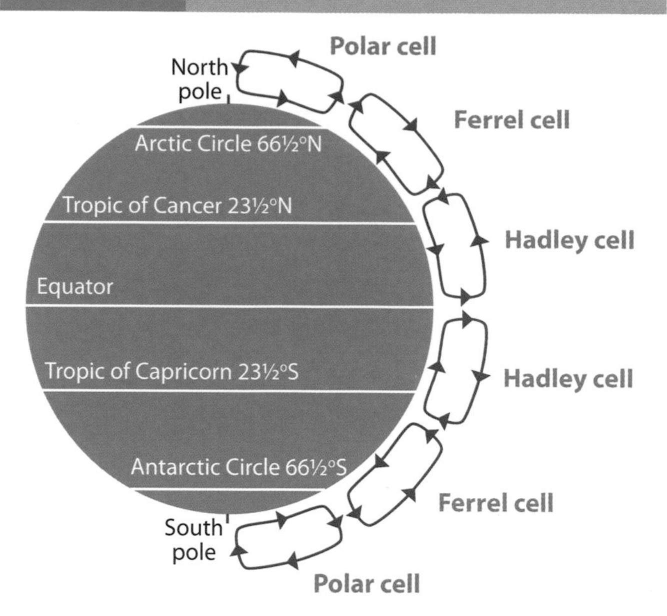

The Tricellular Model

Resource 78

Building on the Hadley cell in the low latitudes, theorists have described a second direct cell around the poles: the Polar cell. This and the Hadley cell initiate a third indirect cell in the mid-latitudes named the Ferrel cell. The cells provide a mechanism for the horizontal exchange of energy from the tropics surplus to the mid and high latitude, and are confined vertically to the Troposphere which, though 16 km in depth at the Equator, is a mere 6 km over the poles.

Hadley Cell

Around the equator the prevailing trade winds meet in an area termed the Inter-Tropical Convergence Zone (ITCZ). Here, in a zone of calm surface conditions, currents of air move vertically in large convection cells that produce cumulo-nimbus clouds stretching as high as 16,000 m up to the Tropopause. At this height the air diverges, flowing poleward to sink again at 30°N and S. As the subsiding air warms, it is able to hold more moisture, resulting in clear skies and dry conditions at the surface, forming the tropical desert climate. From here, some air flows into the mid-latitudes while the trade winds blow towards the equator to complete the Hadley Cell.

Polar Cell

The very cold air over the poles leads to subsiding air, producing persistent high pressure on the surface. From this, surface winds move out into the mid-latitudes where the air rises at the **Polar front** and returns poleward to complete the cycle.

Ferrel Cell

Dominating the mid-latitudes, this cell involves the convergence of cold surface winds from the poles and warm surface winds from the tropics around 60°N and S of the equator. This is the Polar front, possibly the world's most significant zone of energy exchange. Here, convection uplift produces turbulent, wet conditions associated with depressions, and at the tropopause, air diverges away towards the tropics to join the subsiding limb of the Hadley Cell, or poleward to subside at the pole.

Recent modification of the Tricellular model

Research in the upper troposphere has shown that the Ferrel cell is an oversimplification of mid-latitude circulation, and that at this level fast-flowing westerly jet streams are the dominant factor in the circulation and weather of this region (see pages 69 and 70).

Moisture in the atmosphere

Humidity and precipitation

Uniquely, water plays a role in the atmosphere in all its three states as a gas, a liquid and a solid. As noted earlier, the amount of water vapour in the atmosphere varies widely from place to place and time to time. The measure of this water content is termed humidity. Moisture content when expressed as the mass of water vapour (grams) in a given volume of air (cubic metres) is termed the **Absolute humidity** (g/m^3). At any given temperature, air can only contain a certain amount of moisture. At this limit the air is said to be saturated. Cold air, such as that found over the Greenland ice sheet, can only hold a small quantity of moisture. A second useful and commonly stated measure of atmospheric moisture content is **Relative humidity** (RH). This is where the humidity is expressed as a percentage of what the air could contain at that temperature. Air with an RH of 100% is therefore saturated. Relative humidity will change with temperature so that as air cools the RH will increase. If relative humidity reaches 100% it is said to be at **dew point**, at which point the water vapour will start to condense, turn to a liquid, and clouds form. **Resource 79** shows how relative humidity is influenced by changing conditions.

Relative humidity varies with changing conditions ***Resource 79***

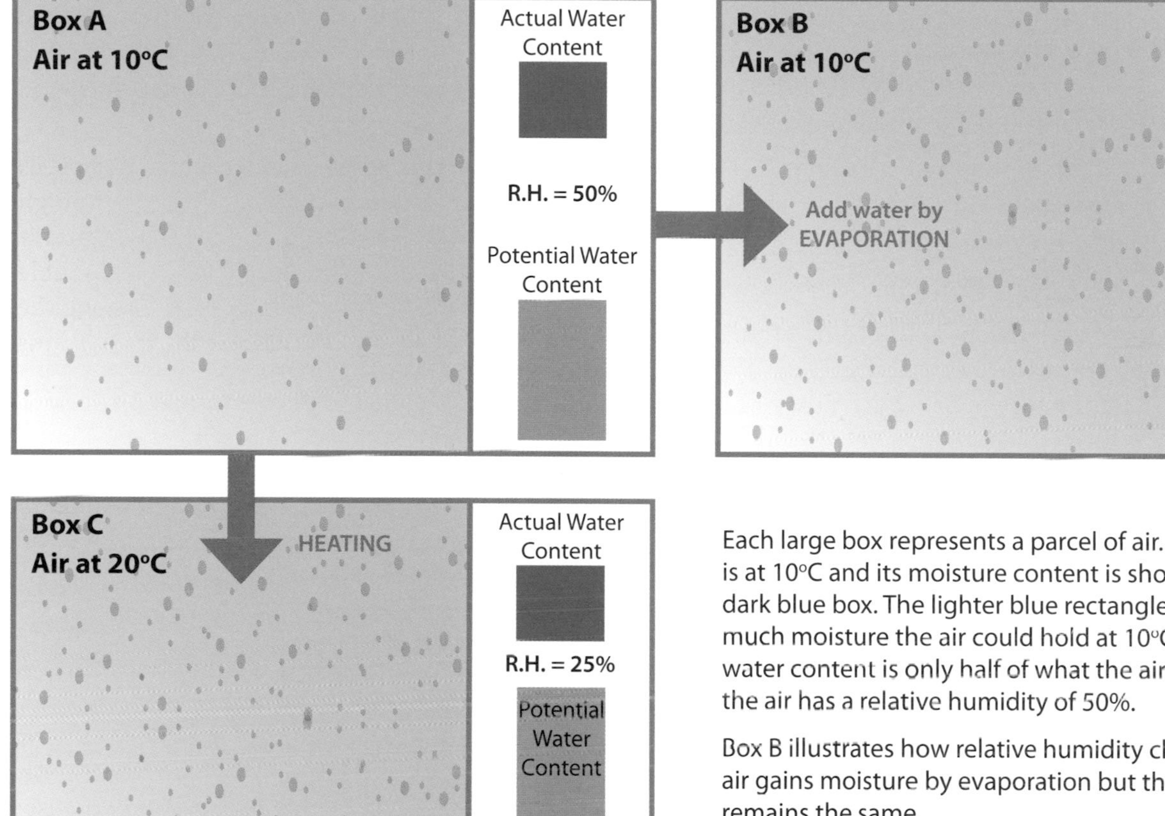

Each large box represents a parcel of air. In Box A the air is at 10°C and its moisture content is shown by the small dark blue box. The lighter blue rectangle represents how much moisture the air could hold at 10°C. The actual water content is only half of what the air could hold so the air has a relative humidity of 50%.

Box B illustrates how relative humidity changes if the air gains moisture by evaporation but the temperature remains the same.

Box C illustrates how relative humidity changes when the temperature rises but the actual moisture content is not altered.

Exercise

Describe what would happen if:

(a) the temperature in Box A fell; and

(b) water was removed from Box C by condensation without any temperature change.

Condensation

This is the process by which water vapour changes from a gas to a liquid. Condensation requires cooling which happens in several ways in the atmosphere and can lead to the formation of clouds and precipitation.

1. Radiation Cooling – overnight and with calm, clear sky conditions the ground rapidly losses heat and so chills the air above. Mist and fog can then form at ground level with dew or, if the temperature is below freezing point, frost on the surface.
2. Advection cooling (advection is the lateral movement of air as opposed to convection or vertical air movement). If warm, moist air from the sea blows onto cooler land it will chill and fogs may develop.
3. Cooling by uplift. Three common mechanisms cause air to rise into the atmosphere and so to expand and cool. Firstly, winds that are forced to rise over mountains (**orographic**); secondly, a body of warm less dense air meeting a body of colder air (**frontal**); and thirdly, where air at the surface is warmed and rises due to its own buoyancy as a thermal (**convection**). In all three cases the rising air will expand and so cool down to reach its dew point temperature. At this point condensation will form cloud.

It is important to note that condensation only occurs where condensation nuclei are available. These are minute particles in the air, often dust or salt crystals.

Precipitation

Precipitation is the term used to describe any form of release of water vapour from the atmosphere by condensation. It includes mist, fog, dew, hoar frost, hail, sleet, snow and rain, the final two being the most common and significant at the global scale. The formation of cloud by condensation is only the first stage in a complex process that leads to precipitation. Clouds are made of very small water particles; a single raindrop could contain millions of them. Two theories are currently used to describe the process by which these tiny droplets grow to form rain or snow:

- the ice crystal mechanism; and
- the collision and coalescence mechanism.

In reality, both of these processes may operate under different conditions in the atmosphere.

Three types of rainfall are linked to the mechanisms of rising air (noted earlier).

Resource 80 *Orographic (relief) rainfall*

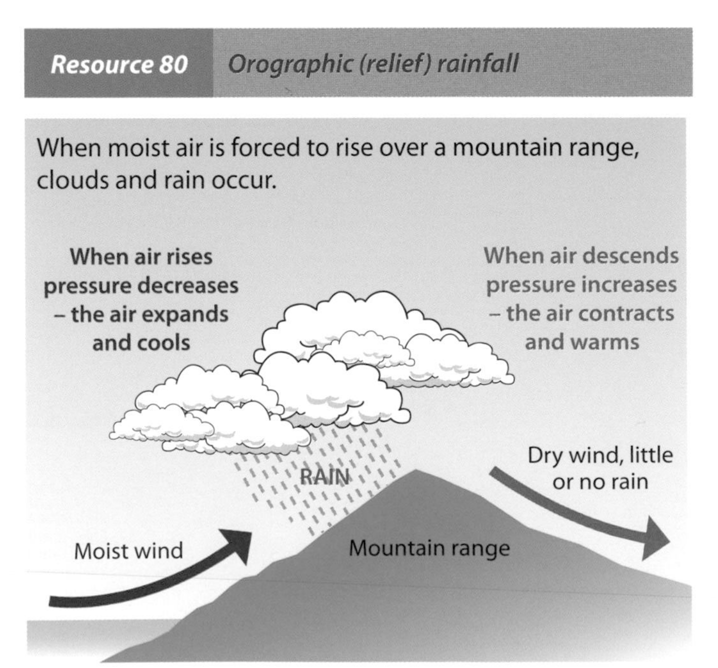

1. Orographic (relief) rainfall

Rainfall totals over even low hills is normally significantly higher than surrounding areas (***Resource 80***). Where a range of mountains lie close or parallel to the coast, such as the Canadian Rockies or the Macgillycuddy's Reeks in South West Ireland, on-shore winds will be forced to rise, expand and cool to produce high rainfall totals. This loss of moisture means that when the air descends on the leeward side it warms, becoming dry and producing an area of lower rainfall – the **rain shadow** effect. This in part explains the contrast in rainfall totals between the west and east coasts of Ireland (1,500 mm–900 mm).

2. Frontal (cyclonic) rainfall

Frontal rain results from the meeting of two air flows. Mid-latitude depressions, common across Western

Europe, form at the Polar Front boundary of contrasting air masses over the Atlantic Ocean. Along the front warm, moist, tropical maritime air is forced to rise by heavier, cold, polar air, producing long wet periods or intense downpours of frontal rain (***Resource 83b***).

3. Convection rainfall

Convection occurs when the ground surface, land or sea, is heated and the air above warms and rises. This is associated with hot conditions in tropical regions and also with long summer days in more temperate latitudes. The rising thermal of air cools to dew point and at that point cumulus cloud development begins. If the release of latent heat keeps the air in the thermal warmer than the surrounding air, it will continue to rise and towering cumulo-nimbus clouds develop. Such a condition is termed instability and it is associated with hurricanes, thunderstorms and intense precipitation such as hail or rain (***Resource 81a***).

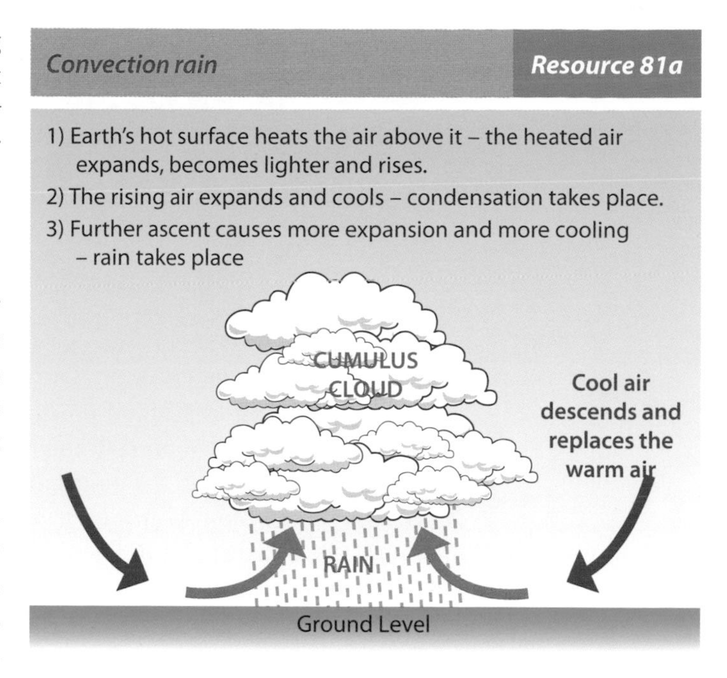

Global distribution of precipitation

Resource 81b shows the global distribution of precipitation. World precipitation totals vary from virtually zero in the Atacama Desert (Peru) to 30,000 mm in exceptional parts of the Himalayan slopes of Northern India. Northern Ireland receives an annual average of around 1000 mm while 250 mm is regarded as the boundary of arid regions.

At a global scale, regions of low pressure, including the Equatorial zone, have higher rainfall totals while high pressure areas such as the Polar Regions and both tropics are associated with low rainfall and arid conditions.

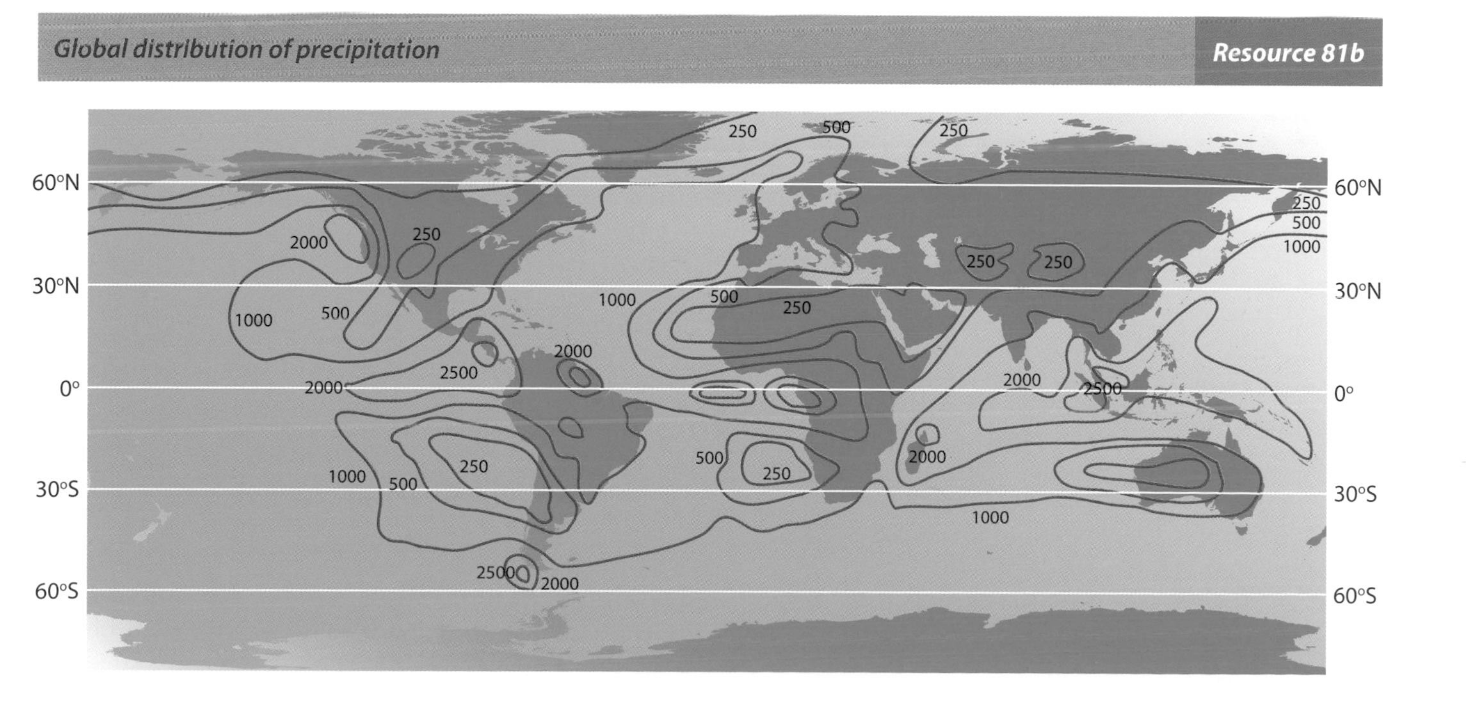

3B MID-LATITUDE WEATHER SYSTEMS

The regions between 30° and 60°N and S of the equator are referred to as the mid-latitudes and they contain the boundary zones between the world's surplus energy budget and its deficit regions. Consequently, the mid-latitude is the region where the critical horizontal energy exchange must occur. At ground level this involves large-scale spiralling weather systems called depressions and anticyclones, which are the creation of large-scale waving westerly air flows in the upper troposphere.

Frontal depressions

The British Isles lie beneath an interface, known as the Polar Front, between the warm moist tropical air from the south and the cooler air from the north.

Air masses

The British Isles are influenced by different air masses. An air mass is a large body of air in which there is great uniformity in its characteristics. Over hundreds of km^2 on the surface, its temperature and moisture content remains similar.

Air masses form when air lies over a region for a long time. Because the atmosphere gains its heat and moisture from the surface below it, air masses reflect these conditions. Air above a hot desert becomes hot and dry, whereas over a polar sea it will be cold and moist.

When an air mass moves location it brings the conditions of its source region to the new location. As **Resource 82** shows, the British Isles is regularly influenced by four major air masses.

Two from further North – **P**OLAR
Two from further South – **T**ROPICAL
Two from the West (Oceanic) – **m**aritime
Two from the East (Land) – **c**ontinental

Resource 82 *The four main air masses that influence weather in the British Isles*

Polar maritime (Pm)
A common air mass.
Cool conditions at any time of year. Heavy showers and storms.

Polar continental (Pc)
In winter very cold, dry air and it may bring snow to the east. It can last for several days. In summer it brings warm air.

Tropical maritime (Tm)
A common air mass in the British Isles. Very mild and wet in winter with thick cloud. Mild to warm in summer.

Tropical continental (Tc)
Hot and dry – normally occurs in summer and is relatively rare

Mid-latitude frontal depressions

The depression is a huge mass of spiralling air up to 2,500 km across, which involves two distinct air masses. More than any other feature, the depression illustrates the exchange of energy by warm, moist air from the tropics with the cold air from the Polar regions. The UK lies in a zone that is dominated by the west to east passage of about 30–35 depressions each year. These powerful features form over the Atlantic and pass, often within 24 hours, over the British Isles. Each depression consists of a body or sector of warm air, from the tropical region of the Atlantic Ocean, (**Tm air**) being gradually surrounded and lifted off the ground by cold polar air (**Pm**). The whole system has surface winds moving anti-clockwise and inwards towards the centre.

The zone between the warm and cold air masses is called a **front**. This term is used because they were first described during World War I when armies faced each other across the Western Front in France. The front is described as being **warm** if, when it passes, you enter warmer air or **cold** if when it passes, you enter cold air. At the front the cold, polar air is undercutting the warm air forcing it to rise. As the air rises it expands, cools, condenses and cloud forms, eventually bringing rain. Fronts are not vertical lines but slope gently upwards; the warm front in advance of the depression, and the cold front behind the system.

Origins

While meteorologists have long recorded depressions and their weather patterns, it is only in the last few decades that it has been shown that their formation is not determined by the conditions on the ground but rather by the flow in the atmosphere 8–10 km above. At this level, fast-moving and waving winds called the Rossby waves or upper westerlies prevail. It appears that where these winds slow down, air moves towards the ground to form high surface pressure and where they speed up (accelerate) they draw air upwards, creating a low pressure system on the surface (***Resource 83a*** and ***83b***).

Weather sequence

While no two depressions are exactly the same many have similar life cycles and bring a common pattern of weather. In the British Isles most travel from the south-west and sweep across the region. Depressions bring cloud, rain and strong winds especially when the Warm and Cold fronts pass over. Depressions last between five to ten days by which time the cold air lifts the warm tropical air off the ground (an **Occlusion**) and the whole system dies away. Depressions rarely come singly; often several follow each other as a 'family' of depressions. The diagrams (***Resource 83b***) show the sequence of weather associated with depressions, including changes in cloud type, coverage, rainfall, and wind speed and direction. This sequence and the reasons for the changing conditions need to be known.

The relationship between the upper westerly winds and surface weather conditions — ***Resource 83a***

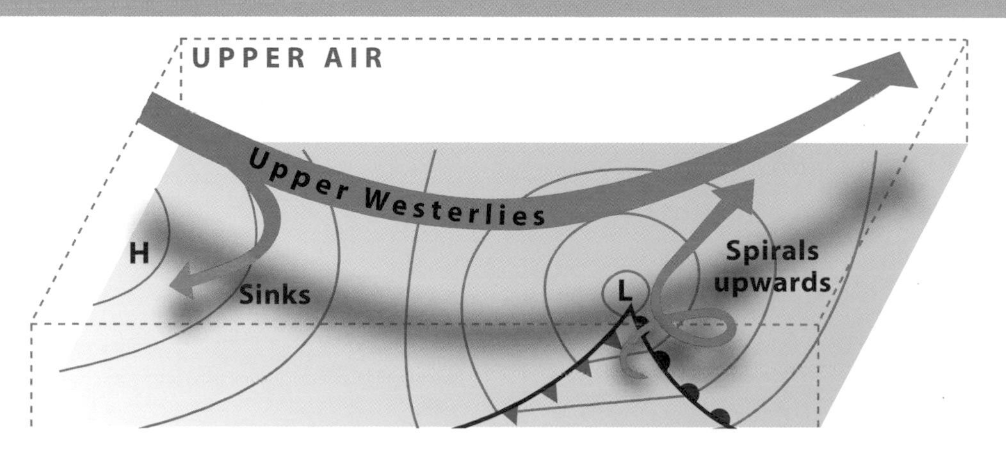

Resource 83b *Weather map showing a mid-latitude frontal depression*

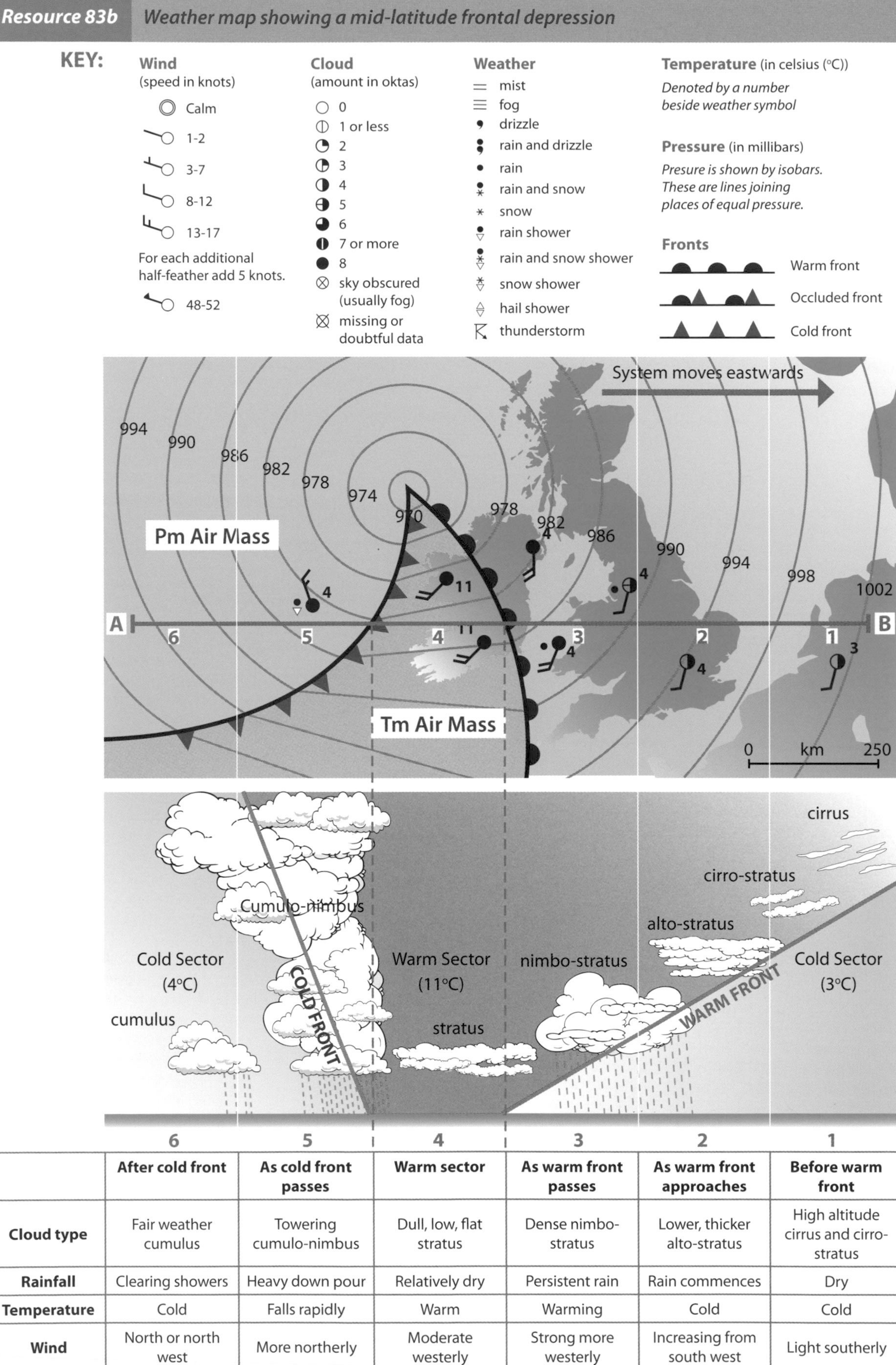

	6	5	4	3	2	1
	After cold front	**As cold front passes**	**Warm sector**	**As warm front passes**	**As warm front approaches**	**Before warm front**
Cloud type	Fair weather cumulus	Towering cumulo-nimbus	Dull, low, flat stratus	Dense nimbo-stratus	Lower, thicker alto-stratus	High altitude cirrus and cirro-stratus
Rainfall	Clearing showers	Heavy down pour	Relatively dry	Persistent rain	Rain commences	Dry
Temperature	Cold	Falls rapidly	Warm	Warming	Cold	Cold
Wind	North or north west	More northerly	Moderate westerly	Strong more westerly	Increasing from south west	Light southerly
Air pressure	Rising	Rising	Steady	Falling	Falling 1002–988	Falling High (1002)

The effects of low pressure systems at the regional scale

CASE STUDY – the February 1994 Storm

Low pressure systems present two possible sources of hazard for people and the economy: intensive rainfall and high winds. On the 2nd February 1994 an embryo low pressure system developed over the Atlantic to the west of Ireland. Over the next two days it moved eastwards and intensified into a deep mid-latitude depression. Its route can be plotted on the accompanying map using the table of central low pressure location and value. The trailing cold front rapidly overtook the system's warm front, lifting it off the ground to form an occlusion. The resulting single band of cloud was the source of intensive sleet or rain, producing over 25 mm for County Down and County Antrim. During the 3rd February much of the island experienced gale to storm force winds with gusts exceeding hurricane force around the coast. On the exposed Donegal coastal promontory of Malin Head gusts peaked at over 90 mph, while on the east coast near Bangor 80 mph winds were recorded (***Resource 84***).

Impacts

Two deaths in Ireland were attributed to the storm – one a farmer electrocuted by fallen high voltage cables and the other the result of someone falling in the darkness of a power cut in County Antrim. However, at sea, the 27 crew of a cargo ship, the Christinaki, were lost when it sank, overwhelmed by 15 m high waves. Across Ulster, from the Rosses of Donegal to the coast of County Antrim a series of events caused the largest failure in the province's electricity supply in over 50 years. Storm events as diverse as falling trees to coastal salt deposition on lines at Kilroot power station meant at least 400 000 homes lost their supply. With almost 70% of Northern Irish homes in darkness, safety and security were threatened. While half the homes had their supplies reconnected within a day, some had to wait up to a week for normality to be restored. Such a sudden failure of the electricity supply is merely an inconvenience to some, for others – the elderly and ill – it is potentially life-threatening and for some businesses, catastrophic. Travel was disrupted by numerous fallen trees blocking roads and rail lines. Ferry links and air travel services to and from the region were seriously reduced on the 3rd and 4th February. The high winds also severely hindered the land-based telephone services in Down and Antrim.

Environmentally too, the storm proved a hazard, with localized flooding occurring on farmland near Newry, County Down, and the undermining of coastal defences south of Dublin by high waves.

The surface synoptic conditions of the February 1994 storm (12 noon Feb 3rd) **Resource 84**

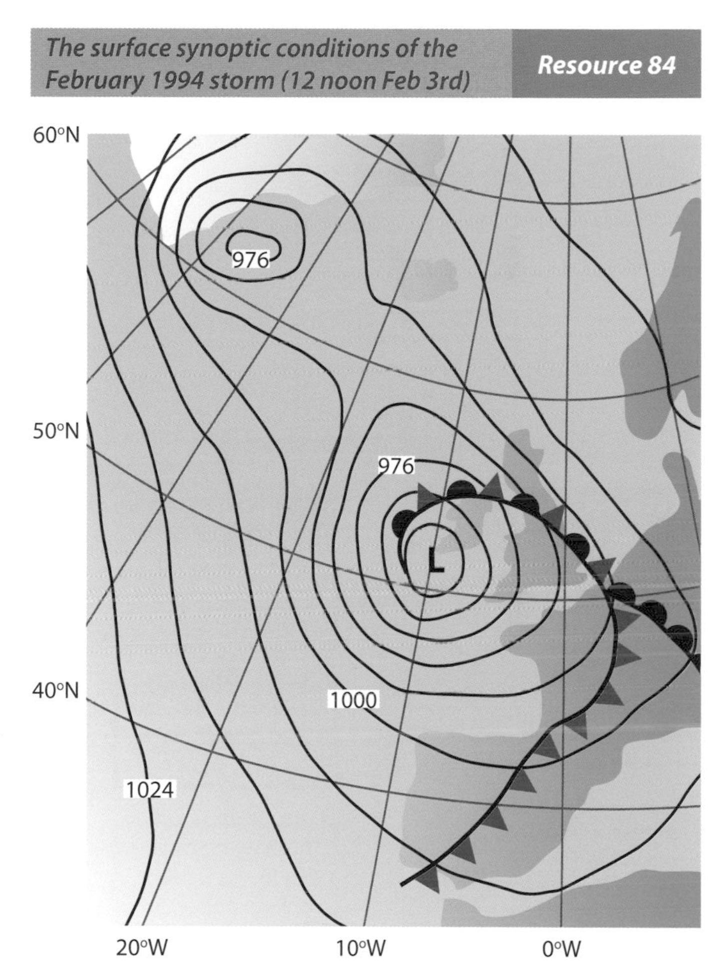

Date and time	Location (latitude and longitude)	Pressure (millibars)
Feb 2nd 0600	45°N 32°W	997
Feb 2nd 1200	47°N 25°W	986
Feb 2nd 1800	48°N 19°W	964
Feb 2nd 2400	49°N 14°W	954
Feb 3rd 1200	51°N 10°W	950

Anticyclones

These are large, 3,000 km in diameter, masses of subsiding air sinking from the upper troposphere, around 8 km high towards the surface. At ground level the air slowly spirals away in a clockwise direction (northern hemisphere). As the air descends it warms, and consequently the air is dry, so clouds and rain rarely form. Anticyclones are larger than depressions, involve only one air mass and move more slowly, lasting for several days or even weeks. The pattern of weather associated with depressions is similar throughout the year but anticyclones produce contrasting surface weather in summer and winter (***Resource 85***).

Resource 85 *Contrasts in summer and winter weather under an anticyclone*

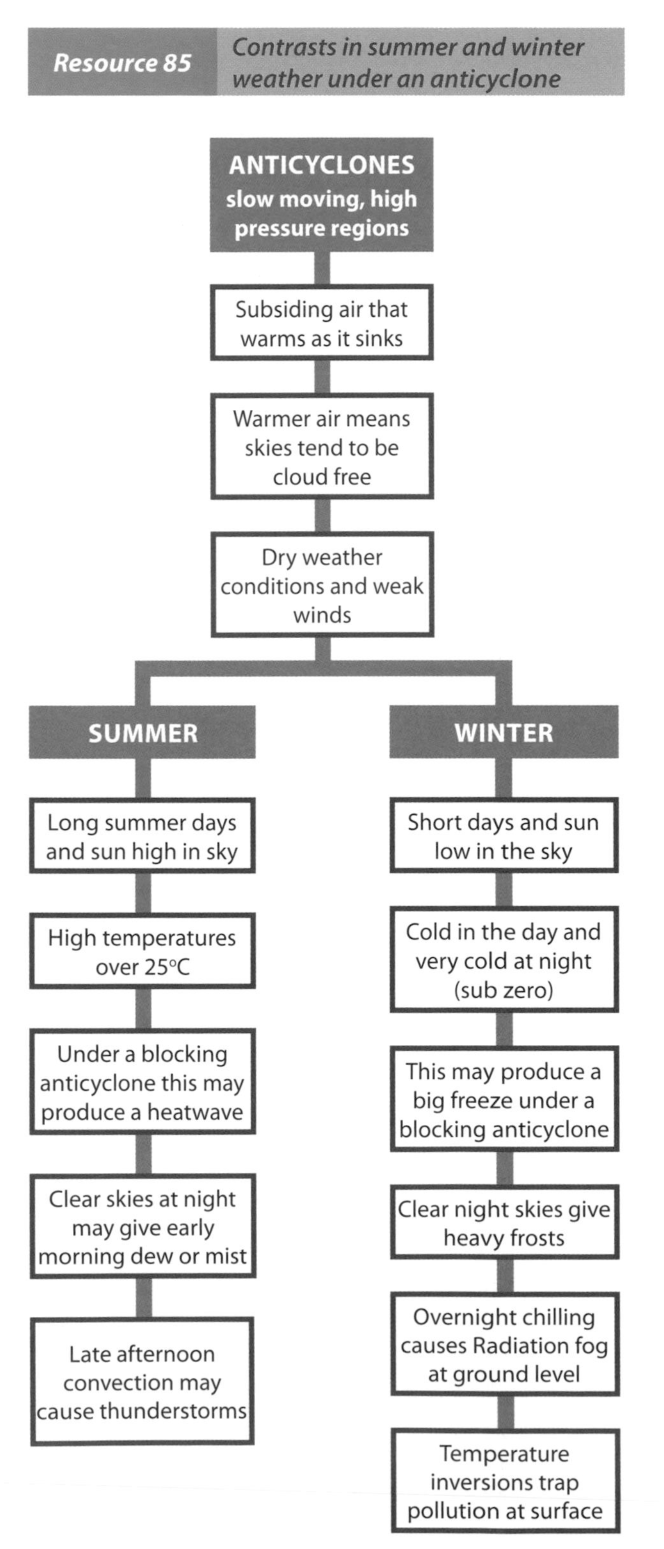

Formation

As mentioned earlier, anticyclones – like depressions – are the product of westerly winds moving in the upper troposphere. Where these air flows decelerate, air is forced to move downwards (subside) to form a surface high.

Summer Anticyclones

In nature summer and winter anticyclones:

- are marked by a central area of subsiding air that warms up, preventing rising air;
- produce calm or gentle surface wind that blows outwards and, in the northern hemisphere, clockwise;
- have clear skies allowing long hours of solar insolation;
- create dry conditions as uplift is restricted by the subsidence;
- may last for several days or even weeks, under a blocking system.

The weather produced in a summer anticyclone involves:

- hot sunny days, 25°C or 30°C in southern Britain;
- overnight cooling of the ground chills the air at the surface, producing dew or mist which normally burns off as temperatures rise the next day;
- local winds can develop including on-shore winds during the day;
- in late afternoons hot air at the surface may rise by convection as a thermal to form vertical cumulus or cumulo-nimbus clouds, causing thunderous rain or hailstorms.

The possible impact of summer anticyclones on people :

1. In a blocking anticyclone situation a large area of high pressure remains in place for days or weeks, forcing the depressions that move in from the Atlantic away to the north or south of their usual path. Under these circumstances and if the air in the anticyclone is Tc (Tropical continental) in origin, high temperatures and rapid evaporation can produce heat waves and drought. Both are at least inconvenient and potentially dangerous to health and the economy. Forest and heath fires can threaten ecosystems and property.

2. The light breeze or calm conditions of blocking anticyclones and the trapping of cool air near the ground can trap pollution. Vehicle exhaust and industrial emissions may react chemically with sunlight to produce photochemical smogs. These and the concentration of ozone at ground levels can be a hazard to the health of young children and people who suffer from asthma or other such ailments.

Winter Anticyclones *(Resource 86)*

The clear sky caused by warming subsiding air produces very different weather conditions in a winter anticyclone. Long nights of radiation heat loss often produce sub zero temperatures, hoar frost (frozen dew), and persistent fogs. These conditions can persist for days under a blocking system as the sun is too weak to warm the land surface. The disruption to everyday life can be severe. Elderly people are at risk from hypothermia as they may not be able to afford adequate heating in their houses. In homes, water in unprotected pipes may freeze, and as the ice expands the pipes break. Then once this ice thaws, homes are flooded. One hundred homes in Eastern Scotland were flooded in this way during the Christmas period of 2000. Icy pavements force people to stay indoors or risk injury. Transport by road, rail and air may be curtailed by both ice or fog. The worst effects are normally seen in inland areas because in coastal regions the proximity of the sea helps moderate temperatures. In both central Ireland and England several days of persistent fog and sub zero temperatures are not uncommon between December and February. The most severe winter on record in the last 300 years was in 1963 when temperatures for two consecutive months remained below zero.

A winter anticyclone over western Europe **Resource 86**

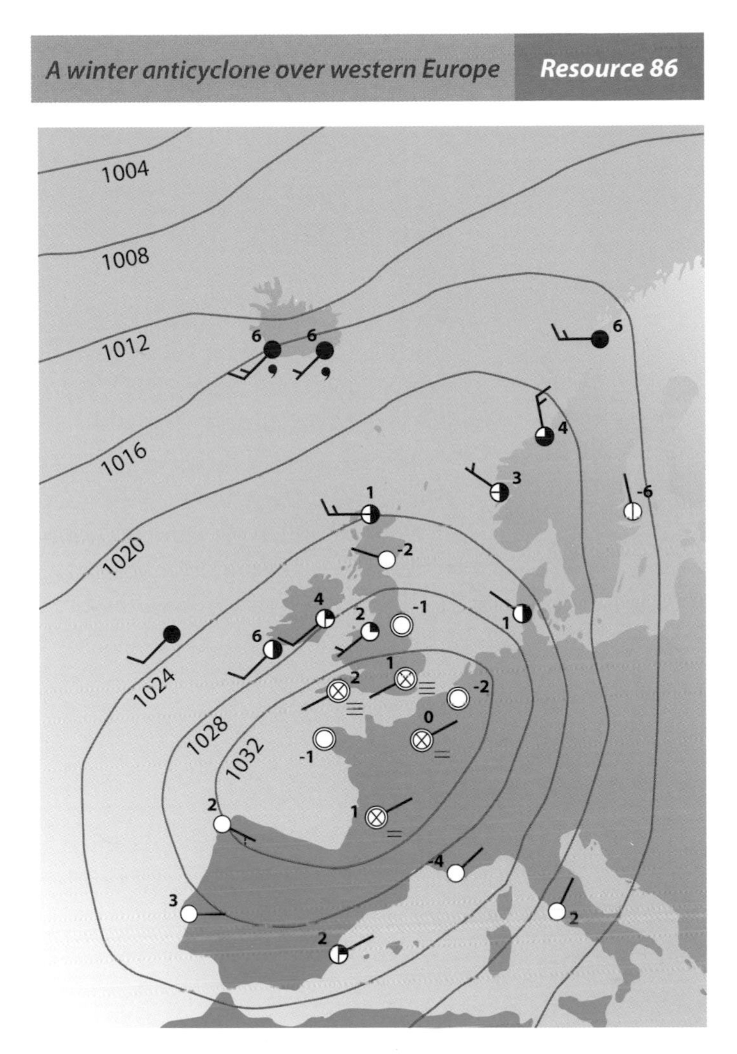

Contemporary reports of the winter of 1963 in Ireland and the UK

"Parts of the River Shannon were frozen over and today Limerick is a city of burst water pipes and frozen oil heaters. Several factories and schools closed down due to oil freezing in the boilers."

"In early February it was reported that coal stocks were seriously depleted in the city and some firms had imposed rationing. Hundreds of homes were without water due to frozen pipes and there have been many homes damaged due to burst pipes. Several tried melting the snow but found it was not suitable for making tea."

"The lowest minimum reported in England was -20.6C at Hereford on the 23rd and then a maximum of only -8C at Ross-on-Wye the next day. There was a snowdrift 25 feet deep on Dartmoor on the 21st. There was much freezing fog on the 24th. For the first time since 1947, there was pack ice on large estuaries such as the Solent, Mersey, and Humber. At Eastbourne the sea was reported as frozen to an extent of 100 feet offshore for a length of 2 miles."

3C EXTREME WEATHER EVENTS

Hurricanes

"In Hereford, Hartford and Hampshire hurricanes hardly ever happen."

This line from the musical 'My Fair Lady' was initially designed to improve Eliza's cockney habit of dropping the letter 'H'. It also happens to be meteorologically accurate as hurricanes do not occur in Southern England.

Hurricanes are severe storm events formed in the tropics, correctly termed Tropical cyclones. Hurricane is the Atlantic Ocean name, elsewhere they are known as Typhoons (Pacific Ocean), Cyclones (Indian Ocean) or even Willy-willies (Australia). They are intense, low pressure systems that develop over warm tropical oceans and are associated with torrential rainfall and high winds. Their central pressure may be as low as 880mb (at least under 970mb) and fierce winds spiral around a calm central zone. They are about 500–1000 km in diameter, with the strongest winds confined to a 2–300 km wide belt. At first glance they may appear similar to intense mid-latitude depressions but they have no fronts and are much more severe.

A world map showing the distribution and path of hurricane tracks reveals some very clear facts (***Resource 87***).

- Tropical Cyclones form within the tropics but not within 5° of the Equator.
- They originate on the surface in east ocean basins
- They track westward and away from the equator
- They only develop where sea surface temperatures exceed 26°C to a depth of 60 m.

A calendar of such events shows they are common in late summer and autumn.

While atmospheric disturbances in the tropics are common, and some develop into storms, only a few, perhaps one in ten, develop into full blown hurricanes.

Resource 87 *World map of common hurricane tracks*

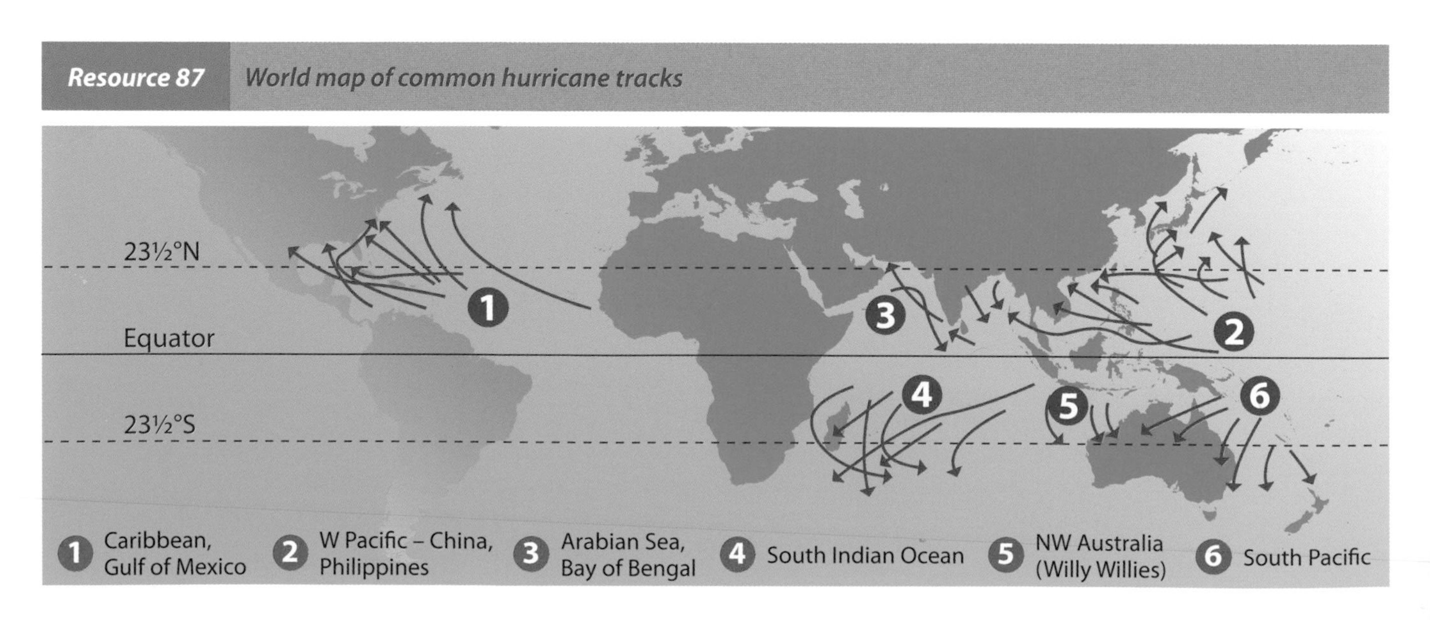

Formation

'Hurricanes have their heads in the clouds and their feet in hot water.'

Common features in the sequence of hurricane formation:

1. Beneath easterly air flows in the upper atmosphere, surface air is drawn upwards;
2. Strong convection (vertical) uplift of hot damp air develops;
3. Beyond 5° of the Equator the Coriolis Force provides the force for air to spiral;
4. Clouds form into spiral bands;
5. The rising air carries stored energy from evaporation, which is released by condensation to form clouds and rain (Up to 200 million tonnes of seawater can be re-cycled each day by a hurricane);
6. This process strengthens the winds until they attain hurricane force – sustained winds over 120 km/h;
7. Cumulo-nimbus clouds extend up to the top of the troposphere, 12 km high;
8. As hurricanes mature they form an unique, distinct feature – the eye, a central zone of light winds and clear skies;
9. In the eye, air is subsiding whereas everywhere else it is rising;
10. At the surface, winds spiral inwards towards the low centre (this is called cyclonic flow – anticlockwise in the Northern hemisphere);
11. Aloft the air spirals outwards and clockwise (Northern hemisphere);
12. Hurricanes only decay and die when their supply of energy fails. Normally this happens when they move over land, or over cooler water away from the tropics (***Resource 88***).

Structure of a mature tropical hurricane ***Resource 88***

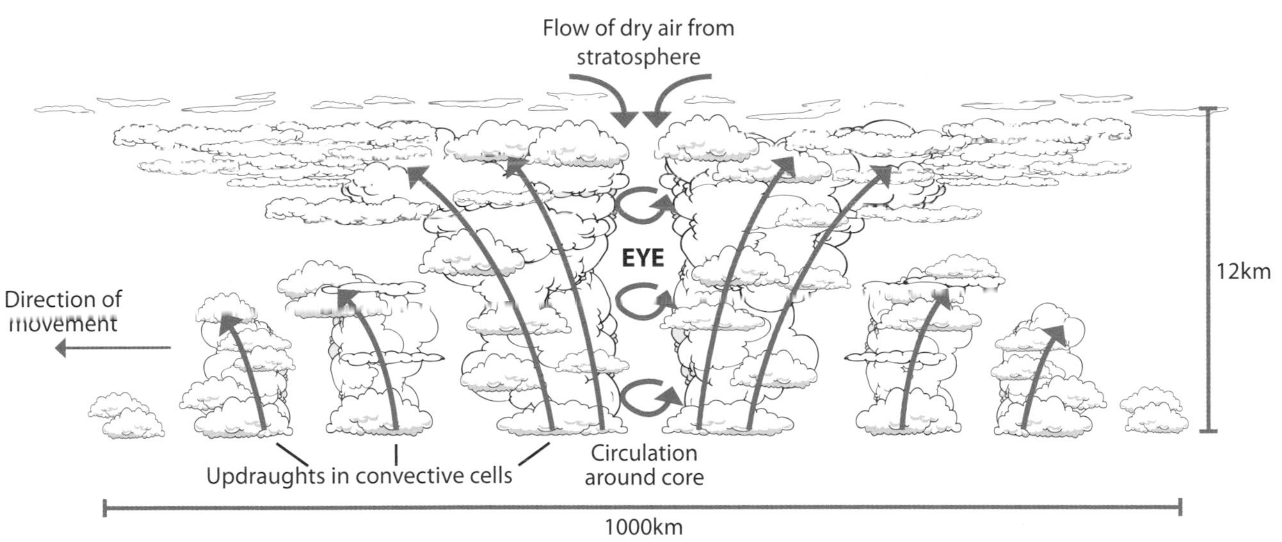

Structure

As the diagrams illustrate, hurricanes are large, dynamic three dimensional features. The key structural features are a series of circular bands of cloud, formed by vertical uplift, the central clear eye of subsiding air and the inward surface air spiral and outward 'exhaust' air spiral in the upper atmosphere.

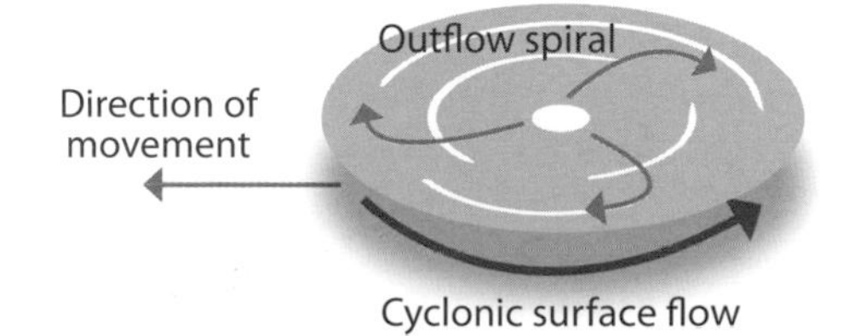

Naming hurricanes

Each season in the Atlantic each hurricane is given a name in alphabetical order. There are six lists of names used in rota. For many years only women's names were used. From 1979 men's names were added and now Hispanic, as well as Anglo-Saxon, names are listed. Names are dropped from the lists after they are given to storms with severe impacts. In the last few years the names Wilma, Stan, Juan, Ivan, Charley and Jeanne have been retired in this way.

The effects of hurricanes on people and property

Tropical cyclones have three distinct destructive elements: intensive rainfall, the powerful winds and, the most dangerous of all, the **storm surge**.

A storm surge is caused when the intense low pressure of a hurricane allows sea level to rise up to 10 metres above its normal level. When this is driven on-shore by the hurricane force winds, and especially if it coincides with local high tide, the outcome may be catastrophic (***Resource 89***). The tropical cyclone named Nargis that made landfall in the Irrawaddy river delta in Burma, May 2008, swept away over 70,000 people, left one million homeless and destroyed food supplies for millions more.

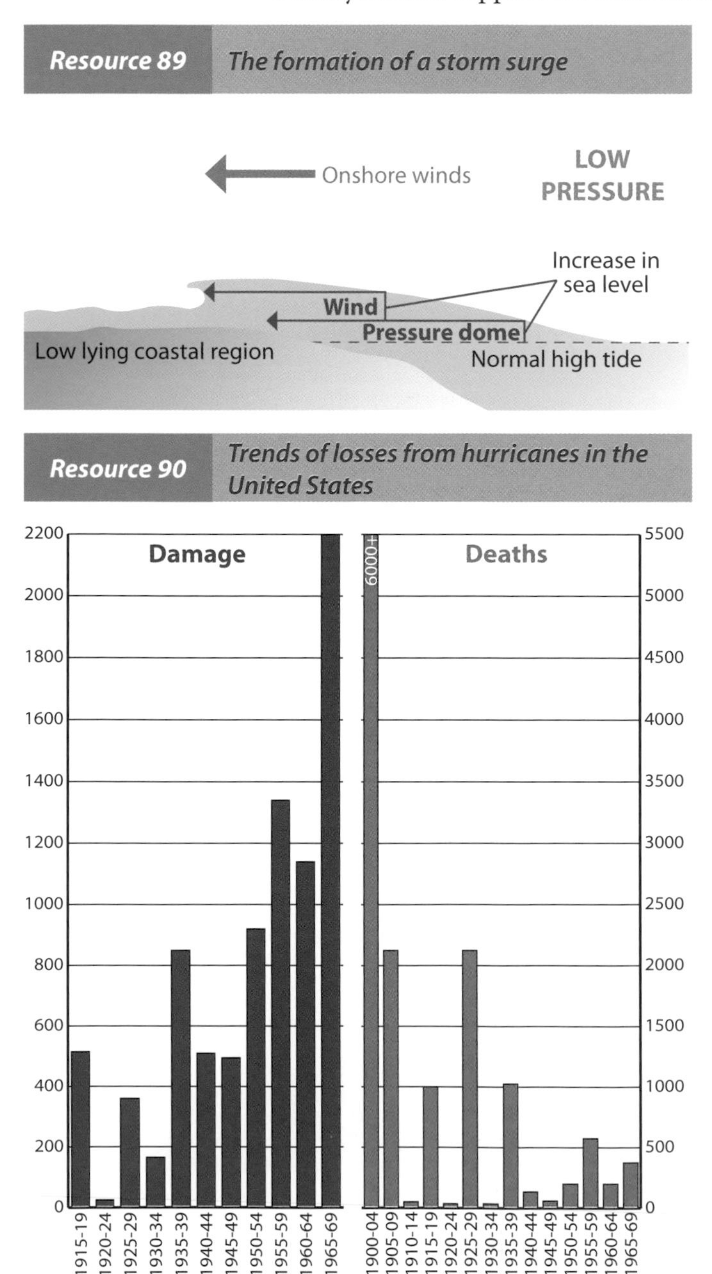

Resource 89 *The formation of a storm surge*

Resource 90 *Trends of losses from hurricanes in the United States*

Hurricanes are classified according to their wind speed with a 5 being the most potent. The intensity of the winds can turn any loose object into a destructive weapon, including dustbins, street furniture, cars and grounded aircraft. Poorly built housing in LEDC nations often suffer; for example, Hurricane Gilbert destroyed the homes of 500,000 Jamaicans in 1988. In places where the storm passes directly overhead, after the battering of winds from one direction and the brief respite of the calm eye, a second period of hurricane force winds from a different direction can demolish weakened structures.

The intensive rains accompanied by storm surges can flood coastal regions and along low-lying coasts – such as Bangladesh or Florida – this may extend for miles inland. As with many natural hazards, their impact depends on both the physical and human geography of the region concerned. Today, due to prediction and evacuation, Atlantic hurricanes that take many lives in the LEDC nations of the Caribbean may have a small death toll in the southern states of the USA (***Resource 90***); however, the property bill may be enormous. Until **Katrina** in 2005, the most destructive and costly hurricane event in the USA had been **Hurricane Andrew** in 1992. Then, despite leaving 250 000 people homeless and a bill for $27 billion, the direct death toll was only 26.

CASE STUDY: Katrina 2005 - Introduction and Effects

Introduction

Around the 23rd August 2005 a small tropical depression formed over the warm waters off the coast of Florida; within a week it had deepened to form a Category 5 hurricane that destroyed the Louisiana city of New Orleans, killed over 1,700 people and caused in excess of $100 billion in damages. It remains the most costly hurricane ever and took more lives than any USA storm since 1930.

As the name suggests, Katrina was the 11th tropical storm that year; a year that produced record numbers of storms, hurricanes and in particular category 5 hurricanes.

Katrina's path took it over the southern tip of Florida (***Resource 91***) but it was over the warm waters of the Gulf of Mexico that the storm deepened to exceed wind speeds of 282 km/h, thereby creating a category 5 hurricane. Despite dropping to a category 3 event before reaching land over the delta of the Mississippi River, Katrina was a huge and powerful storm. The accompanying storm surge varied from 3 to 10 m in height and along with the powerful winds and intense rain, building damage and flooding was extensive.

Path of Hurricane Katrina **Resource 91**

People

An area the size of the UK was impacted by Hurricane Katrina's destruction. Within this it was the city of New Orleans in Louisiana that suffered most and hit the headlines. Built on the floodplain of the Mississippi River and lying mostly below sea level, much of the city rapidly flooded when three protective levees along the river and a lake gave way. Over one thousand of the city's 460,000 residents lost their lives, while the homes of most were destroyed or seriously damaged. Over half a million US citizens became refugees in their own country and half of these said they were unlikely to return to New Orleans as a result of their trauma. It was the poor, working class population, often without insurance, who lost most in the storm. Katrina rewrote the population distribution map of the region; the state population fell by over 8% and within Louisiana, areas outside New Orleans grew in numbers as the city's people fled. In the coastal state of Mississippi 109,000 were made homeless and over 230 died. The Federal Emergency Management Agency (FEMA) later helped 108,000 people with unemployment benefit.

Property

- Housing damage and destruction was widespread and stretched up to 100 km from the storm centre. At least 100,000 temporary homes were needed across the region.
- Services in New Orleans were severely damaged. Even six months after the event the city centre had no functioning sewerage system, and gas and electricity supplies were not available.
- Agriculture suffered heavy losses including the death of nine million poultry in Mississippi, while in the same state the dairy industry lost $12 million.
- The forestry industry in the region was affected; over one million acres of forest was destroyed. Due to Katrina, the total financial loss to the timber industry is estimated at $5 billion.
- The US Corps of Engineers, the group responsible for river management, estimated that in Louisiana there were over 70 million cubic metres of storm debris to be removed.

- In the aftermath of Katrina hundreds of thousands of local residents were left without work. This will have a trickle-down effect with lower income tax paid to local governments. It is estimated that Katrina's total economic impact in Louisiana and Mississippi will exceed $150 billion.

Protective Measures for people and property

Statistics would suggest that in general in the USA, protection from hurricanes is successful in terms of lives, if not property. Despite decades of increasing population in hurricane-prone regions of the nation, essentially the south eastern states, the storm death toll has generally fallen. Katrina was a shock for a country where deaths from hurricanes was now a rarity. Property damage paints a different picture; with an increasing number of inhabitants – and moreover wealthier ones, especially on the storm-surge-prone coasts of Mississippi and Florida – the fixed property at risk has grown. Homes, yachts, cars and other possessions must often be abandoned, though secured as much as possible, during hurricane events. So while prediction and evacuation works well for the population, it is not the answer for their possessions (***Resource 90***).

1. Prediction and Warning

As noted earlier, hurricanes are events normally confined to the tropics and to the summer and autumn seasons. Given these spatial and temporal controls, along with the highly visible nature of hurricane systems, especially from space, they are easily identified and tracked by satellite imagery. Accurate prediction has the potential to save lives if effective evacuation is possible. In the USA, schools often act as hurricane shelters or people move to a safe distance inland or onto high ground. Public evacuation is normally undertaken by private transport; cars use published maps of safe evacuation routes with local police monitoring and clearing housing areas at risk. The low number of hurricane-related deaths in recent decades in North America is largely due to such prediction and evacuation. Prediction also enables homeowners to protect their properties by securing doors and windows, and cutting off potentially dangerous gas and electricity supplies.

The US National Hurricane Centre (NHC) has invested heavily on the most modern equipment and annually spends tens of millions of dollars monitoring and tracking tropical storms. Using ground level observation, along with dedicated weather aircraft flights and satellite imagery, data is constantly gathered.

Case Study - Katrina

In August 2005 Hurricane Katrina was precisely tracked from the initiation to the conclusion of its short but violent life. Katrina's scale, strength and landfall locations, first in Florida and later in the delta region, were accurately predicted.

Despite the years of investment and numerous warnings of the threat that a major hurricane posed to the people of New Orleans, evacuation plans proved inadequate. Following the event, a Congressional Report described the government's response to Katrina's impact as a national failure, stating that "... clumsiness and ineptitude ... characterised behaviour before and after this storm." Part of the problem was that over 25% of households in New Orleans did not own cars and had no easy way of evacuating the city. Residents also refused to leave their homes, fearing looting in their absence; some paid for this inaction with their lives.

2. Education

Knowledge about, and an accurate perception of, the threat of hurricanes are keys to helping people save themselves and their possessions. Local government uses schooling, internet sites and hurricane-preparedness events to raise the public's perception of how they can prepare for and reduce risk before, during and after the storm.

Case Study – Katrina

It appears that despite the official awareness of the risk to New Orleans, most residents had little fear that the city would be badly impacted.

3. Building codes

One approach to the threat of hurricanes along low-lying coasts is Zoning. This is when specific 'at risk' areas are set aside to be left undeveloped, while in other areas any development that is permitted is constructed and designed to cope with the hazards. In the USA such building codes are well developed for both the public and private sectors.

Case Study – Katrina

In New Orleans the city's situation – much of it lies below sea level – meant that the flooding was so severe that even the strict building codes did little to reduce the impact. In the aftermath some people have suggested that the city should not be rebuilt on such a dangerous site. Others have suggested that the worst affected areas should not be redeveloped but that a smaller population should be catered for, and flood-prone regions given over to low risk land uses such as parkland or sports pitches.

4. Coastal and river engineering

In many regions of the coastline of south eastern USA the authorities have attempted to use hard engineering defences and soft engineering beach feeding to reduce the impact of storm surges. These may be effective but, in common with all such developments, they are designed with a risk level in mind. Sea walls can deflect and reduce the impact of storm-surges, while wide sandy beaches are an effective method of using up wave energy on the shore.

Case Study – Katrina

The levees and defences along the Mississippi and canals in New Orleans were built to withstand a hurricane up to a category 3, anything beyond that would cause flooding. The cost and practicality of protecting the city and the rest of the region from the worst possible event is prohibitive. Currently the US Corp of engineers, who have the task of rebuilding New Orleans defences, is restoring them to their previous height while the city's mayor wants them raised and improved to deal with category 5 hurricanes.

5. Insurance

Personal and household insurance cannot prevent death or property loss but it can provide compensation.

Case Study – Katrina

Most households in New Orleans could not afford the premiums required to cover hurricane damage and this explains why in the hardest-hit, working class regions of the city many of the residents have chosen not to return.

Resource 92 *The Saffir-Simpson Hurricane magnitude scale*

The Saffir-Simpson Hurricane Scale is a 1–5 rating based on the hurricane's intensity. This is used to give an estimate of the property damage and flooding expected along the coast from a hurricane landfall. Wind speed is the determining factor in the scale, as storm surge values are highly dependent on the geomorphology of the local coastline.

Category One Hurricane:

Winds 119–153 km/hr, Storm surge generally 1–1.5 m above normal. No real damage to building structures. Damage primarily to unanchored mobile homes, shrubbery, and trees.

Category Two Hurricane:

Winds 154–177 km/hr, Storm surge generally 1.6–2.5 m above normal. Some roofing material, door, and window damage to buildings. Considerable damage to shrubbery and trees with some trees blown down. Damage to mobile homes, poorly constructed signs, and piers.

Category Three Hurricane:

Winds 178–209 km/hr, Storm surge generally 2.6–4 m above normal. Some structural damage to small homes and buildings. Damage to shrubbery and trees with foliage blown off trees and large trees blown down. Flooding near the coast destroys smaller structures with larger structures damaged by battering from floating debris. Land under 1.5 m above mean sea level may be flooded for up 13 km or more inland.

Category Four Hurricane:

Winds 210–249 km/hr, Storm surge generally 4–6 m above normal. Shrubs, trees, and all signs are blown down. Complete destruction of mobile homes. Extensive damage to doors and windows. Low-lying escape routes may be cut off by rising water 3–5 hours before arrival of the centre of the hurricane. Land lower than 3 m above sea level may be flooded, requiring massive evacuation of residential areas as far inland as 10 km.

Category Five Hurricane:

Winds greater than 249 km/hr, Storm surge generally greater than 6 m above normal. Complete roof failure on many residences and industrial buildings. Some complete building failures with small utility buildings blown over or away. All shrubs, trees, and signs blown down. Complete destruction of mobile homes. Major damage to lower floors of all structures located less than 5 m above sea level and within 500 m of the shoreline. Massive evacuation of residential areas on low ground within 8–16 km of the shore may be required.

Only three Category 5 Hurricanes have made landfall in the United States since records began: The Labor Day Hurricane of 1935, Hurricane Camille in 1969, and Hurricane Andrew in August, 1992. Hurricane Katrina, a category 5 storm over the Gulf of Mexico, was still responsible for over $100 billion of property damage when it struck the USA Gulf Coast as a Category 3.

Websites for research concerning Tropical Cyclones (Hurricanes)

General

For background, archive and current material the National Hurricane centre (NHC) site:
www.nhc.noaa.gov

An alternative location for Hong Kong typhoons:
www.hko.gov.hk/informtc/informtc.htm

Case study related sites

Katrina's storm surge:
www.wunderground.com/education/Katrinas_surge_contents.asp

Hurricane preparedness week 2008:
www.nhc.noaa.gov/HAW2/english/intro.shtml

Images of new Orleans flooding:
http://www.nytimes.com/packages/khtml/2006/08/26/us/nationalspecial/20060827_REVISITED_FEATURE.html
or search in www.nytimes.com

Other Atmosphere resources

Geofile 530 – Hurricanes Katrina and Rita – The after effects (September 2006)
Geofile 500 –The 2004 Hurricanes season in the Caribbean and South USA (September 2005)
Geofile 493 – Depression Rainfall and its effects on the British Isles (April 2005)
Geo Active 342 – Weather patterns associated with depressions (September 2006)
Geo Factsheet 162 – Hurricanes – A Predictable Hazard? (2004)
Geo Factsheet 146 – Depressions (2003)
Geo Factsheet 100 – Anticyclones – A Potential Hazard (2001)

Websites for animations of General Circulation

http://wxpaos09.colorado.edu/atoc1060/overheads/23_WeatherPat.html
http://wxpaos09.colorado.edu/atoc1060/overheads/34_Coriolis.html
http://wxpaos09.colorado.edu/atoc1060/overheads/25_JetStream.html
http://wxpaos09.colorado.edu/atoc1060/overheads/01_EarthSun.html
http://wxpaos09.colorado.edu/atoc1060/overheads/04_GlobalWind.html
or http://esminfo.prenhall.com/science/geoanimations/middle.htm

AS 2 HUMAN GEOGRAPHY

1A POPULATION DATA

Population Geography involves the study of all aspects of the spatial interactions between man and his surroundings. In particular, Population Geographers are concerned with the dynamics of population change both in terms of growth and distribution. The patterns that human populations have created vary greatly over time and space. In order for Geographers to describe and explain these variations there is a need for accurate and reliable data. Two main sources available to Geographers are:

- The National Census
- Vital Registration

The National Census

The word 'census' comes from a Latin word meaning 'tax' but it has come to refer to a count of all of the population and those social and economic characteristics that can easily be counted. Censuses have been taken for many centuries as Governments have sought to collect information for many reasons. The ancient Egyptians collected information on their population numbers so that they could plan to build the Pyramids. The Romans carried out regular censuses throughout their empire in order to plan the regional distribution of power. It was one of these censuses ordered by Caesar Augustus that caused Mary and Joseph to travel to Bethlehem before the birth of Jesus.

History of census in the UK

In England, the Domesday Book, produced in the reign of William the Conqueror in 1086, was the first real attempt at conducting a national count of population, land and property. Again, the main reason for collecting this data was to inform the new king of the extent of his conquests in England. There were other piecemeal attempts at surveys of population from then on, but the idea of a national census to count all of the people did not come about until the end of the eighteenth century. This was a time of considerable social change in Britain – agriculture had improved, industry was developing, towns were increasing in size due to rural-urban migration and new advances in medicine were resulting in increased knowledge about sanitation and nutrition. There was some debate about the impact of these changes on population numbers but most agreed that there was a significant increase in population size.

It was against this background of change and uncertainty about the future that Thomas Malthus published his theory on 'The Principle of Population' where he stated that population growth, if left unchecked, would soon exceed the available food supply and resources. In this situation, Britain would face famine, disease and war. It became increasingly clear that a national census was required. Initially there were objections, as some feared that a census could reveal too much information to the enemies of Britain, but parliament passed the **Census Act** in 1800 and the first census was carried out in Britain in 1801 and in Ireland in 1821. This first census was conducted by door-to-door collectors and it was not until 1841 that an official registrar nominated a specific date when each head of household would complete a form, and this has remained the format of the UK census to the present time.

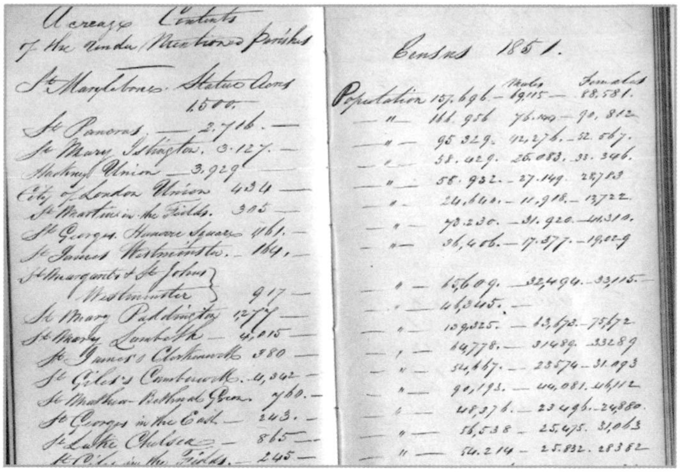

Since its introduction in 1801, a census has taken place in the UK every ten years, with the exception of 1941 during the Second World War. The information collected provides a unique snapshot of the social, economic and demographic conditions of all of the people at a specific time. This information enables central and local government, health and education authorities to target their resources more effectively and to plan housing, education, health and transport services for years to come.

The last census taken in the UK was in 2001 and already plans are well underway for the 2011 census. In England and Wales the census is carried out and processed by the Office for National Statistics. In Northern Ireland, the Northern Ireland Statistics and Research Agency performs a similar function. For a census to be worthwhile it has to be reliable and fully inclusive. The UK, like most MEDCs, has succeeded in gathering accurate and comprehensive information.

Success of UK census

Some reasons for the success of the UK census are as follows:

(1) The Government has made completion of the census legally binding and undertakes an expensive publicity campaign in the media to explain the importance of reliable and comprehensive information.

(2) The rationale behind each question is carefully explained and there is a helpline where further information can be obtained.

(3) Confidentiality of the information collected is assured. All of the information is processed in secure conditions and confidentiality is guaranteed by legal protection.

(4) Every effort is made to ensure full participation in the census and there is a substantial fine imposed on anyone refusing to complete the census form.

(5) There are helplines for those who have difficulty completing the census, especially those who are not fluent in English, and forms are provided in Braille for the visually impaired.

(6) The distribution of the census is carefully planned and every household in the country will receive its census package well in advance of the completion date.

(7) The census package comes with a detailed set of instructions.

In spite of all of this, there is still some opposition to the completion of the census forms. Amongst the reasons for opposition are suspicions of government trying to find out too much personal information, doubts about the rationale of the census, and the nature of the questions asked on the census form. Over time some of the questions have been modified and others have been added to more accurately reflect the changing conditions in society. For the first time the Northern Ireland census of 2001 asked people to state their nationality and religion, though neither of these questions was compulsory.

2011 census

Plans for the 2011 census are well under way and include a pilot test of the data collection process in 2007 and a pilot test of the census form in 2009. The 2007 test will use a number of additional questions to those used in the 2001 census. Some examples of the new questions include:

- National identity – to allow respondents to record their English,

Reference

A copy of the 2001 census form and details of plans for the 2011 census can be obtained at the following website:

www.statistics.gov.uk/census

Welsh, Scottish, Northern Irish, Irish or other identity

- Income – to collect level and sources of income
- Language – to collect information on proficiency in English, Welsh, and other languages. In Wales, people will be asked about the frequency of their use of the Welsh language.
- Second address – to identify the number of people with a regular second address and the purpose and frequency of its use
- Month and year of entry into the UK – to collect extra information about international migration

Modified questions to be tested include:

- Illness and Disability – expanded to collect information on the nature of illness and disability
- Marital or civil partnership status – expanded to include civil partnership equivalent for each marital status
- Each question included in the 2007 Census Test will be evaluated to assess the quality of information that can be collected and the public acceptability of the topic.

Following the test there will be a census rehearsal in 2009, before the proposed questions are put forward to Parliament later that year.

Censuses in LEDCS – lack of reliability

In LEDCs, census-taking has been much less successful both in terms of the reliability of the data collection process and in the frequency of the census itself. Some countries have not had a regular census (***Resource 1***).

There are many reasons to account for this lack of reliability:

(1) In many LEDCs their first experience of a census occurred during colonial times when censuses were carried out by the colonial rulers for the purpose of tax collection, conscription to the army or simply to gain information about the number of their subjects. As such, the census came to be seen by the indigenous population as a symbol of foreign power and domination. The local population sometimes reacted by providing inaccurate information.

(2) The inaccuracy of the census information is often an attempt to gain political advantage in areas where political power, parliamentary seats or budget allowances are determined by information gathered in a census. During the 1941 census in pre-partition India there was a tendency for the two main ethnic groups – Hindus and Muslims – to exaggerate their numbers because it was commonly believed that the country would be partitioned on ethnic lines. More recently, in the state of Kashmir, a boycott by some ethnic groups threatened to obstruct the operation of the 2001 census.

(3) Gross per capita income is often used to determine the amount of international aid allocated to LEDCs and therefore a large population could be seen as an

Resource 1 *Countries that have not taken a Census since 1990 or earlier*

Country	Year of last census	Census taken or planned
Burundi	1990	August 2008
Uzbekistan	1989	*
Somalia	1987	*
Dem. Rep. of Congo	1984	2010
Eritrea	1984	2009
Myanmar	1983	*
Togo	1981	2009
Afghanistan	1979	August 2008
Angola	1970	2010 or 2014
Lebanon	1970	*
Liberia	1984	April 2008
Western Sahara	1970	*
Djibouti	1960	*

* It is expected that a census will be held during the decade (2005-2014).

Sources: United Nations, Population and Vital Statistics Report (2007); and UN Statistics Division, 2010 World Population and Housing Census Programme: Census Dates for All Countries (http://unstats.un.org)

advantage. Several incidents of over-estimations of total population size have been reported across Africa. Gabon refused to accept the results of its census for its total population, which the government considered to be too small, even though it was ratified by the United Nations.

(4) In many ethnically diverse countries there is often a delicate balance between the numbers in each group and the allocation of power, and this has led to manipulation of the results in favour of the ruling group. In Nigeria the 1962 census was so disputed, and the results were deemed so unreliable, that a new census was called the following year, but it fared little better and it too was scrapped. In Lebanon no census has been taken since 1970 for fear of results-tampering and the resultant unrest that might follow between Muslim and Christian groups.

(5) Census-taking is expensive – the 2001 census in the UK cost £255 million. Central government considered such expenditure worthwhile in order to gain accurate and comprehensive information. Few LEDCs can afford such large sums of money.

(6) In LEDCs, literacy problems mean that many censuses are collected by door-to-door interviews. This can result in some people being omitted or inaccurately recorded. In certain cultures women may not be interviewed by men, resulting in further misrepresentation.

(7) The nature of the data collected is often hotly debated. In Nigeria, where census-taking is very infrequent, the last census was delayed for over a year because the Muslim leaders in the north of the country mounted strong opposition to the inclusion of questions on ethnicity or religion. At the same time, the Christian leaders in the south east of the country mounted opposition to the omission of such questions. In the end the questions on ethnicity were omitted leading to incomplete information.

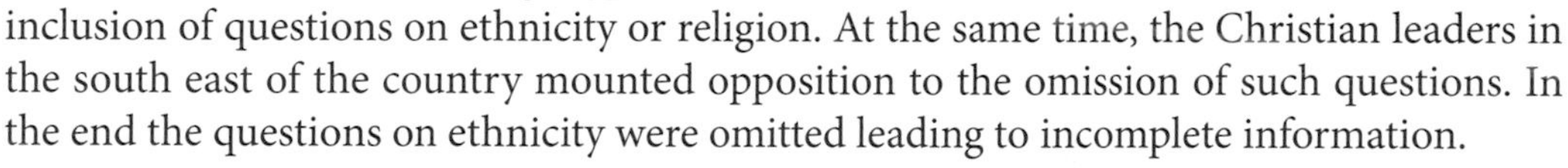

(8) There are many examples in LEDCs where the transport and communications are poor, resulting in some people being omitted from the census altogether.

(9) Wars and civil unrest are common occurrences in LEDCs and this can result in a census being postponed, cancelled or incomplete.

Informal settlement in Tripoli, Lebanon

Additional information can be found on the following websites:

www.prb.org/articles/2007/ObjectionsOverNigerianCensus.aspx

www.prb.org – *a number of very useful short articles on the 2006 census in Nigeria*

http://news.bbc.co.uk/1/hi/world/africa/4853562.stm – *Thumbs up for Nigeria's census*

http://news.bbc.co.uk/1/hi/world/africa/4512240.stm – *Nigeria's Counting Controversy*

www.iussp.org/Brazil2001/s30/S35_PO2_Ramachandran.pdf

www.statistics.gov.uk/census

www.prb.org/articles/2008/liberia.aspx – *Liberia's first census in 24 years*

Vital registration

Vital Registration is the other important source of population data. Vital Registration is the official recording of all births, including stillbirths, adoptions, marriage and civil partnerships, and deaths. In Scotland there is also information on divorce. Vital Registration has been compulsory in England and Wales since 1837 and in Northern Ireland since 1864.

The information collected in this way gives a continuous record of population, whereas the census gives a snapshot image of population and its characteristics at one given time. The management of the information collected is the responsibility of the General Records Office in England and Wales. There are equivalent offices in Northern Ireland and Scotland.

In non-census years the total population can be estimated using the information held by these offices. For example, the total population for Northern Ireland in 2002 (one year after the census) was the total population recorded in the census plus the new births, minus the deaths, plus or minus the net migration in that year (see question 4 below).

Exercise

1 Distinguish between national census-taking and vital registration in terms of population data collection. (CCEA January 2006) (3)

2 Discuss the ways in which population data can be collected and the problems associated with this task. (CCEA January 2007) (6)

3 Study the information on the Nigerian census below. Use this information to help you explain why census data is unreliable in many LEDCs. (6)

Nigeria's Population, 1991 and 2006

Year	Census	Government Estimate
1991	89 million	112 million to 123 million
2006	140 million	120 million to 150 million

Source: National Population Commission, 1991

4 (a) Study the table below, which shows data published by the Office for National Statistics. Make a copy of the table and complete the missing values for 2004–2005. (2)

Population change in the UK 2003–2005

	2003-04 (thousands)	2004–05 (thousands)
Existing population	59 553	59 834
Births	707	717
Deaths	603	
Natural change	+104	+126
Net migration	+177	
New population total	59 834	60 209

(b) Describe the contribution of migration to population growth in the UK, as illustrated by these figures. (3)

1B POPULATION STRUCTURE

Components of Population Change

Geographers are interested in how and why populations change spatially and over time. In order to do this effectively it is necessary to understand how population change is brought about through the interaction of the three components of population change – births, deaths and migration.

This interaction can be represented as a systems diagram, as shown below.

Births, deaths and migration **Resource 2**

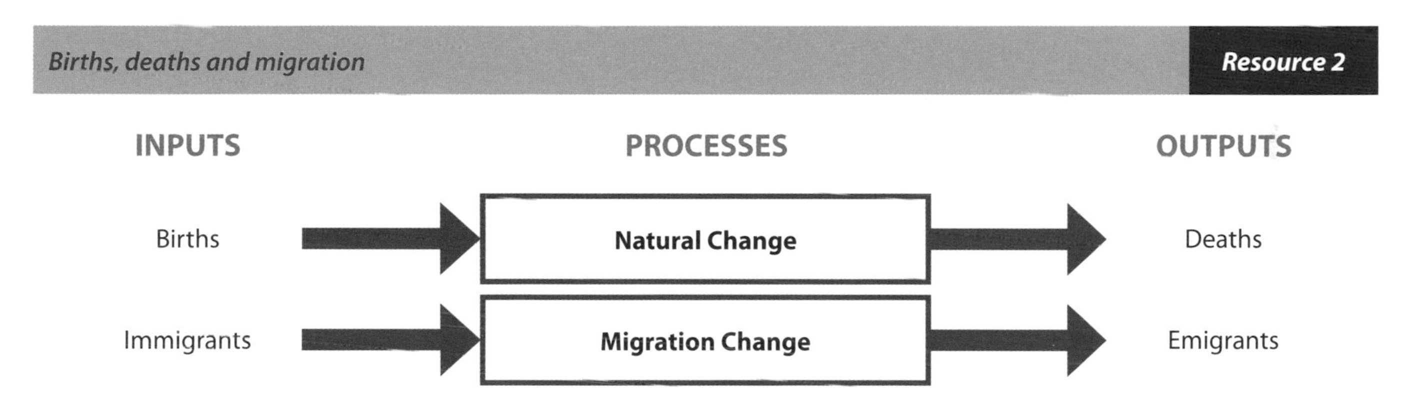

Natural population change refers to the balance between birth rates and death rates. There are several ways of measuring these but the most commonly used are:

Crude Birth Rate

The Crude Birth Rate (CBR) is defined as the number of live births in a year per 1000 of the mid-year population. Birth rates are usually higher in LEDCs than in MEDCs.

Crude Death Rate

The Crude Death Rate (CDR) is defined as the number of deaths in a year per 1000 of the mid-year population. High death rates are more characteristic of LEDCs although these are falling in most LEDCs at present. MEDCs have lower death rates but these are rising in some countries due to ageing of the population.

If the difference between these two measures is positive, there is an increase in population and conversely if the difference is negative, there is a population decrease.

Migration Balance

Few countries live in isolation and most will have in-migration and out-migration. The balance between these two movements has to be calculated to determine whether there is a net gain or a net loss in population (***Resource 3***).

Resource 3 *Components of population change for selected countries*

Country	CBR (per 1000)	CDR (per 1000)	Natural change %	Net migration (per 1000)	Growth rate %
Afghanistan	46.21	19.96	2.625	0	2.625
Bangladesh	29.36	8.13	2.123	-0.66	2.056
Cameroon	35.07	12.66	2.241	0	2.241
Belgium	10.29	10.32	-0.003	1.22	0.12
Germany	8.20	10.71	-0.251	2.18	-0.33
Japan	9.24	9.38	-0.014	0	-0.014
UK	10.67	10.09	0.058	2.17	0.275

LEDC MEDC

Source: US Census Bureau, International Data Base

Population Pyramids

One important aspect of population studies is the proportion of people in each age group. This is made possible through the study of population pyramids. These are diagrams that show the proportions of males and females that occur in each of the five-year age bands. Usually, but not always, percentage figures are used, thereby making comparisons between regions of different sizes possible. The population pyramid can be used to compare spatial differences in population structure between areas such as rural and urban, or between MEDCs and LEDCs. They can also be used to study changes in population structure over time, such as previously high birth rates or the reduction in births during a period of war followed by a post-war baby boom.

Resource 4 *How to analyse spatial differences in population pyramids*

Shape	Feature	Typical example
Narrow base	Low birth rate	MEDC
Broad base	High birth rate	LEDC
Straight sides	Low death rate	MEDC
Narrowing bars	High death rate	LEDC
Broad top	Long life expectancy	MEDC
Bulges	Immigration/in-migration	Urban areas
Waists	Emigration/out-migration	Rural areas
Narrowing base	Falling birth rate	MEDC

The following resources show comparisons between the population pyramids of different areas. The percentages on the pyramids represent the percentage of 'all males' (to the left) and the percentage of 'all females' (to the right) that are in that age group.

Comparison between an LEDC (Afghanistan) and an MEDC (UK) — Resource 5

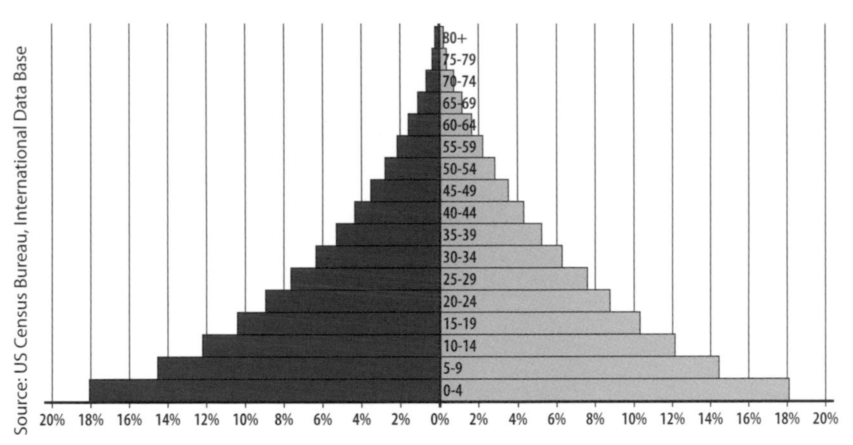

Afghanistan in 2000

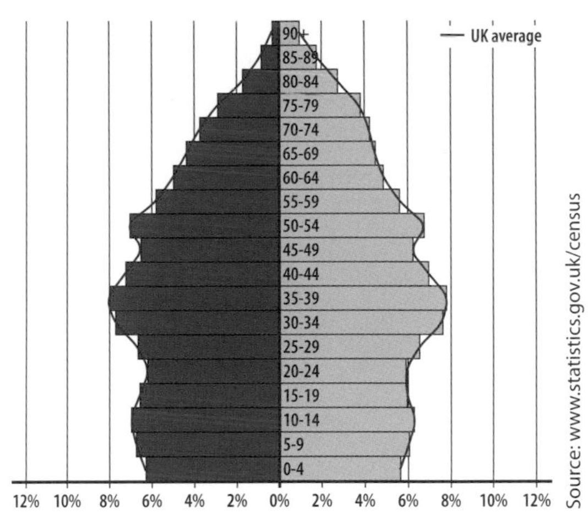

United Kingdom in 2001

Regional comparison in the United Kingdom — Resource 6

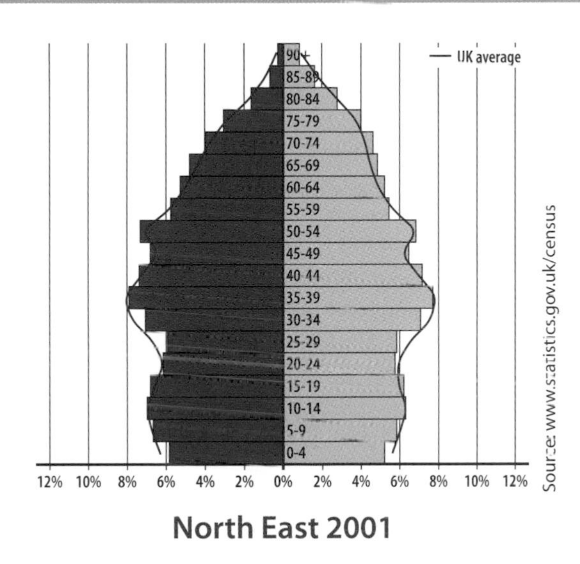

North East 2001

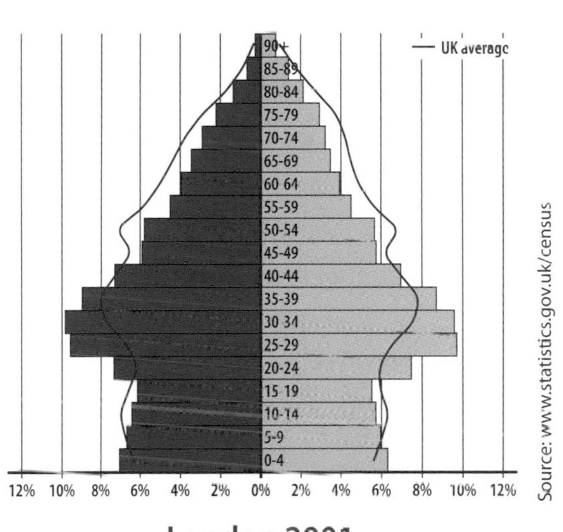

London 2001

Comparisons between an inner London borough (Kensington and Chelsea) and an outer borough (Barking and Dagenham) — Resource 7

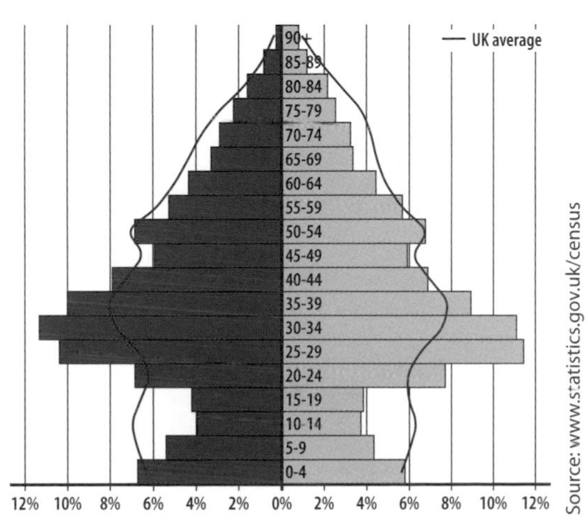

Kensington and Chelsea 2001

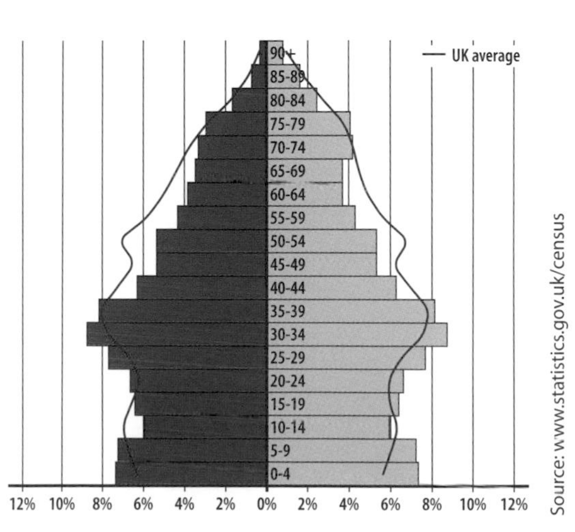

Barking and Dagenham 2001

Resource 8 *Comparison between Northern Ireland and the United Kingdom*

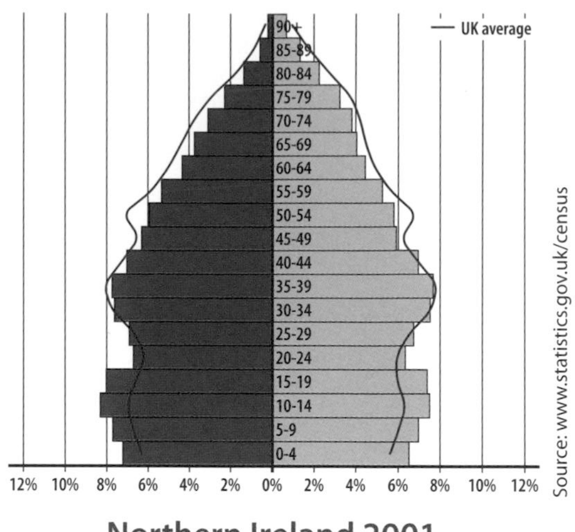

Northern Ireland 2001

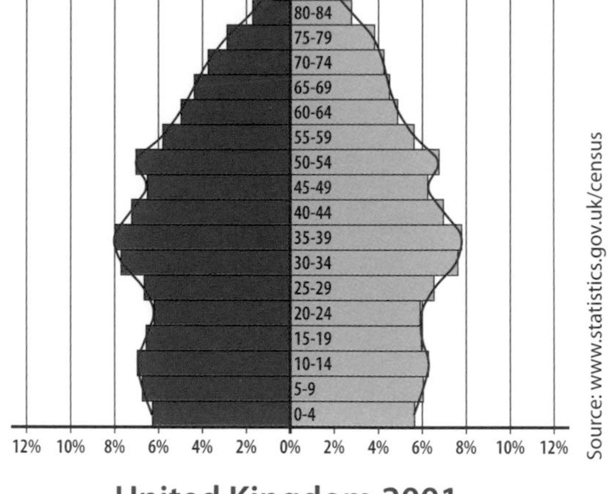

United Kingdom 2001

Resource 9 *The population structure (in millions) in Afghanistan in 1980, 2000 and the projected structure for 2020*

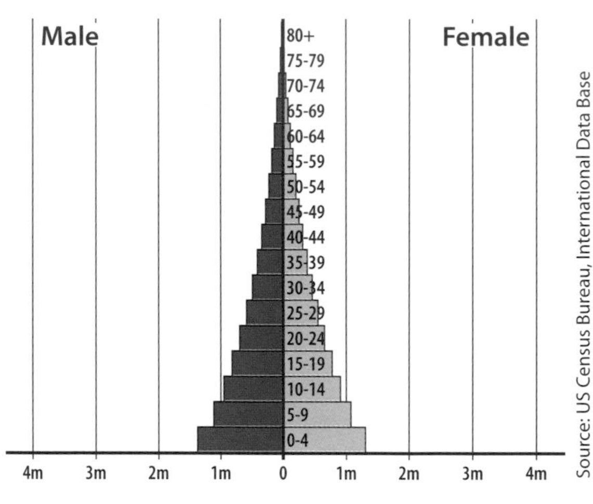

Afghanistan in 1980

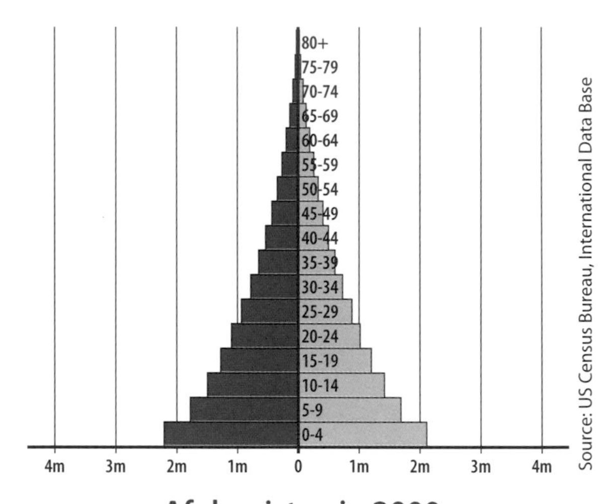

Afghanistan in 2000

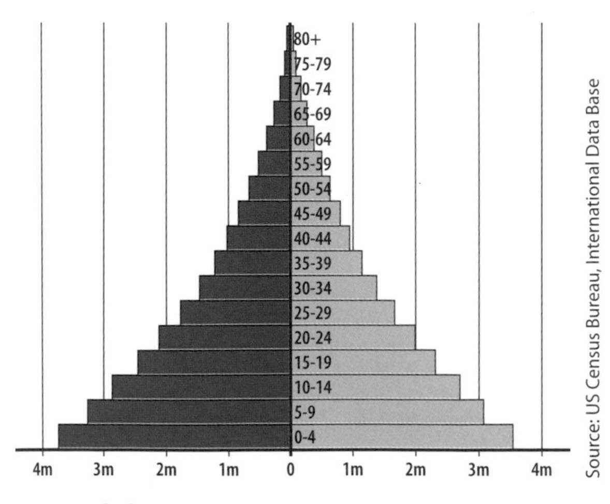

Afghanistan in 2020 *(Projected)*

Exercise

1. Using the following website, select two countries from different levels of development and annotate each with appropriate labels from **Resource 4** – www.census.gov/ipc/www/idb
2. Compare and contrast the population pyramids shown in **Resource 5** and **6**.
3. Suggest possible reasons for the increase in the proportions of the population in Kensington and Chelsea between the ages 20–34.
4. **Resource** 7 includes the population structure for an outer London borough. Why do you think this area has a birth rate greater than the national average?
5. Compare and contrast the Northern Ireland population structure with that of the UK shown in **Resource 8**.
6. Describe the effects of migration on the shape of a population pyramid. (2)
7. **Resources 5–8** use percentage scales on their horizontal axes while **Resource 9** uses actual numbers. Explain why these different methods have been used. (4)
8. *Question from CCEA February 2005*
   Study the Figures below, which show population pyramids for three countries in 2000.

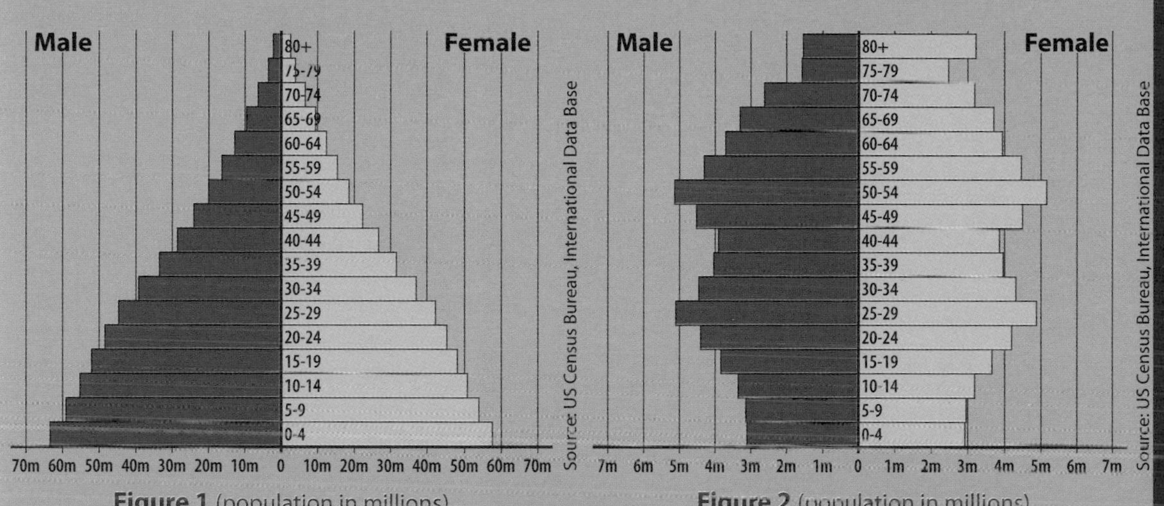

Figure 1 (population in millions)

Figure 2 (population in millions)

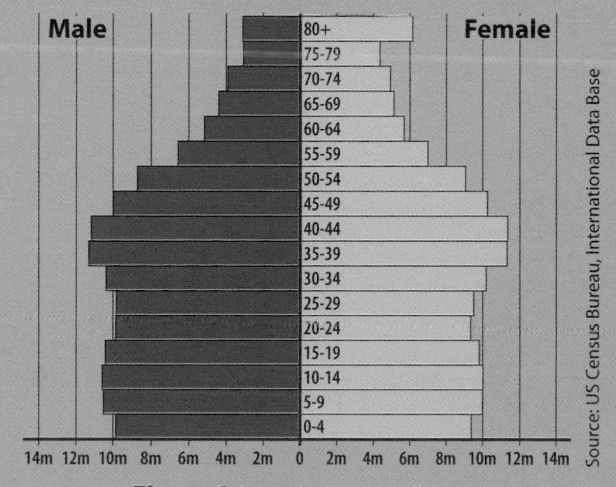

Figure 3 (population in millions)

i) Complete a copy of the table overleaf by matching the correct pyramid with the correct country. (3)

Exercise continued

Country	GNP per capita (US dollars)	Infant morality (per thousand live births)	Death rate (per thousand)	Birth rate (per thousand)	Pyramid
India	2540	60	9	23	
USA	37 600	6	8	14	
Japan	28 000	3	8	9	

ii) With reference to the population pyramids, explain your reasons for **one** choice you made. (2)

9. *Question from CCEA June 2005*
Study the Figures below, which show three population pyramids for the UK: in 1881, 1991 and a projection for 2050. Source: adapted from British Census data and data from the US Census Bureau

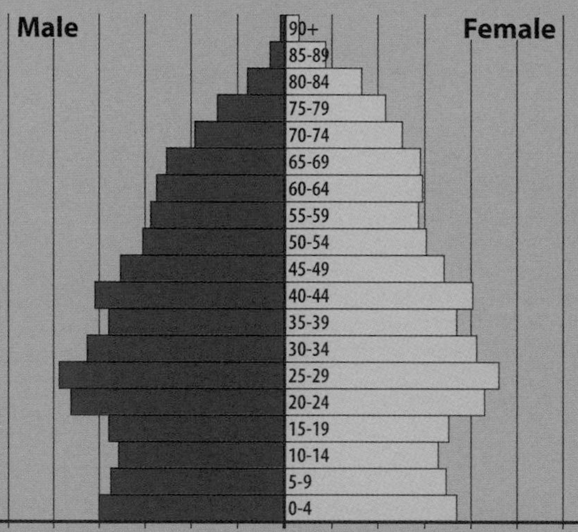

Figure 1 (population in millions)

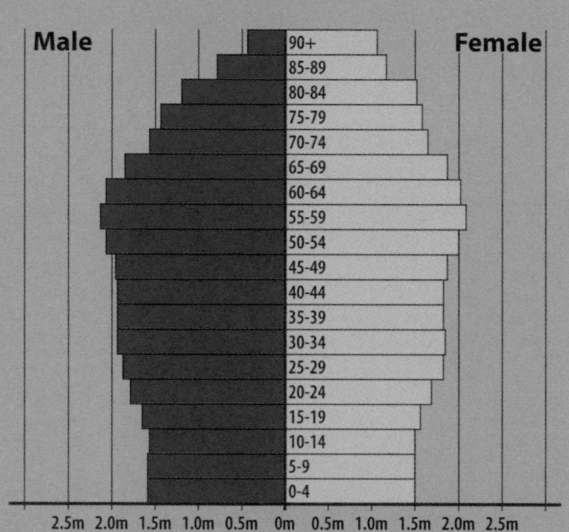

Figure 2 (population in millions)

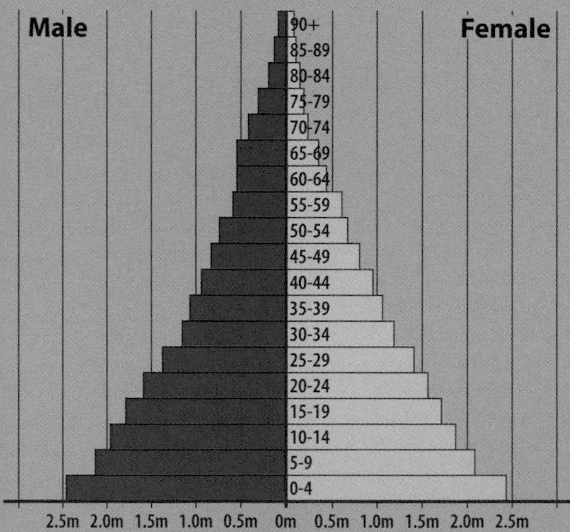

Figure 3 (population in millions)

i) Complete a copy of the table by matching the correct pyramid with the correct date. (3)

1881	
1991	
2050	

ii) Using the resource above, describe and explain how either changes in birth rates **or** changes in death rates have affected the shape of the pyramids from 1881 to 2050. (3)

Reference

Additional reference material can be obtained at the following website:
www.statistics.gov.uk/census

NATIONAL CASE STUDY: Population change in Britain 1700-2000

Reliable population figures are only available for England and Wales since 1801 and slightly later in Scotland. This makes it difficult to be precise about anything much earlier than this. However, it is generally believed that population growth in Britain was slow throughout the eighteenth century and birth and death rates were both high, fluctuating between 31–36 per thousand. Total population increased from 7 million in 1700 to 10.5 million at the time of the first census in 1801. Social and economic conditions were poor at this stage and at times the death rate exceeded the birth rate, resulting in a slight fall in total population. Towards the end of the eighteenth century a number of important social and economic changes began that would dramatically transform Britain over the next one hundred years. The two most important of these changes were the improvements in agriculture, and industrialization.

Agriculture and industrialization

Once agricultural output had increased, a growing number of workers could be freed from food production to seek work in the new factory-based industries in the towns. Initially, towns were very unhealthy places and according to some, death rates may even have increased due to the increased risk of infection from large numbers of people living in high density and in unsanitary conditions. At this stage the government did little to improve the situation and it was left to some enlightened individuals to provide the necessary funds to improve the living conditions for the working classes. The founding of voluntary hospitals and the provision of elementary outpatient care had an immediate impact on the death rate.

Improved healthcare

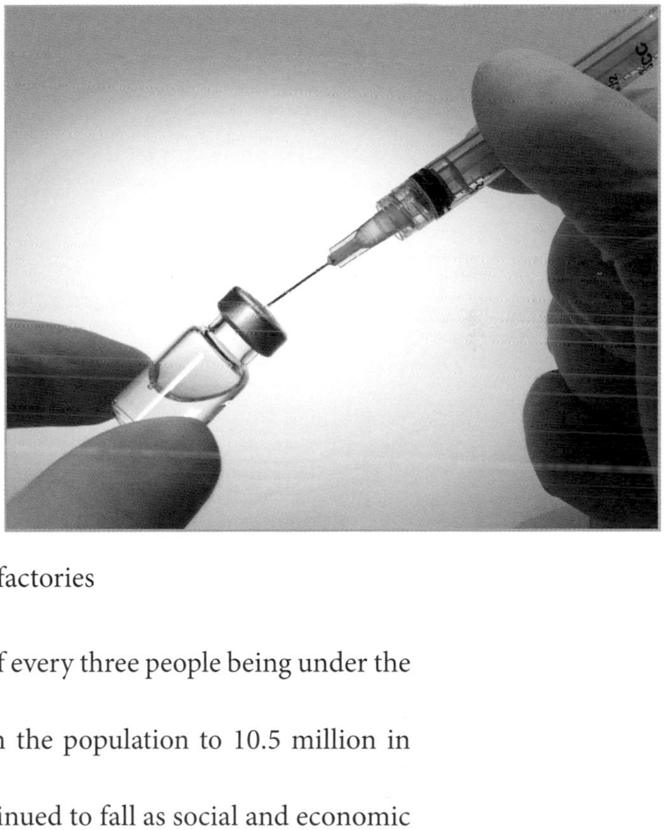

The discovery of a vaccination against smallpox was one of the most significant medical developments towards the end of the eighteenth century. Alongside these medical advances and the significant improvements in agriculture was an increase in disposable income for many. Food consumption increased and, together with a better understanding of nutrition, resulted in a healthier population and a significant decrease in mortality rates in general, and infant mortality rates in particular. While death rates fell throughout the latter half of the eighteenth century, to 25 per thousand in 1801, the birth rates were still high at 37 per thousand. The high birth rates at this time were due to a number of factors including:

- Lack of birth control and family planning
- The need for large families to provide a workforce in factories
- Infant mortality was falling but still high
- The population structure was youthful with two out of every three people being under the age of 30.

The end result of this was an increased rate of growth in the population to 10.5 million in 1801.

Throughout the nineteenth century the death rate continued to fall as social and economic conditions saw further improvements. Government began to pursue policies and fund projects that would ultimately improve medical and sanitary conditions. The Public Health Act of 1848 went a long way towards providing clean water, which saw a considerable reduction in the number of deaths from diseases such as typhoid. A general improvement in hygiene caused a decrease in the occurrence and spread of infectious diseases. Consequently, the death rate fell steadily throughout the nineteenth century reaching 17 per thousand in 1901. Life was steadily improving for the working classes. There were many opportunities for employment in

Resource 10 *Population structure in Britain in 1840 and 1880*

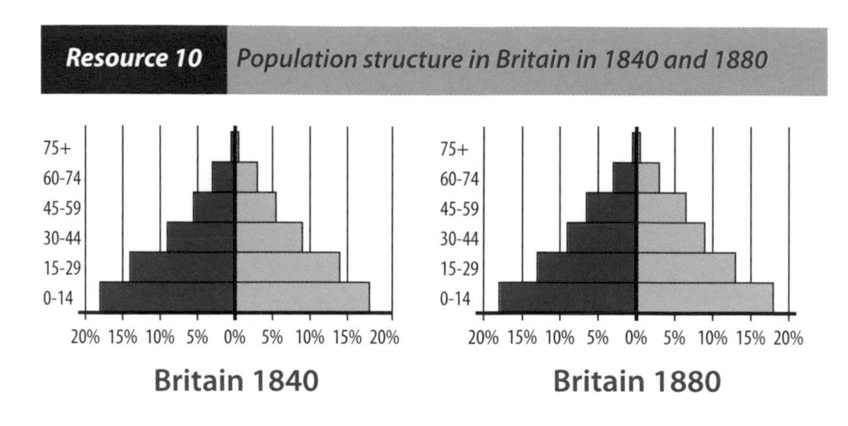

factories, shipyards or in the growing service sector. Disposable incomes were rising, fewer children died in infancy, child labour had been abolished in 1833 and people began to see the material benefits of having smaller families. Certainly, many of the reasons for large families had been removed. Towards the end of the nineteenth century there was a marked fall in the birth rate to 26 per thousand. In spite of this, overall numbers continued to increase due to the youthful structure of the population. In 1901 total population was 36.2 million.

Slower rate of population growth

In the twentieth century, population growth continued but at a slower rate than in the previous century. In 1939 the total population in Britain was 47.5 million. The main reason for this slower rate of growth was the rapid fall in the birth rate. By 1940 the birth rate was 14 per thousand and apart from the post-war baby boom in the period after the end of the Second World War the downward trend has continued. The economic benefits of smaller families, the increased participation of women in the workforce, along with the greater availability of contraception and the changing role of women in society were the main reasons for this fall in the birthrate. The death rate steadied somewhat as the population began to age. In fact, there are concerns that Britain may be nearing the situation where the death rate could exceed the birth rate, resulting in an overall decline in population. The figures from the 2001 census reveal that Britain's total population is now 57 million. For the first time, there are now more people aged 60 and over than there are children aged under 16. Children make up 20.2 per cent of the population in England and Wales while those aged 60 and over make up 20.9 per cent. In Scotland, the figures are 19% under 16 and 16% over 60.

Resource 11 *Population structure in Britain in 1920, 1940 and 2001*

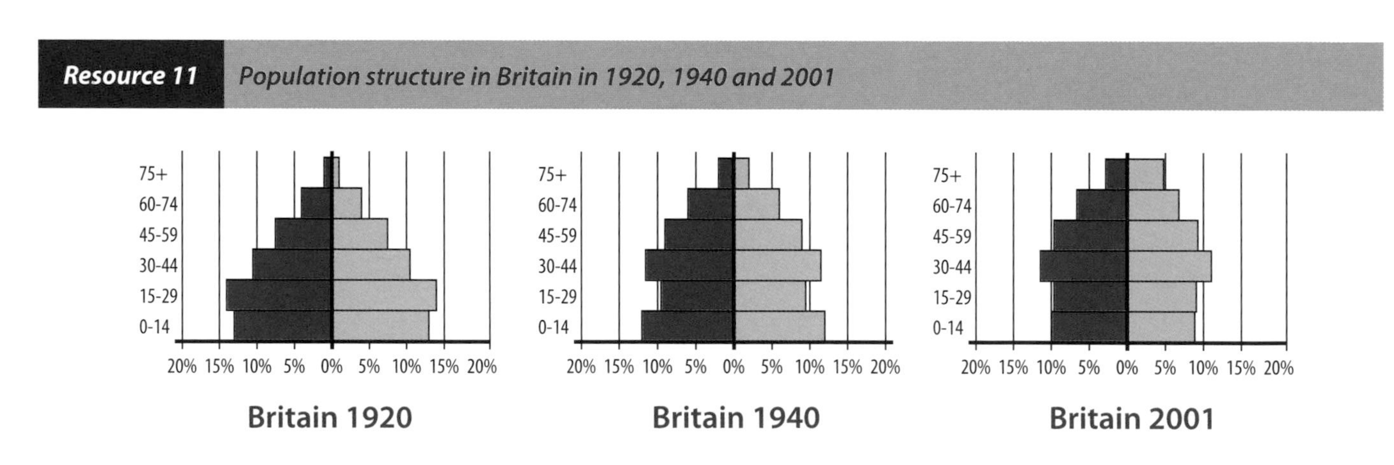

Migration

International migration has played a part in the population growth of Britain since the 1950s, but in recent years the importance of migration has increased. Many of these migrants originate from those countries that joined the EU in 2004. As most of these migrants are in the younger working age groups, they are likely to contribute towards a higher birth rate in years to come.

Regional variations

As well as changing over time, the population of Britain also displays distinctive regional variations. Rural areas, particularly the more remote areas, have significantly larger proportions

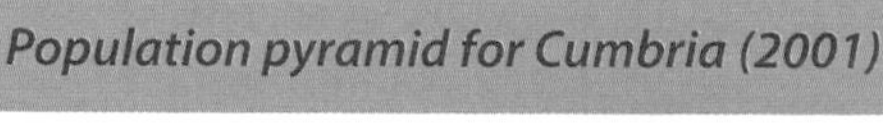
Population pyramid for Cumbria (2001) **Resource 12**

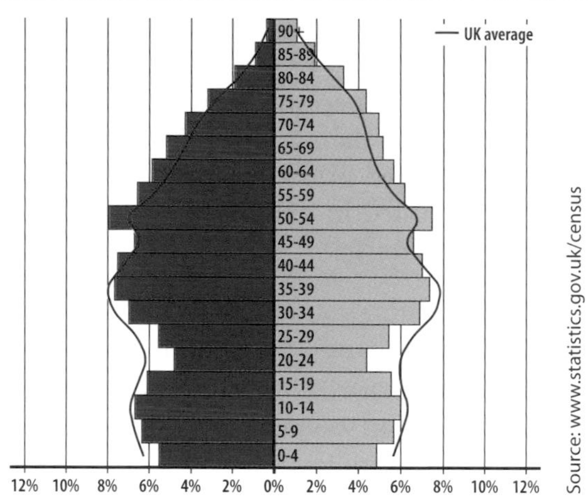

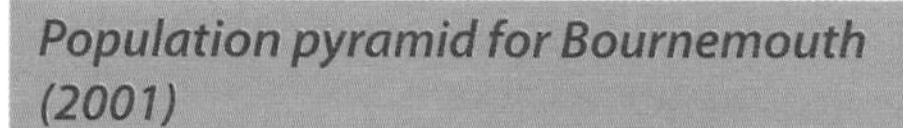
Population pyramid for Bournemouth (2001) **Resource 13**

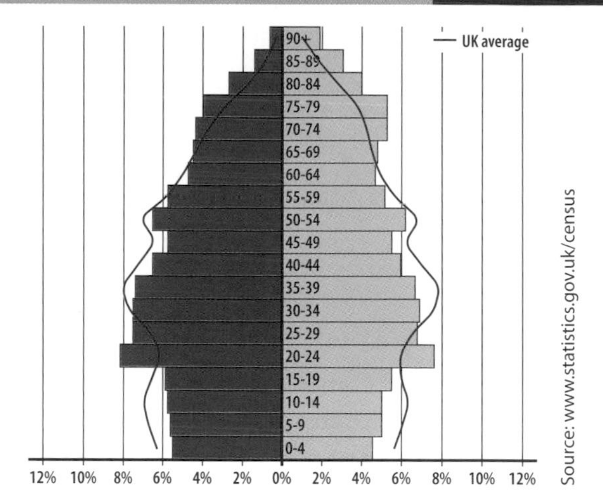

of elderly. The main reason for this is the out-migration of the younger working age population. The Highlands of Scotland, Wales and Cumbria all have above the national average of their population in the over-65 age groups. In many cases, these regions have experienced economic decline and are isolated from the main core of economic development. Consequently, they offer few attractions or opportunities for investment and the working population has little alternative but to seek employment in the surrounding urban areas. A population pyramid for these regions will show a distinct narrowing in the working age groups between 20-45 and a bulging in the over-65 age groups.

Above-average concentrations of elderly are also found in the south coast resorts of England such as Brighton and Bournemouth. These areas have attracted retired groups for some time now. A combination of a number of factors, such as a mild climate, a pleasant and scenic environment, along with the availability of facilities specifically designed to cater for the needs of those over 65, have acted as a magnet for many retired people.

Younger age groups are found in greatest numbers in the urban areas of the Midlands, Lancashire and in the new and expanding towns along the M4 corridor west of London. These towns and cities are close to the economic core of Britain and as such provide excellent employment opportunities for the economically active age groups. Along with the usual range of educational and recreational facilities, they have attracted substantial numbers of young migrants. In some of these urban areas the young age structure has resulted in a higher than average proportion of children. In Milton Keynes 30% of the population is under 15 compared with 25% for Britain as a whole.

Exercise

1. Use the following website to find population pyramids for some of the places mentioned in the text and annotate their main features.
   www.statistics.gov.uk/census2001/pyramids/pages/UK.asp

2. With reference to your national case study, describe how and explain why a country's population structure has changed over time and space. (12)

8.5 per thousand. This will result not just in a reduction in the rate of growth of the population in India but there will also be significant changes in the population structure. The changes will be most significant in the proportions under 15 and those over 65 (***Resource 16***).

Regional variations

The patterns of fertility and mortality decline in India show marked regional variations. **Resources 19** and **20** show distinct differences in life expectancy and fertility rates between the northern states of Uttar Pradesh and Rajasthan and the southern states of Kerala and Tamil Nadu. In general, birth rates and death rates are lower in the southern states and in the more affluent urban areas in the north. Life expectancy can vary by anything up to 20 years between the northern and the southern states. These variations are due to the better education and health services in the southern states. The southern states have always been the more advanced in terms of development and attitudes towards the role of women in society and their entitlement to education. In Kerala and Tamil Nadu fertility levels (average number of children per woman) are similar to those in MEDCs at 2.0 but in the northern states fertility levels are close to or over 4. There are also marked differences in fertility levels among the different ethnic groups in India. Muslims generally have larger families than Hindus. Although Muslims are a small minority nationally, there are some states where numbers are significant enough to ignite historical animosities between them and the Hindu majority.

Resource 19 *Total Fertility Rate (average number of children per woman) for selected states in India, 1971 and 2003*

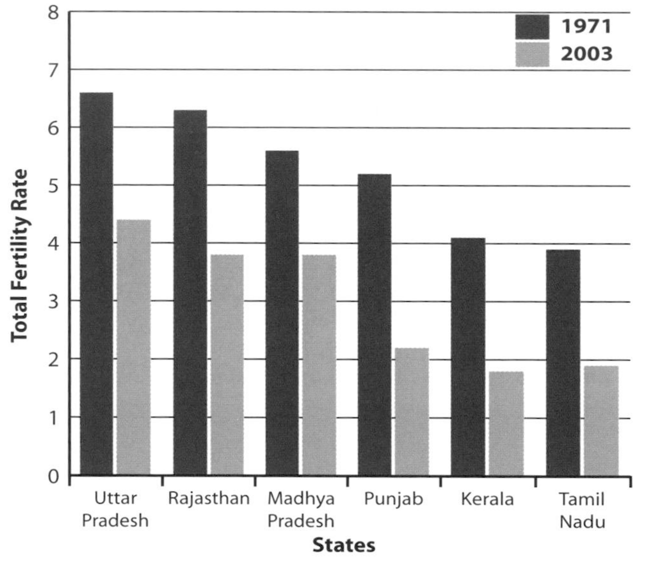

Original data from Population Reference Bureau 2006 by Carl Haub and Sharma

Resource 20 *Average life expectancy in selected states in India 1999-2003*

State	Male	Female
Uttar Pradesh	59.6	58.7
Rajasthan	60.7	61.8
Madhya Pradesh	57.2	56.9
Punjab	67.6	69.6
Kerala	70.9	76.0
Tamil Nadu	64.3	66.5

Northern States Southern States

Original data from Population Reference Bureau 2006 by Carl Haub and Sharma

Exercise

With reference to your national case study, describe how and explain why a country's population structure has changed over time and space. (12)

Reference

Additional reference material can be obtained at the following website:

www.prb.org/Articles/2006/ChinasConcernOverPopulationAgingandHealth.aspx

The Implications of Dependency Ratios

One aspect of Population Geography which is of particular interest to government planners is the relative proportions of economically active and inactive populations. It is common practice to regard those over 15 and under 65 as the economically active and those under 15 and over 65 as economically inactive. A widely used measure of these proportions is the dependency ratio. Dependency ratios are calculated by expressing the number in each group as a percentage of the total population:

$$\frac{[\text{Total number (0-14 + 65 and over)}] \times 100}{[\text{Total number 15-64}]}$$

UK example:

Age groups	0-14	15-64	65+
Totals	7 604 852	40 628 085	10 556 257

$$\frac{18\ 161\ 109 \times 100}{40\ 628\ 085} = 48$$

This means that for every 100 people in the age group 15–64 there are 48 dependants.

Governments are interested in this dependency ratio because they need to plan for future demands in services. In MEDCs the population is ageing as the birth rate falls and life expectancy increases and it is the relative size of the over-65 group that is of most concern. In LEDCs it is the relative size of the under-15 group that is most significant. Two other ratios are used to reflect these differences in dependency between LEDCs and MEDCs:

Aged Dependency is calculated:

$$\frac{\text{Total number over 65}}{\text{Total number 15–64}} \times 100$$

Youth Dependency is calculated:

$$\frac{\text{Total number 0–14}}{\text{Total number 15–64}} \times 100$$

Calculate the youth and aged dependency for the UK using the figures in the table from the example.

Exercise

Aged Dependency in MEDCs

In most MEDCS standards of living have improved dramatically over the last half century. Practically everyone in MEDCs is literate, has access to state health care, and has adequate food and nutrition. There is widespread information on health care and disease prevention. In general, people are reasonably well informed on the requirements for a long and healthy life. It is not surprising therefore that life expectancy has risen and at present the average life-expectancy is 74 years for men and 80 for women. This increase in life expectancy, along with the fall in the birth rate, has resulted in a disproportionate number of the elderly relative to the younger age groups and gives the characteristic top-heavy shape to population pyramids for MEDCs. In Europe those over 60 will make up 40% of the population by 2050 and a significant number of these will be in the oldest category, the over eighties. This increase in the proportion of elderly in society is set to have a major impact on service provision in the future and has become an issue for governments and planners.

Resource 21 *Predicted proportions of population over sixty in selected European countries*

Country	2000	2020	2050
Germany	22.9	30	41
Belgium	22.1	30	38
Denmark	19.9	28	36
Spain	21.8	28	44
France	20.7	29	38

Source: Eurostat

Advantages of an aged population

There are considerable advantages in having an aged population structure. Many senior citizens have considerable disposable incomes. Mortgages are paid off and children have left home so outgoings are substantially reduced. Many will be in receipt of private pensions as well as the state pension so they have significant purchasing power. In the UK there is a whole raft of services geared towards the demands of the 'grey' population. SAGA is a company that provides holidays at home and abroad for the over 50s as well as insurance packages for health care and personal belongings. Housing Associations and private developers have set up nursing homes and other accommodation suited to the elderly who have sufficient funds to pay for their care.

As the present generation of pensioners enjoy better health, many of the more affluent have chosen to move away from the busy towns and cities, preferring to live in smaller and quieter areas. The seaside towns of the south coast of England are good examples of this type of development where social and recreational services target the senior citizens. In these towns many of the younger population will find employment in the provision of services for the elderly residents. It is also claimed that a society with a large proportion of elderly should be more stable as the younger generations can benefit from the experience and advice from the senior members of society. However, the importance of this factor will vary from one country to another and will depend on how older people are perceived in society.

Disadvantages of an aged population

Increasingly it is the problems of an ageing population that challenges planners and politicians. There is no doubt that an increasing number of pensioners create an economic problem for the health service and so for the government. The elderly will at some stage require medical care and as their numbers increase the financial cost will also increase. More hospital beds and specially trained staff will be needed just to keep up with present standards of health care.

Medical care

According to government figures, care for the elderly alone is calculated at £1.6 billion pounds. The cost of health care in the UK has always been covered by the contributions to national insurance that employees pay from their salaries. These contributions are used annually and there is no reserve of money left for future years. This means that healthcare provision is funded through the contributions of those still in employment. If that sector of the population has to fund the health care of a growing elderly population, it is easy to see that the health service will not be able to cope.

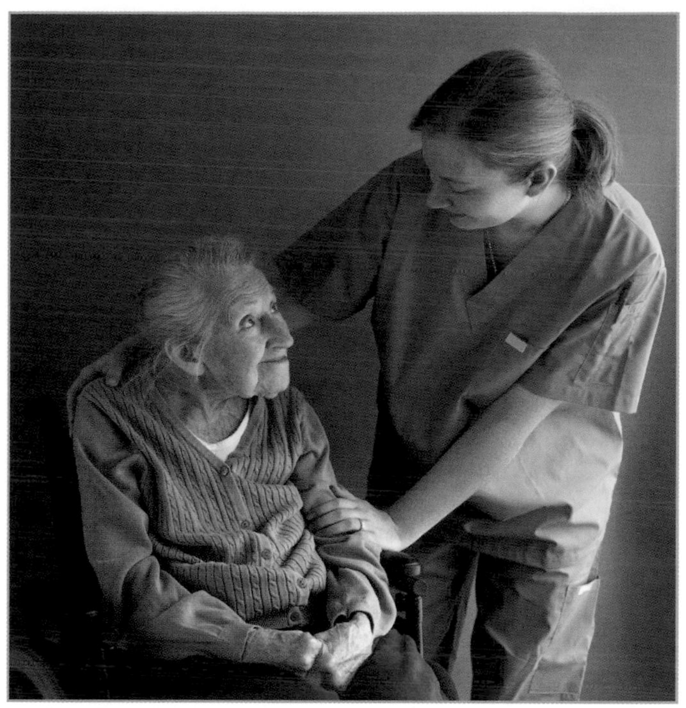

Care for the elderly is calculated at £1.6 billion annually in the UK and puts a massive strain on the health service.

State pension

All pensioners are entitled to a state pension, which is funded from the national insurance contributions paid throughout their working lives. When this scheme was introduced in the 1950s life expectancy was lower but now that people are living longer the financial burden on government has risen dramatically.

The 2001 Labour government commissioned an enquiry into this problem and made substantial changes to pension provision for the future. At present the normal age of retirement and state pension entitlement is 60 (for women) and 65 (for men). This has increased to 68 for both men and women for those entering the workforce since 2005. The government has also encouraged individuals to take out a private pension that they will contribute to throughout their working lives. At one stage government considered means-testing the state pension. This would have meant that only those pensioners with no alternative source of income would have received a state pension. This became a highly emotive issue and the proposal was scrapped. Pensioners are a significant group in society and the government does not want to lose the 'grey' vote. It is clear however, that state pension funds are stretched and for those who depend on it solely for an income there is considerable hardship.

Residential care

Due to the prohibitive nature of residential care, the government was forced to take action; the cost of health care and particularly long-term health care in a residential or nursing home must be paid from the patient's own funds where possible. In some cases this has meant that the patient's home will have to be sold to meet the costs of nursing care. This has brought considerable hardship to many, but it highlights the problem of the increased financial burden of an ageing population.

The future: insufficient workforce

The provision of a future workforce is the other major problem facing an ageing population. With a falling birth rate and an ageing population some MEDCs will fall below replacement level. In other words their populations will decrease. In the UK at present there are insufficient workers to fill all the skilled and unskilled manual jobs. Whilst this shortage is due partly to the unpopularity of these jobs, it nevertheless highlights another problem facing an ageing society. Currently these shortages in the UK are being met by the large number of migrants from Eastern Europe. For many this is only a short-term solution and in some countries such as France and Sweden the governments are pursuing policies to entice younger people to have more children.

Exercise

1. Study the extract on Ageing Europe and describe the social, economic and political problems brought about as a result of increased aged dependency in MEDCs. (12)

Ageing Europe is unprepared

Two-thirds of the inhabitants of the village of Cersosimo in southern Italy are over 65. Improved healthcare and changing lifestyles mean people are living longer, but local women are marrying later and seem increasingly reluctant to have children. "We've had four babies this year," the mayor told me morosely, "in the same period we've had 14 funerals." Things are so bad the village school has combined all of its classes to maintain a quorum of pupils. The mayor is planning to turn the redundant classrooms into an old people's centre – if he can find the money. But with the population of the village down from over 1,000 to just 850, his local tax income is going down too.

Cersosimo's finances are still reeling from last year's effort to turn back the tide of depopulation. The mayor decided to offer 2,500 euro to any family having a baby in the village. Even though there were only half a dozen recipients of the handout, it was an experiment that the village cannot afford to repeat. Besides, there is no evidence that it changed anyone's mind about the merits of procreation.

Rafaele and Marisa Lofiego, owners of the village bar, received the windfall when they had a baby boy, Vincenzo, 14 months ago. Did it make them feel any different about having children? "No way," said Rafaele. "The fact is, children cost too much."

Across Italy the average number of children a woman can expect to bear in her lifetime is now down to 1.2. Yes Catholic Italy, now has the lowest birth rate in Europe. Demographers calculate that by 2050 the current population of 56 million could have dwindled to 40 million. Towns and cities will be left with thousands of unwanted apartments, schools may well be half empty and whole swathes of the countryside could be depopulated. And, naturally the proportion of old people within the population will continue to rise.

By mid-century there may be one pensioner for every one productive worker in Italy, which begs a simple, devastating question: how on earth is Italy going to maintain its pensions system? Either the next generation of workers will have to pay unthinkably high levels of tax, or the current, relatively generous benefits will have to be radically scaled back.

This is not just Italy's problem, it is Europe's problem. Spain, Germany, Austria and Greece all have disturbingly low birth rates.

Story from BBC News (http://news.bbc.co.uk)
2003/08/02

2. Discuss possible options that governments might consider to address the problems of ageing societies. (5)

3. *Question from CCEA January 2007*
   Study **Figure 1** (opposite) which shows the percentage of world population in three groups from 1970 with projections to 2100.

   i) Compare the trends for those under 15 with those aged 60 and over between 1970 and 2100. (2)

   ii) There is a projected increase in the percentage of the population aged 80 and over. Explain **one** social **or** economic consequence of this trend. (3)

Exercise continued

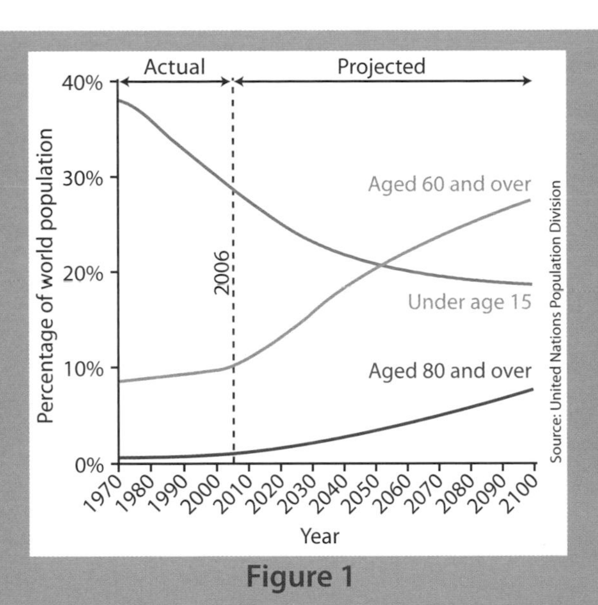

Figure 1

4. Study **Figures 2–5** showing predicted population structures to 2025. Describe and explain how the patterns emerging in the MEDCs differ from those in the LEDCs. (10)

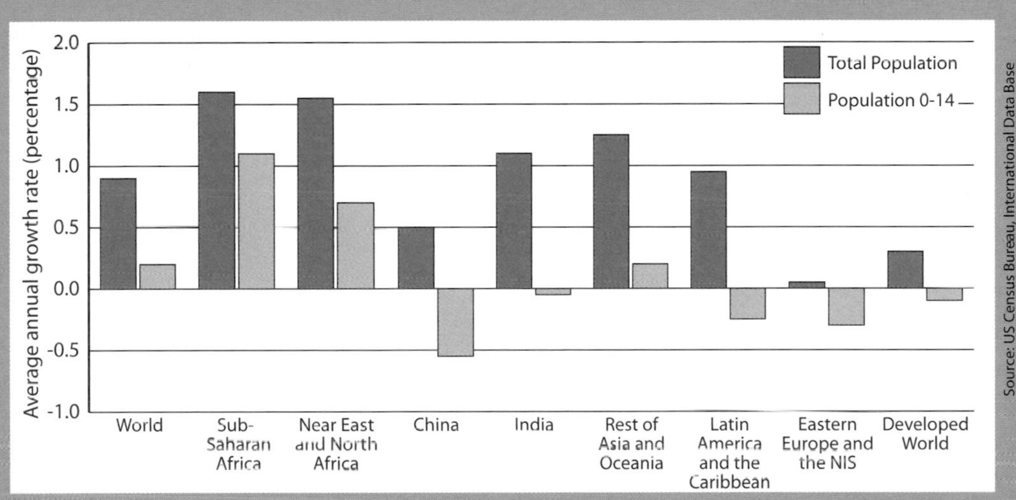

Figure 2 – **Average annual growth rates in the child population by region: 2002-2025**

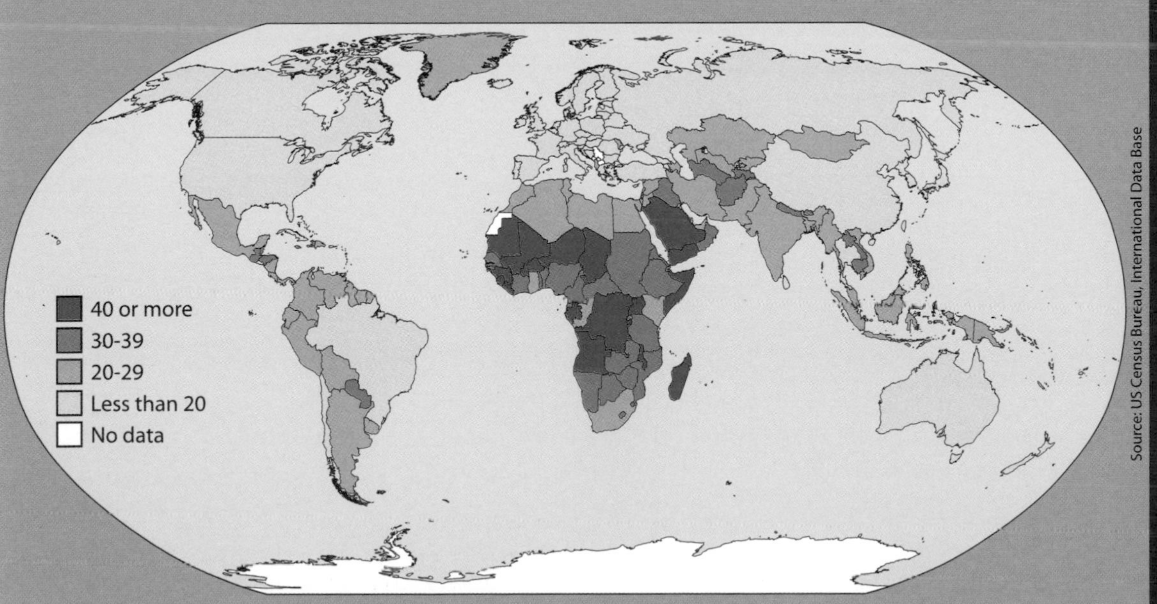

Figure 3 – **Children (ages 0-14) as a percentage of the total population by country: 2025**

Exercise continued

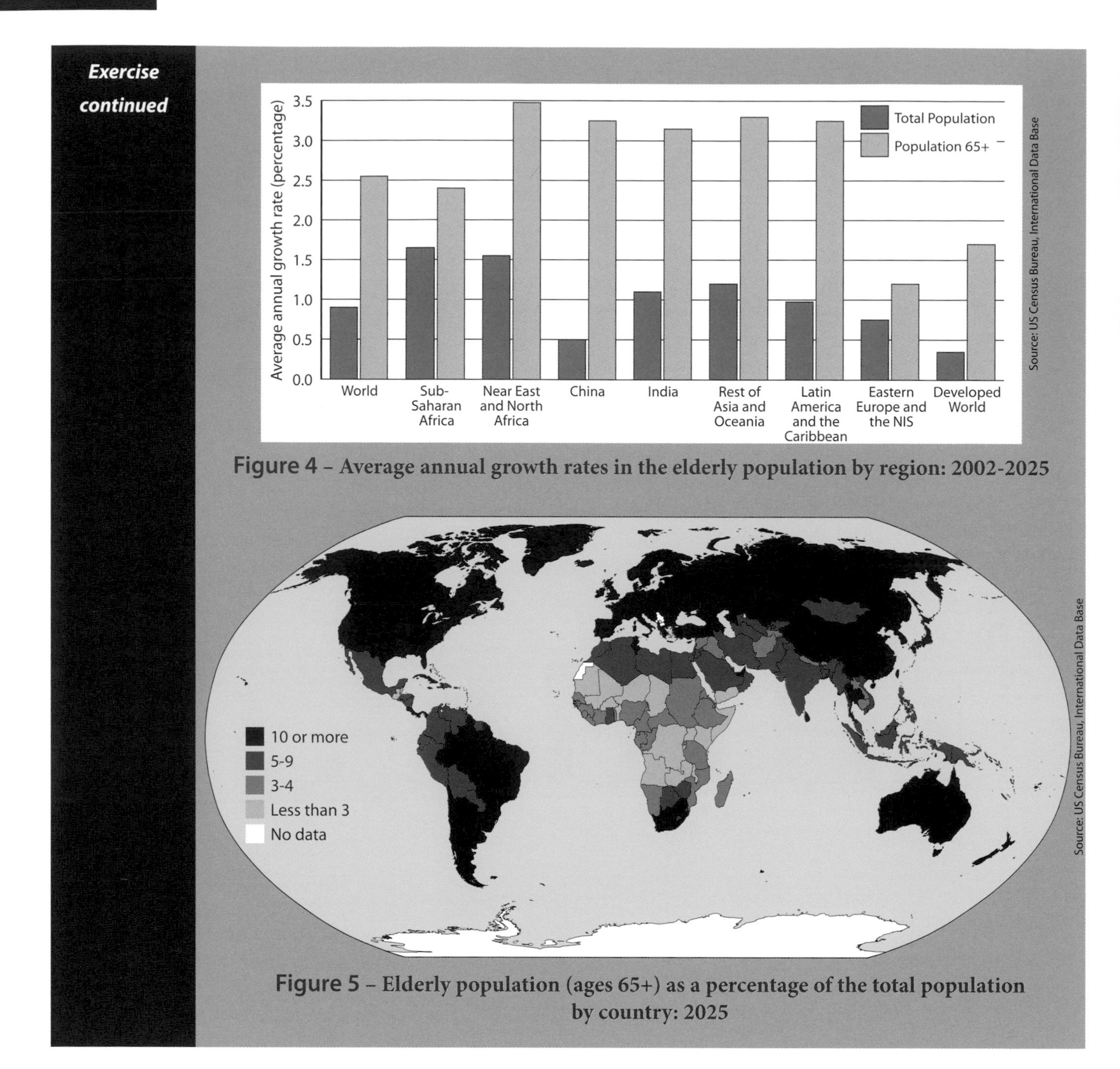

Figure 4 – **Average annual growth rates in the elderly population by region: 2002-2025**

Figure 5 – **Elderly population (ages 65+) as a percentage of the total population by country: 2025**

Additional references

http://news.bbc.co.uk/1/hi/business/2452683.stm – *Italy's pensions timebomb*

http://news.bbc.co.uk/1/hi/business/2248531.stm – *Ageing Europe's growing problem*

http://news.bbc.co.uk/1/hi/world/europe/3092045.stm – *French parliament backs pension reform* (http://tinyurl.com/4fucsf)

http://news.bbc.co.uk/1/hi/world/europe/3008527.stm – *France wrestles with pension reform* (http://tinyurl.com/3fe8q6)

http://www.prb.org/Articles/2006/FullTimeWorkAmoungElderlyIncreases.aspx

The Globalisation of Ageing – Geo Factsheet number 196

Ageing Society in Japan – Geography Review November 2001

Youth Dependency in LEDCs

In LEDCs population trends differ greatly from those in MEDCs. Death rates have been falling for some time now due to elementary improvements in health care and vaccinations against some childhood diseases. The picture is not the same everywhere but a marked fall in the death rate has been recorded for most LEDCs. It would be optimistic to assume that LEDCs will now follow the pattern of the MEDCs, such as Britain, where the birth rate fell some time after the reduction in the death rate. There are fundamental differences between the situation operating in Britain in the nineteenth century and that currently operating in most LEDCs. In Britain, industrialisation, followed by social and economic advancement, made small families seem an attractive option. In many LEDCs the economy is still largely agricultural and children are viewed as an economic asset. Consequently, birth rates remain high in many LEDCs and the population structure shows a large and growing proportion of persons under 15. Indeed, in the next decade as these people become parents themselves there may be an increase in the proportion under 15 even if the average family size is reduced.

A growing young population should provide the opportunity for an adequate labour supply and market but the provision of services for the current population places an enormous financial strain on the governments of LEDCs. It has made these countries attractive to multinationals that follow low wage labour forces. Multinationals are major employers in LEDCs and many see this as a first stage in the development of the LEDCs.

Countries with a large youth-dependent population will need to invest large amounts of money into education and health care. In LEDCs this is very difficult and as a result there are usually inadequate educational opportunities – many will not even receive primary-level education. This lack of education will be a considerable disadvantage as an educated and skilled workforce is necessary if a country is to progress. A similar situation arises in relation to the provision of housing, food supply and medical care. Often the demand exceeds supply of even the most basic of essential services and many people will be left well below the poverty line.

Some countries with a large youth-dependent population were seriously concerned that they would be unable to support their populations in the future and their governments had to make plans to deal with this impending crisis. The Chinese government undertook a very stringent one child policy whereby permission to have a child had to be obtained from the local authorities and only one child per family was permitted. In Indonesia, the threat of a serious imbalance between available resources and a growing population prompted the government to undertake a transmigration policy, which involved moving large numbers of people from overcrowded islands to less crowded areas, often over several thousand kilometres. The example of Mauritius is a less severe example and involved a two pronged approach of encouraging smaller families and at the same time carrying out economic reform.

The Chinese government's one child family may create problems for the future, including a gender imbalance in favour of boys, as families prefer to have a male heir.

Exercise

1. Use the following website to find information to allow you to calculate the youth dependency ratios for two LEDCs:
   www.census.gov/ipc/www/idb

2 'LEDCs with a large youth-dependent populations are often unable to provide even the most basic of services for their populations'. Research one LEDC to find facts and figures that would allow you to discuss this statement.

3. Compare and contrast the economic, social and political problems of dependency in MEDCs with those in LEDCs. (12)

1C POPULATION DISTRIBUTION AND RESOURCES

Population distribution or variations in the density of population are very closely related to the resources of a region. Resources are used in the widest context to include not only physical or natural resources but human resources as well. Physical resources include mineral wealth, fertile soil, climate and relief. Human resources include an appropriate labour force, stable government, financial and trading assets. The nature of the economy and the opportunities available for people to earn a living are very strong influences on the number of people that can be sustained within an area. Areas which have favourable resources will have more people than areas with few resources.

CASE STUDY: The relationship between population distribution and resources in France

According to the 2006 census, France had a population of just over 61 million – the twenty-first largest population in the world and third largest in the European Union.

With an average population of 107 inhabitants per km^2, France is relatively densely populated in global terms but in Europe, France is among the less densely populated countries, coming ninth on the list of EU countries, a long way behind the Netherlands (at 460 inhabitants per km^2), and the United Kingdom (240 inhabitants per km^2). However, average density is not a particularly useful indicator since population distribution is highly uneven. Half of the population occupy just over 10% of the surface area. This includes the Paris area, the lowland basins, the Mediterranean and Brittany coasts and the industrial areas in Lorraine and Nord-

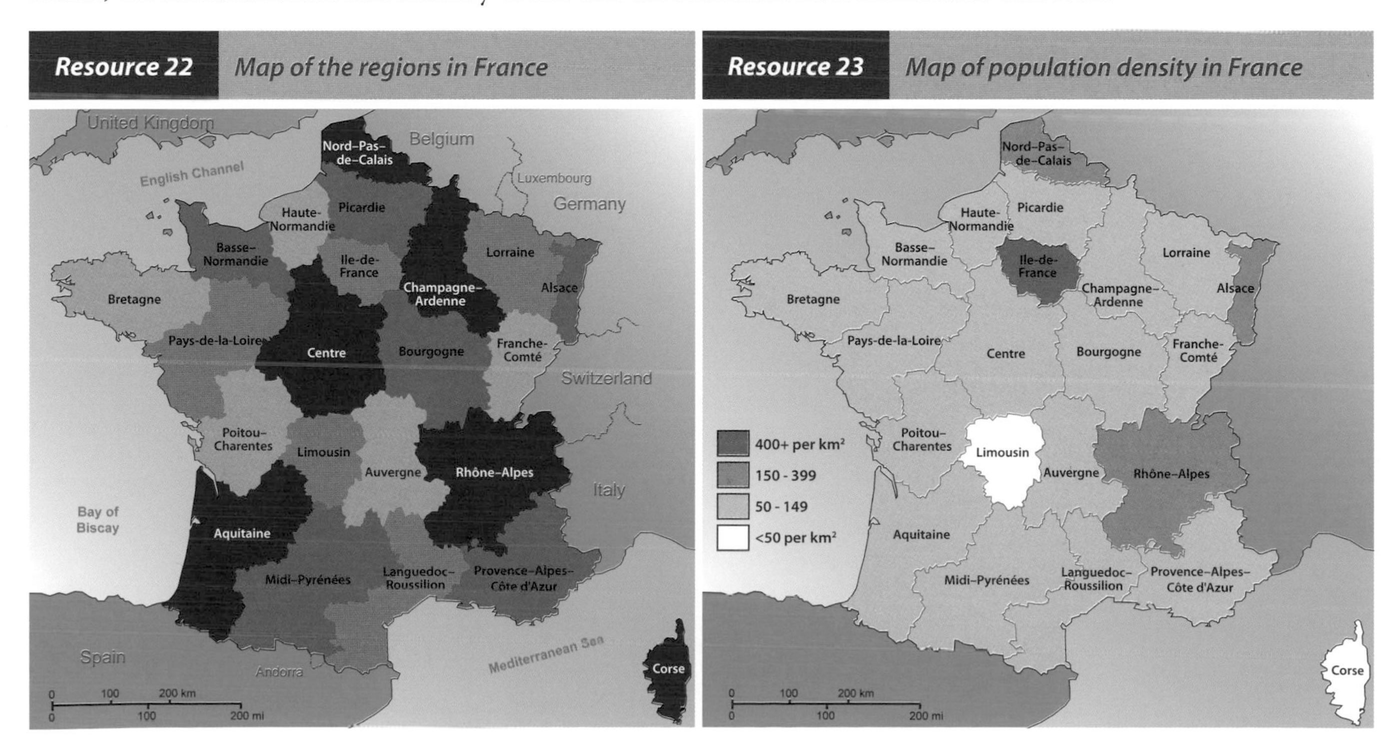

Resource 22 *Map of the regions in France*

Resource 23 *Map of population density in France*

Pas-de-Calais. These areas have the densest populations, but it is Paris which holds the record, with 20,000 inhabitants per km^2. Meanwhile, large areas of countryside are sparsely populated – sometimes with fewer than 20 inhabitants per km^2. These sparsely populated areas include the mountainous zones of the Alps, the Massif Central, the Pyrenees and Corsica. How can this pattern be explained?

1. Physical factors

There are three main physical regions in France:

(i) Lowland Basins

These are mostly in Northern France and include the Paris Basin and the Aquitaine Basin, and the Garonne and Loire valleys. These are the most fertile areas in France and have the best agricultural soils in the country. Climate is characterised by mild winters and warm summers. Farming is modern and intensive and these regions have a moderate population density between 51–150 per km^2.

(ii) The Central Massif and the Armorican Massif

These areas have poor infertile soils and are of little use for agriculture. Pastoral farming is the main agricultural activity. Climate is cool and often wet. These areas are remote from the main centres of economic activity in France and have long been associated with out-migration of the young and economically active. Population densities are low, typically less than 100 per km^2. The most remote part of the Massif Central has the lowest population density in France. This is the district of Limousin with a population density of 44 per km^2 .

(iii) The High Alpine Mountains and the Pyrenees

These mountains are over 3000 m in height. They have cold mountain climates and the land is unsuitable for agriculture so consequently population densities are very low.

Resource 25 *Climatic regions of France*

Resource 24 *Physical map of France*

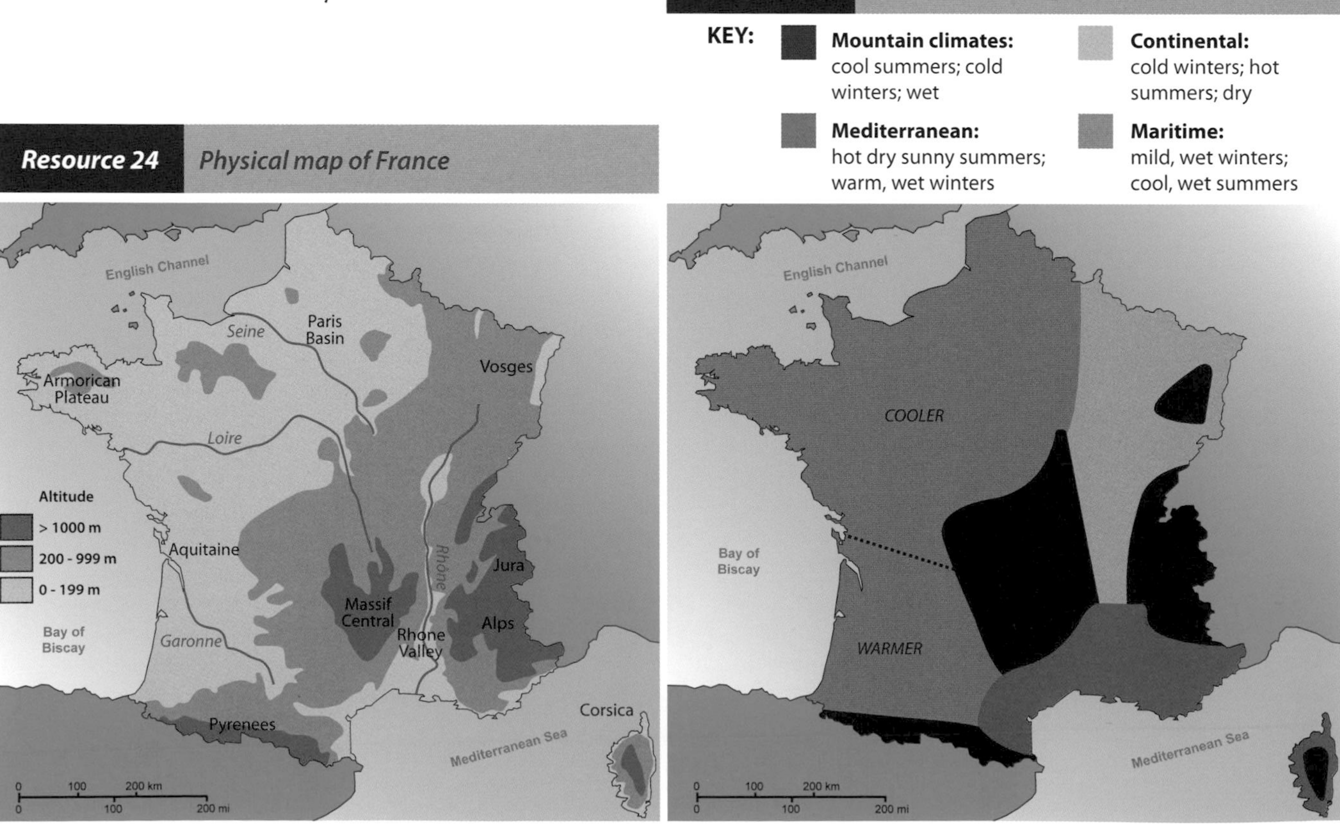

2. Human factors

(i) Urbanisation

Urbanisation associated with industrialisation occurred later in France than in some other European countries, such as Britain and Germany. It was only in 1930 that the urban population overtook the rural population. From the 1950s, France started to catch up and in 1999, 75.5% of the population were living in the 361 urban areas. The largest urban area is Paris, whose conurbation is home to 9.6 million people – over 20% of the total number of city-dwellers in France and a population density exceeding 20,000 per km^2. It is estimated that the population of the Ile-de-France (Paris area) will increase by 16% between 1990 and 2020, when it will reach 12 million. Since the 1970s the populations of most city centres and inner suburbs of the larger conurbations have decreased in size while the outer suburbs and rural districts around them have experienced a significant rise in the number of inhabitants. These outlying suburbs now house a total of 9 million people, an increase of nearly 800 000 between the 1982 and 1990 censuses, whilst the populations of town and city centres decreased by a total of over 700,000. In the rural areas surrounding Paris, population densities have increased significantly in the last ten years. Future population growth will be even greater in the Provence-Côtes d'Azur and Languedoc-Roussillon regions, with respective increases of 30 and 37%. During the last two decades, these have been the areas where population has grown fastest, along with Rhône-Alpes, the Centre and Aquitaine which are all associated with modern electronics industries. Conversely, the coalfields of the north, especially in Alsace and Lorraine, which were the focus for earlier French manufacturing industry are now in decline partly due to migration of younger people.

A view of Paris from the top of Notre Dame Cathedral

(ii) Government influence

The French government has tried to encourage new industry to locate away from the Paris region to the peripheral regions, for example the cities of Rennes, Bordeaux and Toulouse.

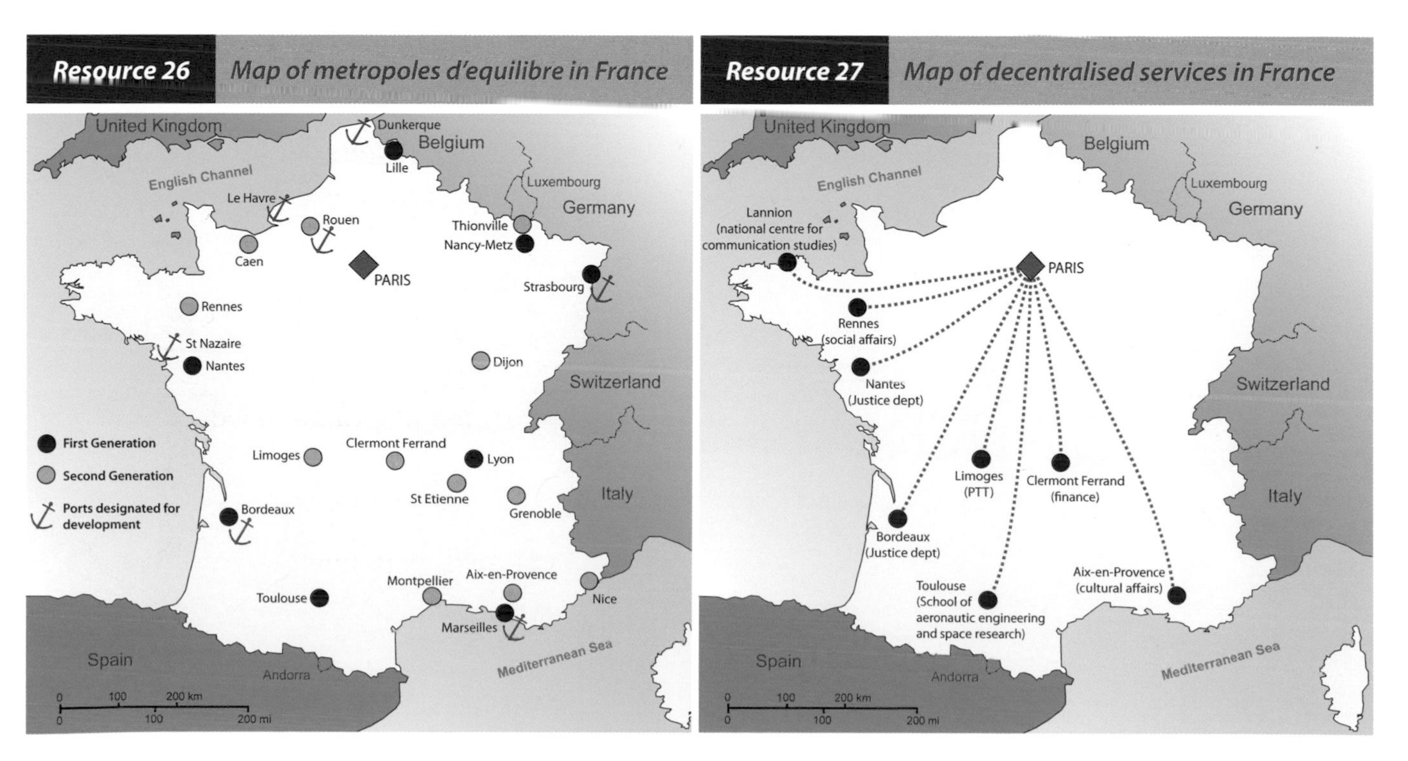

This has been carried out in a number of ways. The government made it difficult to get planning permission to build factories in the Paris region and charged high rates of tax on profits. They also set about providing incentives for industry to relocate to the peripheral regions by improving infrastructure in these areas, building industrial estates and decentralising a number of public service jobs. In addition, a number of towns were targeted for growth. These towns, or 'Metropoles d'equilibre', were given financial assistance to develop industry in the hope that people would leave the Paris region in favour of them. These plans have been fairly successful as there has been considerable growth away from Paris, but as noted earlier, the greater Paris region still contains the highest densities of population in the country.

Exercise

1. Make a larger copy of the table below, and using information from the maps as well as the text, add examples from this case study of France.

Factor	Low population density	High population density
Relief	High land and steep slopes	Low lying or gently sloping
Soil	Infertile soils	Fertile soils
Climate	Hot wet, arid or extremes of temperature	Mild temperate climates
Mineral wealth	No mineral wealth of economic importance	Accessible reserves of economically viable minerals
Distance from the sea	Areas in centre of continents	Coastal areas

2. Discuss the importance of human resources on the population distribution in France. (6)

3. (a) What is the name of the mapping technique used to show population distribution in France in **Resource 23**? (1)

   (b) State two advantages and two disadvantages of using this technique. (4)

4. *Question from CCEA January 2007*

   Study **Figures 1** and **2**, which shows population change in Canadian provinces between 1996 and 2001.

   (a)(i) Identify the region which has experienced the greatest percentage change in population in Canada between 1996 and 2001. [1]

   (ii) Using **Figures 1** and **2**, describe how one **human** factor may have affected population change in Canada. [2]

   (b) Describe how one **physical** factor can affect population distribution. [2]

Exercise continued

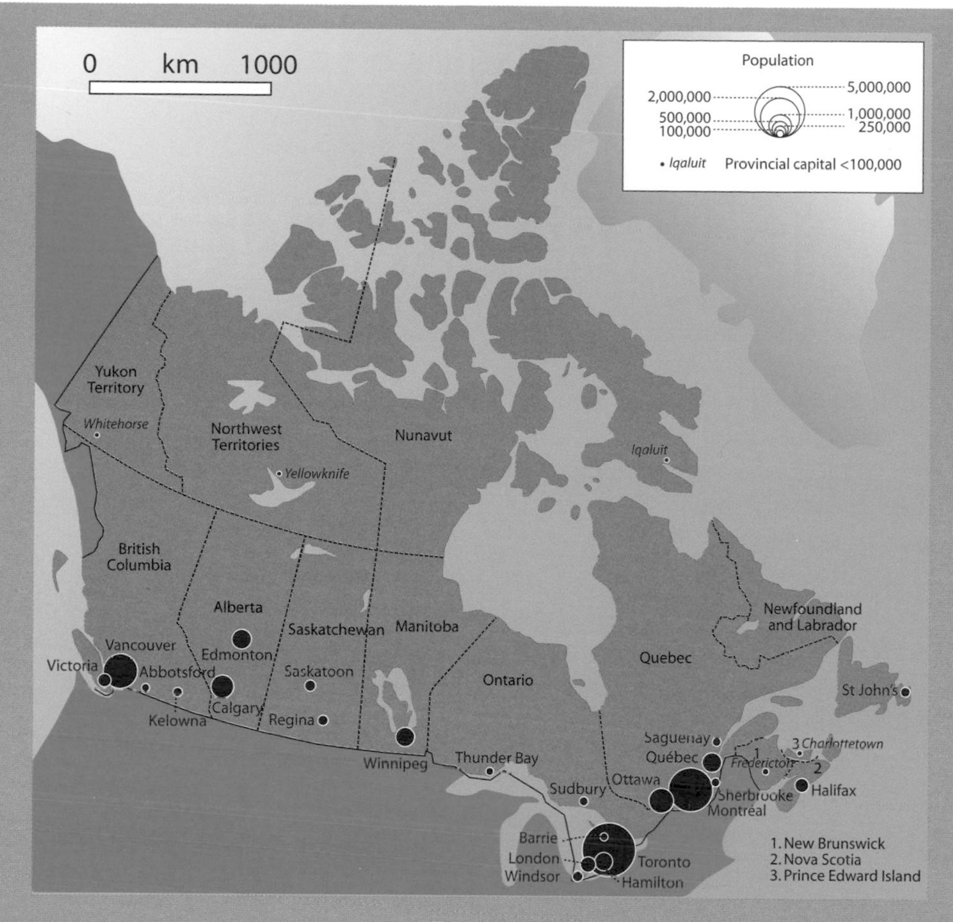

Figure 1

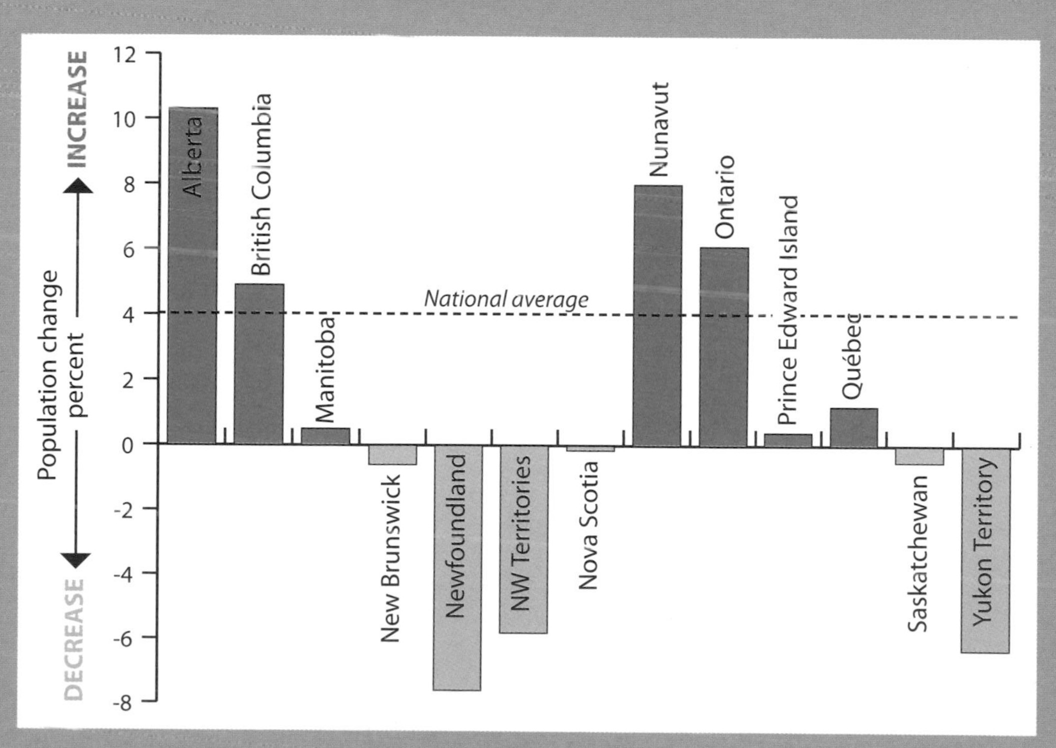

Figure 2

2A CHALLENGES FOR RURAL ENVIRONMENTS

Before embarking on a study of the challenges facing rural environments it is necessary to distinguish between rural and non-rural environments. Numerous attempts have been made to do this using criteria such as population size, population density, land use and employment structure. To date, no single and absolute definition has been achieved. The 2001 census used a number of descriptors, including population size of 10,000 plus, to distinguish an urban area from a rural area. Based on this classification approximately 65% of Northern Ireland's population is urban and 35% is rural-based.

Other common approaches employ a number of criteria such as Cloke's index of rurality, which distinguishes a gradual change from extreme non-rural to extreme rural. The Council for the Preservation of Rural England (CPRE) employs the concept of tranquil/non tranquil areas.

Resource 28 *Indicators used in Cloke's index of rurality*

Indicator	Description
Occupancy rates	The proportion of dwellings that are occupied
Commuting	The percentage of residents who commute to work
Female population	The proportion of the population who are female and aged 15–44
Amenities	The percentage of households that have inside bath and toilet
Population density	The number of people per km^2
Agricultural employment	The proportion of the workforce employed in agriculture
Elderly population	The percentage of the population aged 65 or over
Remoteness	The distance from the nearest settlement of more than 50,000 people

Source: Essential AS Geography by Ross, Morgan and Heelas published by Stanley Thornes 2000 ISBN 0-7487-5175-0

Issues in the Rural-Urban Fringe

If a clear distinction between urban and rural is complex, their physical separation is even more complex. As urban areas increase in size they encroach on former rural environments and it is where urban and rural meet that the major challenges occur. The zone where urban and rural meet is referred to as the **rural–urban fringe**.

(1) Suburbanisation

Urban areas increase in size through urban sprawl or **suburbanisation**, a process that refers to the decentralisation of people, services and industry to the edge of the existing urban area. Cities grow outward from the centre in a series of stages and with each outward advance some rural environments are transformed into urban areas. The division between town and country is not a clear one and there will always be a zone of transition between true urban and true rural. In the case of Belfast, the first phase of suburbanisation occurred as early as the late 1920s and 1930s when wealthy merchants and factory owners moved away from the densely populated and unhealthy inner parts of Belfast to the slightly higher and better drained lands at Stranmillis, Malone and Ormeau in the south, Stormont in the east and along the Antrim Road in the north. This marked the first separation between place of work and residence. Suburbanisation has always depended on the ability of suburban dwellers being able to access their work with relative ease. In the case of Belfast's earliest suburban dwellers that meant living close to the new public transport routes of trams and railways. The outer limit of suburbanisation was determined by the extent of the public transport network. Over the next decades the process of suburbanisation continued to expand as public transport improved, but large scale suburbanisation did not occur until private transport became generally available.

During the 1950s and 1960s two major developments led to rapid suburbanisation and to the first serious challenge for the surrounding rural environments.

(i) At this stage the condition of many of the working class housing areas had deteriorated to an unacceptable level and some people were rehoused in public sector housing at the edge of the city in what were called greenfield sites (rural land previously used for rural use only). Examples of such public housing included Castlereagh, Cregagh and Sydenham.

(ii) There was a general increase in the standard of living in the 1950s and people became more affluent, which enabled many to buy into the private housing market, and suburban housing developments continued to be built at the edge of the city.

The result of these developments was a sprawling city with more and more people leaving the inner areas in favour of the more attractive suburbs. In a reversal of previous trends, new industry followed the people to the purpose-built industrial estates such as Castlereagh Industrial Estate surrounded by the public housing estates of Cregagh, Castlereagh and Braniel. This in turn was followed by retail suburbanisation. The present site of Forestside in south Belfast was occupied by Supermac – one of the earliest edge-of-city shopping complexes, built in 1964 on what was then a greenfield site.

(2) Counterurbanisation

It appeared that the process of suburbanisation was now posing a significant threat to the surrounding countryside. Former villages such as Newtownbreda were absorbed by the growing suburbs. In an attempt to curtail the growth of the suburbs the Belfast Regional Plan (the Matthew Plan) of 1963 implemented a containment policy by placing a stopline or limit of growth to Belfast. Beyond the stopline, future growth was to be directed to a number of new towns away from the Belfast area. An area of open and controlled space – a greenbelt – would separate Belfast from the surrounding smaller towns. The stopline was successful to an extent in that suburbanisation was curtailed for a while, but after considerable pressure, permission was eventually granted to build public sector housing across the stopline in the west of the city at Poleglass. Once the stopline had been breached for public housing, the private sector mounted increased pressure for the release of land in the greenbelt for private housing. Eventually land was released to a number of private developers notably at Cairnshill and Four Winds in the south east.

The Belfast Urban Area Plan 2001 took additional

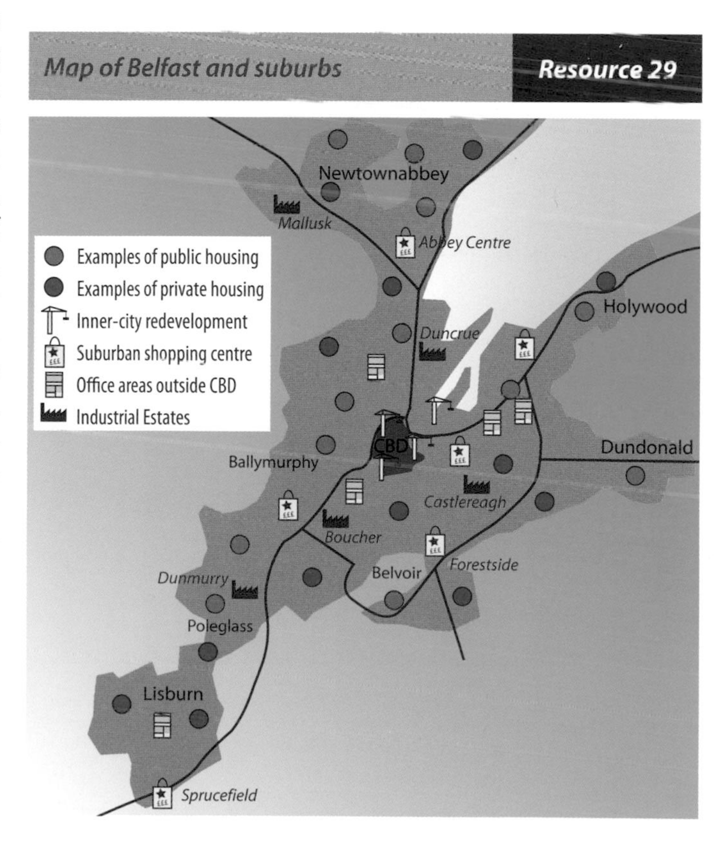

steps to prevent further suburbanisation, but if these containment policies prevented urban sprawl, another process was challenging the rural areas. This was **counterurbanisation** – a movement of urban workers to rural towns and villages within commuting distance of the city. Towns such as Comber, Carryduff, Saintfield and Hillsborough have all experienced considerable growth of private housing in greenfield sites in the last 20 years. It is not just the increase in population that challenges these rural towns but the character of the towns is often changed significantly. The new urban migrants often have little connection with their adopted living place. They work, shop and use the leisure facilities of the urban area. House prices rise as the demand for housing in towns that allow a rural living place close to an urban work place becomes highly desirable. This can be an advantage for the original inhabitants if they wish to sell property or land for development but a considerable disadvantage to the rural dweller who wants to buy into the housing market.

Resource 30 *The Belfast Metropolitan Area*

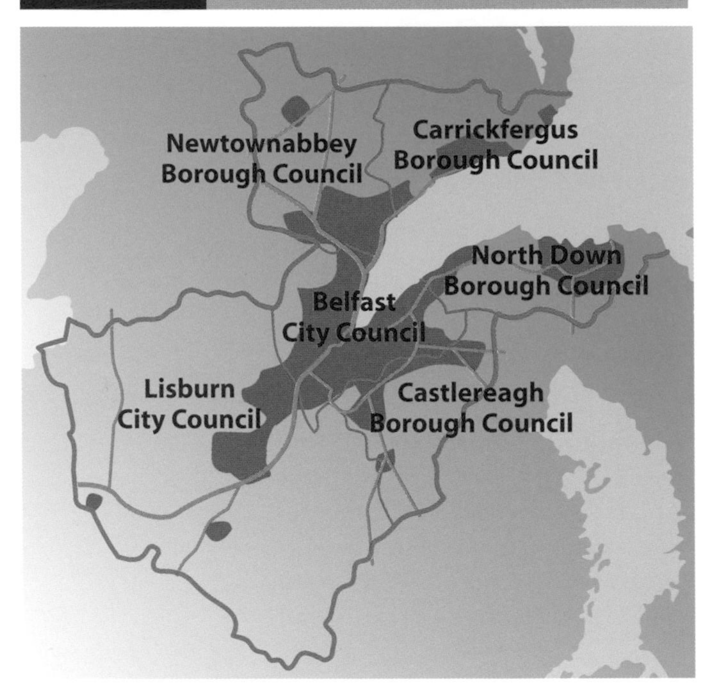

Resource 31 *Traffic Growth in the Belfast Metropolitan Area*

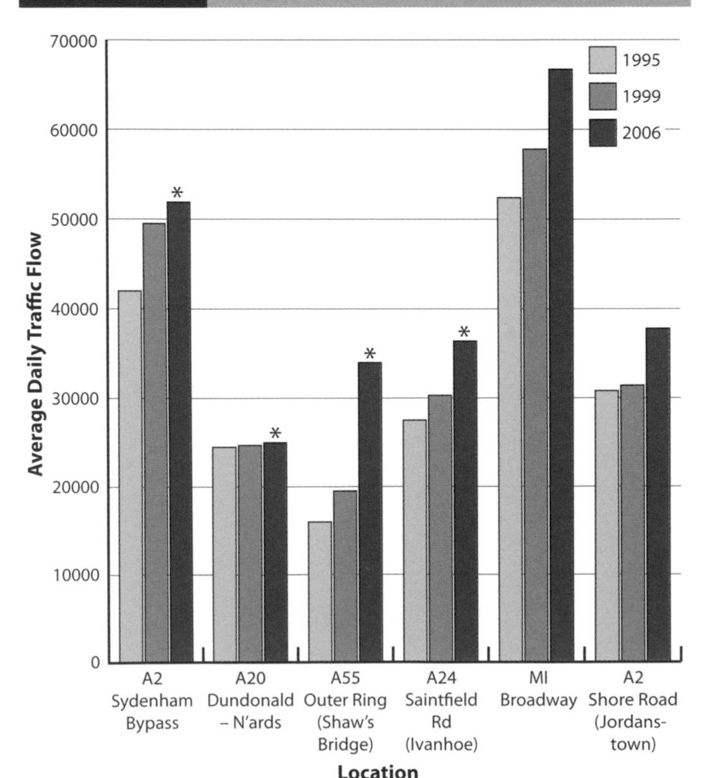

Source: 1995 and 1999 - Annual Traffic Census Report 1999, 2006 - DRD *adjusted to 2006

(3) Transport infrastructure

Greenbelt policies have also led to an increase in the commuting distance and added to the congestion on the main arterial routes to the city centre (***Resource 31***). The Belfast Metropolitan Area Plan 2015 cites a 20% increase in traffic on the major arterial routes out of Belfast. General increases in road traffic leads to congestion and longer journey times, which reduces economic competitiveness. Many road projects have been undertaken to deal with this increase. A major Ring Road skirts the outer edge of Belfast with links to the M1 and M2. Bypasses have been built around several towns, including Holywood, Comber and Hillsborough. Other roads have been widened to add additional lanes. Often these developments have taken over farmland. Occasionally where the planned road developments would encroach on land of scenic or recreational value there has been considerable public protest. A notable example of successful opposition occurred over plans to build over parts of Belvoir Forest park. More recently planners have become aware of the pollution caused by increased traffic and attempts have been made to make other forms of transport more attractive through the use of bus lanes and bicycle lanes.

(4) Greenfield developments

Another major issue is the building of retail centres in the rural-urban fringe. Greenfield developments such as Sprucefield (close to Lisburn) and Abbeycentre (in Newtownabbey, close to Belfast) have attracted criticism from local residents because of the increased traffic, and from local retailers for the loss of income. Once established these retail centres attract new retailers. A plan by the John Lewis organisation to build a store at the Sprucefield centre was successfully opposed by local residents.

Despite such opposition to new developments in greenfield sites, the process of counterurbanisation and the challenges it poses for rural areas will continue and planners have to formulate policies to deal with the increased demands for housing in the rural-urban fringe. The Labour government has allowed four million new homes to be built on greenbelt sites in the south-east of England. In Northern Ireland the recently published Belfast Metropolitan Area Plan (BMAP) has outlined future strategies and demands in the commuting hinterland of Belfast, which includes this rural urban fringe. It is estimated that an additional 51,000 houses will be required in the BMA, 9,000 of which will be built on greenfield sites. A major issue for the planners is how they can reconcile this need against their stated aim to maintain the rural character of the settlements in the outer parts of the BMA through very strict planning controls on new buildings. Most of the 9,000 new homes will be in designated settlements including Carryduff, Ballyclare and Moira. These new housing developments on greenfield sites will be subject to strict planning policies. The intention is to create balanced communities by building a mix of housing tenures including owner-occupied, public sector or social housing and specialised housing for the elderly and people with disabilities. In this way they hope to prevent the building of uniform housing developments that characterised the earlier commuter settlements.

Exercise

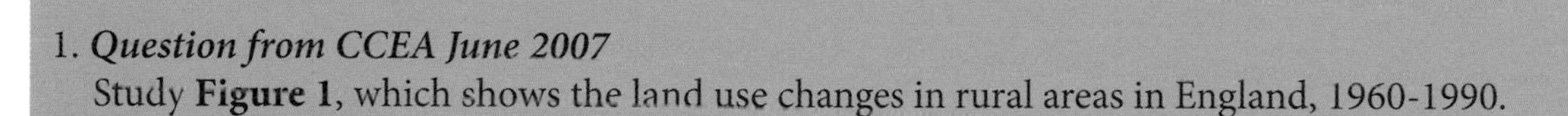

1. *Question from CCEA June 2007*
   Study **Figure 1**, which shows the land use changes in rural areas in England, 1960-1990.

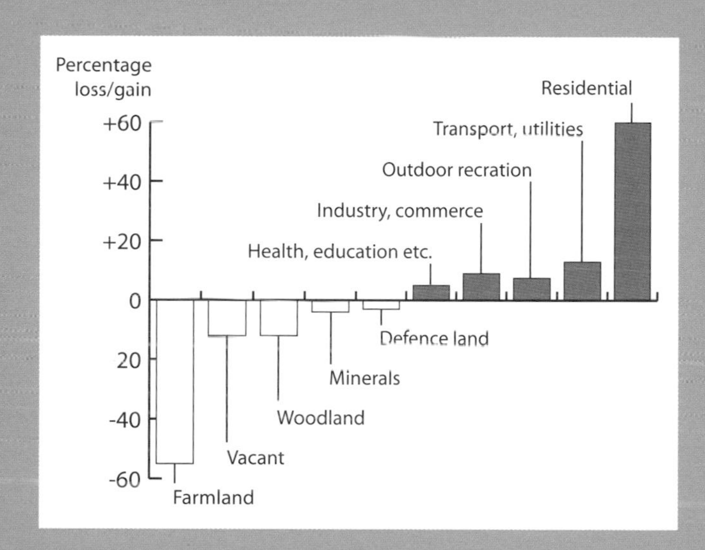

Figure 1

   (a) Identify the main changes to rural land use in England. (3)

   (b) Discuss the possible impacts these land use changes may have had in the rural areas in England. (8)

2. *Question from CCEA June 2005*
   Study **Figure 2** overleaf, which shows average journey to work travel times for commuters to selected areas in the USA in 1990 and 2000.

   (a) With reference to **Figure 2**, describe the change shown. (2)

   (b) Discuss how this change could create problems for urban environments. (3)

Exercise continued

Areas	Average travel time in minutes		Percentage
	1990	2000	
Raleigh	20.2	24.9	23
West Palm Beach	20.9	25.7	23
Charlotte	21.6	26.1	21
Atlanta	26	31.2	20
Miami	24.1	28.9	20
Greensboro	18.8	22.4	19
Las Vegas	20.3	24.1	19
Orlando	22.8	27	18
Providence	19.6	23.2	18
Jacksonville	22.6	26.6	18

Figure 2

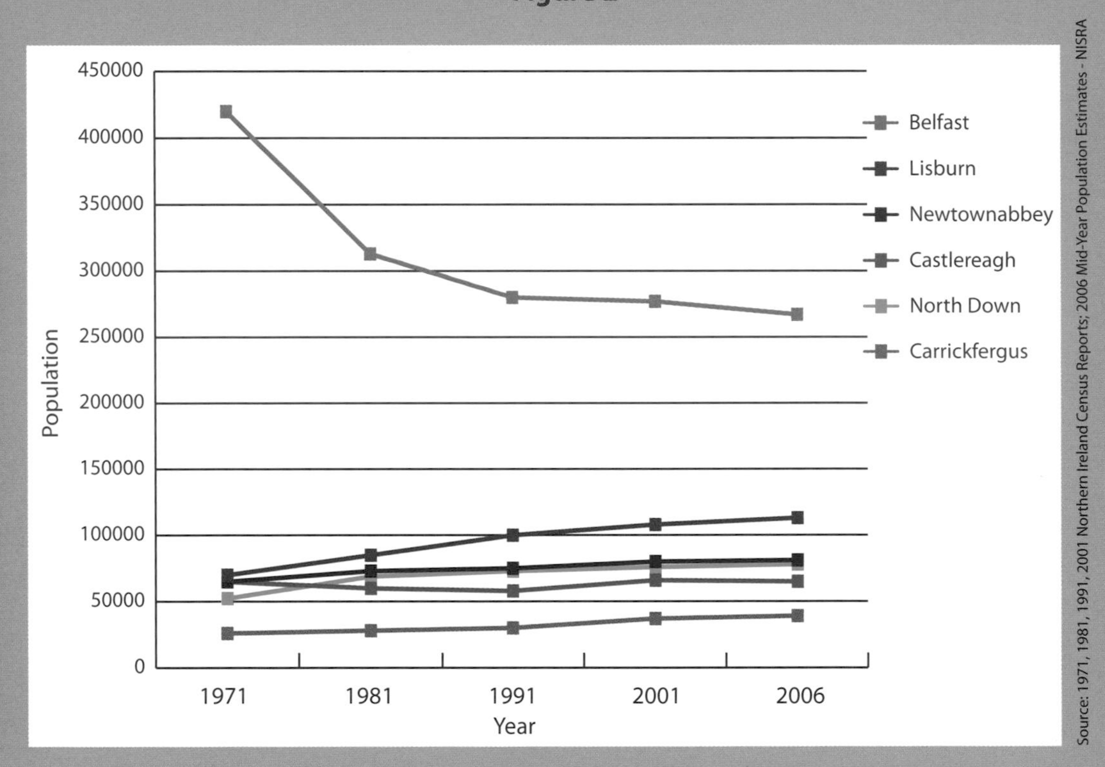

Figure 3

3. Study **Figure 3** above, which shows population change within the BMA 1971–2000.

(a) Describe the patterns of population change illustrated. (3)

(b) Identify and explain the processes that would explain these changes. (6)

References

Belfast Metropolitan Area Plan available online at: http://www.planningni.gov.uk/AreaPlans_Policy/Plans/BMA/IssuesPaper/BMAP2015.pdf

Change and conflict in the rural urban fringe Geo Factsheet no. 165

www.geographyinaction.co.uk

Issues in Remote Rural Environments

A remote rural environment refers to an area sufficiently distant from major urban areas so as not to be affected by suburbanisation or counterurbanisation. Such regions are often described as peripheral – both in terms of their location and of their economy. The 2001 census recorded 600,000 people living in remote or very sparsely populated areas. It is generally assumed that these areas will share some of the following characteristics:

Out-migration and ageing populations

In the UK the development of transport and communications has facilitated the out-migration of large numbers of people from rural peripheral regions including the Highlands and Islands of Scotland, Cumbria, Yorkshire, Cornwall and Wales, to urban areas. The perception of rural areas as old-fashioned, dull and lacking in economic and social opportunities contrasts strongly with the modern, young and vibrant perception of urban areas. Meanwhile, economic hardship is a reality in many remote rural areas. In the Highlands and Islands of Scotland there are few economic opportunities outside of farming and fishing. Small-scale manufacturing and service industries are concentrated in the main towns on the east around Inverness. Tourism does offer some employment but much of this is seasonal. Over the years a combination of an inhospitable climate, mountainous terrain and poor communications has provided sufficient impetus for many of the younger generation to move away.

Not only are the present numbers of young people affected, but so too will future population growth. With fewer people moving into the reproductive age groups, the population gradually ages, and this in turn can lead to accelerated out-migration. In recent times some of these peripheral regions have experienced an in-migration of second-home owners and retired people. These developments often lead to an increase in house prices beyond the reach of the original inhabitants and the character of the settlement can change. Either way, the influx of second-home owners and retired people have done little to curb the out-migration of the younger age groups. A government report on 'The State of the Countryside 2007' noted that the proportion of young people in rural areas aged 15-24 had fallen 6% in the last 20 years. It was also noted that whilst the median age for much of rural England is currently 45, in the most remote areas it can be as high as 62 years (East Devon).

Travel times from Penzance **Resource 32**

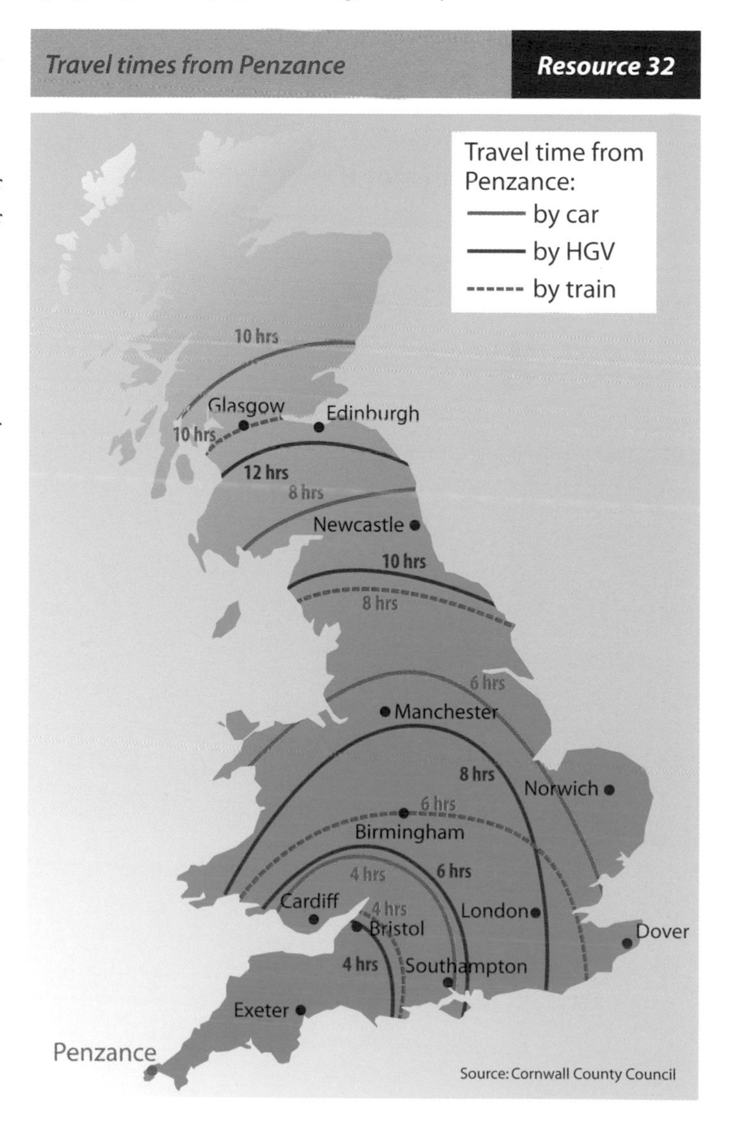

Source: Cornwall County Council

Inaccessibility and lack of economic activity

As the term 'Remote Rural' suggests, these areas suffer from inaccessibility. This may be the result of physical difficulties as in the Highlands and Islands of Scotland or simply due to physical isolation as in Cornwall. **Resource 32** shows the time taken to travel to various parts of Britain from Penzance. Inaccessibility makes these areas unattractive to modern industry because although most industry is now footloose, most investors will want to be near the large urban markets to keep transport costs low. Many rural areas have a limited range of economic activities and wage levels are significantly lower than the national average. In the most remote areas in Wales GDP per capita is almost 20% below the European average and in Cornwall average weekly earnings are 25% below the UK average. In addition, unemployment levels have risen in recent times. In the last 20 years some 65,000 job losses have occurred in farming and a similar number has occurred in the rural collieries since 1984. For many there is little alternative to finding work closer to the main urban areas.

Poor service provision

The decline in population has had an adverse impact on the provision of services in remote rural areas. Services are there for the people and if the population numbers fall, the services become non-viable and eventually they will be withdrawn. Primary schools are among the first services to be affected by an ageing population and this example demonstrates the downward spiral that can follow. The unemployment that follows from the closure of the primary school will result in less spending-power for local shops and entertainment facilities and a reduction in their income. The area then becomes less attractive for those living there and further out-migration can result, leading to a downward spiral effect.

Improvements in transport, which increase the accessibility of larger settlements, can also adversely affect service provision in these remote areas. Recent government policies have seen some services, such as hospitals, government offices and postal services, updated and centralised in the larger settlements with the closure of smaller services in more remote locations. This policy is driven by economics and is claimed to be more efficient but it has a negative impact on those settlements where closures have occurred. The growth of supermarkets and large shopping centres has had a similar impact on local village shops.

The majority of people nowadays are car owners so demand for public transport is less. In Great Britain public transport is managed by competing companies, which are concerned with making a profit. In such circumstances these companies will not operate in areas where the demand is falling. Therefore, in extreme rural areas the service is at best infrequent and at worst has been removed. Often it is in areas where there is most need for public transport that the poorest service is found.

Regarding road transport two scenarios are possible:

- The most remote areas are served by poor quality roads, which increase their relative isolation.
- Some rural settlements are by-passed by new road developments. Whilst this will preserve the rural character of the settlement the local businesses will lose potential custom.

Exercise

1. Read the three extracts below relating to rural service provision.

   (a) Describe and explain the underlying reasons for the decline in rural services.

   (b) Discuss the impacts of the closure of rural services on the local community.

2. Suggest strategies that governments might use to improve the social and economic problems facing rural areas.

Government poised to axe 2,500 post offices
By Brendan Carlin, Political Correspondent

The Government will today be accused of ripping the heart out of rural communities across the UK by condemning 2,500 post offices to closure. Despite massive protests, Alistair Darling, the Trade and Industry Secretary, is expected to confirm that more than one in six local post offices will shut over the next 18 months. The cutbacks – to take place across many countryside areas – will reduce an already dwindling post office network, now about 14,500, to roughly 12,000.

Exercise continued

It also means that Labour will have presided over the closure of over 7,100 local post office branches since 1997. In a statement to MPs today, Mr Darling will seek to sugar the pill by promising investment worth £1.7bn in the Post Office by 2011 – including £150m to keep the least viable branches open.

Susan Kramer, the Liberal Democrat spokesman, accused Labour of preparing to "gut rural communities that have already lost local shops and local schools". Mervyn Kohler, from Help the Aged, urged Ministers to understand that many people over 65 neither used a car nor went on the internet for transactions.

However, Mr Darling will repeat his view that the current size of the network is unsustainable and that drastic action is now needed. Whitehall sources last night stressed that post offices overall currently lose £4m a week despite an existing annual subsidy of £150m.

Story from Daily Telegraph
19/5/07

Tories and SNP back rural fight for services

By Kate Devlin, Scottish Political Correspondent

Communities across Scotland have fought to save local schools and hospitals in recent years. More than 100 rural schools have been closed over the last decade. In one of the latest developments, parents in Roybridge, Inverness-shire announced last December that they had raised £1 million to buy their local school, which currently has about 30 children. Local authorities say that smaller pupil rolls and out-of-date buildings are often behind decisions to close rural schools. Concerns over loss of services have also hit the health service. Last August the local health authorities in Grampian unanimously voted to scrap maternity services at four of its rural hospitals.

Protection of local services including schools, post offices and transport links were included as part of a formal statement by rural and farming organisations launched in Edinburgh yesterday. The document also calls for all legislation passing through the Scottish Parliament to be "rural proofed". Keith Arbuthnot, from the Scottish Rural Property and Business Association, said that it was "crucial" that the impact on the rural and farming communities was taken into account before any new laws were passed. The document calls for all politicians to sign up to an agenda that also includes a reduction in red tape, an increase in investment in renewable energy sources including biofuels, the promotion of a "buy Scottish campaign" to help Scottish food producers and the highlighting of the economic benefits of country sports.

Story from Daily Telegraph
21/3/07

Quarter of villagers cannot walk to post office

By Graham Tibbetts

While all city dwellers live within walking distance of a post office, only three-quarters of villagers are similarly blessed, a report says today. A post office is considered nearby if it is no more than a mile and a quarter away. According to a study by the Commission for Rural Communities, the people worst served are those in more remote hamlets where just 45% are within easy reach of a post office.

The report by the Government advisory body, entitled State of the Countryside 2007, also says that almost a quarter of a million people live in a "financial services desert" – several miles from the nearest bank, building society or cash point.

Exercise continued

Other key findings include:

- Less than half of people in the most isolated areas have access to a nearby GP.
- Over the past decade car use for food shopping has risen by a quarter.

Between 2000 and 2007 the percentage of rural households with access to a post office declined from 90 to 87, while the number within reach of supermarkets increased by 3.4%.

Story from Daily Telegraph 17/7/07

References

Rural Deprivation in Cornwall Geo Factsheet No. 187

The State of the Countryside July 2007

CASE STUDY: The Highlands and Islands Enterprise

The Highlands and Islands of Scotland is a remote peripheral region in the UK and Europe. As the name suggests it is a mostly mountainous region including the North West Highlands and Grampians. Included within the region is Ben Nevis, the highest mountain in the UK, standing at 1,300 metres. The Islands are also mountainous and even more remote and are totally dependent on ferries for linkage to the mainland.

Resource 33 *Map of the Highlands and Islands Enterprise Network*

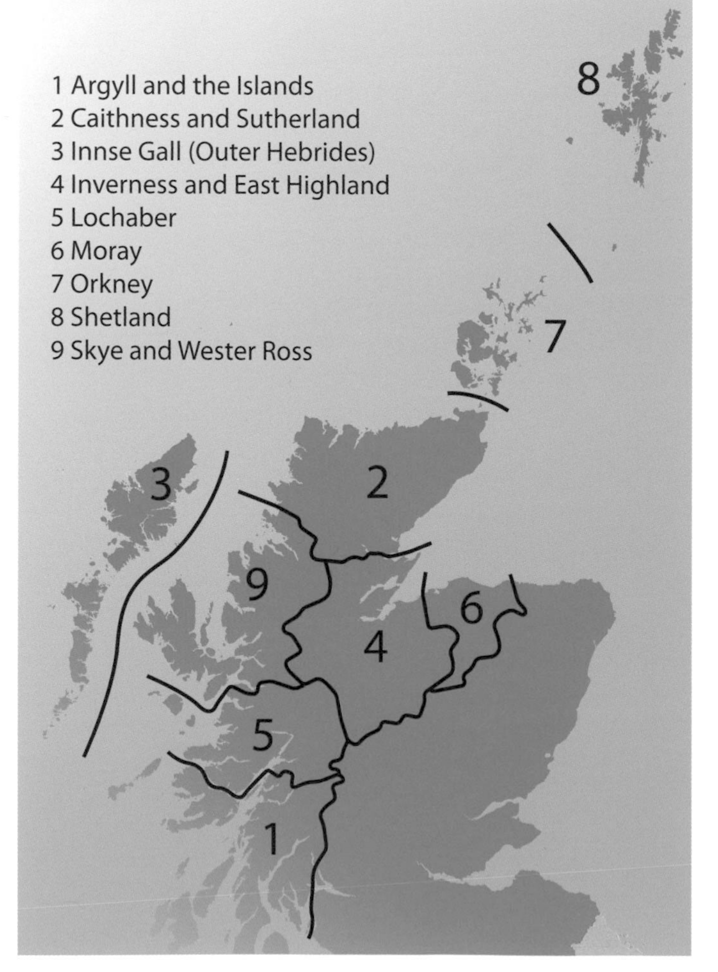

Climate

Climate is affected by latitude and altitude, resulting in lower than average temperatures in summer and winter. Rainfall totals exceed 2,500 mm in the highest areas and these same areas also have up to 60 days of snowfall annually and only five to six months of a growing season. Soils are either peat bog or podsol and therefore have limited agricultural potential. There are only scattered lowland areas along the coast.

Raw materials

Raw materials are lacking in this region with the exception of oil in the North Sea. There is potential for tourism in the snow-covered Cairngorms and the scenic highlands and islands.

Farming

Farming is difficult due to the high mountains and poor, infertile soils. About 40% of the land is unsuitable for any agriculture, 50% is only suitable for rough grazing and a mere 1.5% is suitable for arable farming. The area qualifies as a Less Favoured Area within the EU and is eligible for grants. Most farms are for hill sheep. This has led to the development of a textile industry, albeit a small scale craft industry producing Scottish tweed (Harris). Fishing, fish farming and fish processing are also developed with approximately 50% of Scotland's total catch coming from here. The Highlands and Islands have also world famous whisky distilling industries.

Traditional farms or crofts used to cover a much larger part of the Highlands and employed more people. During the nineteenth century large landowners forced many of the crofters off their land because they wanted to create large open pastures for herds of sheep. As a result of these Highland clearances, many people moved to the towns of Southern Scotland and England where they hoped to find work in the new factories. Rural depopulation has continued and some of the islands have no settlements – only 16 of the 100 islands that make up the Shetland Islands are inhabited.

Scrabster harbour, Caithness, Scotland

Industry

Large-scale industry is limited. Labour supply is inadequate and local markets are small. There is some industry associated with oil around the Moray and Cromarty Firth regions and Inverness. A combination of North Sea oil, timber and water resources have attracted a number of firms to this area. On the Orkney and Shetland islands industry is heavily reliant on the oilfields. In addition, unemployment has risen significantly in the Caithness area because of the continuing decommissioning of the nuclear plant at Dounreay, and the restructuring of some RAF bases in the Morar region.

Regeneration

It is against this background that the Highlands and Islands Development Board and now the Highlands and Islands Enterprise was set up to attempt to regenerate the economy of this very remote rural region. Similar development agencies exist in other remote rural areas of the UK.

The Highlands and Islands Enterprise (HIE), consists of a main body based in Inverness, and nine Local Enterprise Companies (LECs). Each LEC is distinctive and has a particular set of difficulties to deal with regarding economic development. One overriding problem is the continued out-migration of the younger age groups. If the HIE is to be successful it must address this issue by attracting developments that will discourage further out-migration. In order to do this a considerable amount of funding is required to attract modern industry to this region. The HIE has obtained funding from several sources including the Scottish Executive, the Department of Trade and Industry at Westminster, and the European Union; this region qualifies as an Objective 5b status area meaning that financial help is available to assist the restructuring of the local economy. Each LEC has attempted to adapt modern technologies in order to use local raw materials and skills more efficiently.

Highlands and Islands Enterprise are campaigning to have transport improved throughout the region. Their plans would create 100 new jobs in the construction phase and could bring as much as £400 million to the economy over the next 30 years as a result of increased accessibility. Poor communications is one of the main obstacles to development in this region as a whole.

Highlands and Islands Enterprise continues to promote developments within this part of Scotland that will hopefully encourage more young people to remain in the area. Developments such as Distance Lab should provide employment for young people with the appropriate skills. The new educational establishments, including the UHI, will go some way towards providing these skills. In addition, the remoteness and difficulty in providing services has been addressed through the initiatives to increase the use of home working and telemedicine as well as the

Resource 34

LEC	Action taken
Orkney	• £12 million invested in a European Marine Energy Centre working on a scheme to harness wave and tidal energy.
Argyll and Islands	• A Marine Science Biotechnology research centre has been built close to Argyll College and will eventually become part of the proposed University of the Highlands and Islands (UHI).
Caithness and Sutherland	• This area has had high unemployment due to the decommissioning of the nuclear plant at Dounreay, which resulted in the loss of 2,500 jobs. £12 million has been invested to help regenerate this area. • An off-shore wind energy project is planned and a wind farm has been built at Sutherland. • A Business park has been built at Wick. This will include two educational research facilities funded by the European Regional Development Fund (ERDF) and HIE. • A marine energy research facility has been funded at Thurso. • Money has also been made available for the development of tourism, which is predicted to increase its revenue by 30% over the next ten years.
Inverness and Eastern Highland	• £5.5 million has been invested in a Northern Ireland-based firm to create a combined heat and power plant and a wood fuel factory at Invergordon. This will create 38 new jobs and will produce 5MW to the National Grid as well as reducing oil demand by approximately 6 000 litres annually. • A new theatre complex at Inverness has been built with money from HIE, ERDF and the Scottish Executive. • A Business Park has been built close to Inverness airport. It is hoped that this will attract modern engineering industries and already it has attracted a firm that services the European jet. An airport hotel is also planned. • A Higher Education Centre for Health Science at Inverness, which will eventually become part of the UHI.
Skye and Western Ross	• £3 million has been invested in a new creative and cultural centre on Skye to promote Gaelic culture and craft industries.
Moray	• A Distance Lab has been built at Forres. This is a high-tech research centre, which plans to develop products that will overcome some of the problems caused by remoteness. Developments in telemedicine are also being explored whereby patients in outlying districts can access a GP by video link. • Other developments include the provision of hand-held computers and specially adapted mobile phones for tourists travelling in the Highlands to provide specific detail about their exact location. This virtual tour-guide will enhance the visitor's experience and stimulate further tourism. • There are plans to increase access to broadband that will enable people to work from home. • The restructuring of two RAF bases in the region has caused a rise in unemployment but it is hoped that the building of two new business parks at Buckie and Keith, along with the new skills developed, will encourage outside investors. Already, there are new food-processing industries installed, and plans for a renewable energy plant are underway.
Lochaber	• A new timber-processing plant at Fortwilliam has been constructed. • Existing industries, including tourism and other outdoor activities, as well as fish farming and fish processing are being supported and modernised.
Innse Gall (western islands)	• State-of-the-art technology is being used to bring Hebridean culture to tourists through the use of hand-held computers and satellite navigation. • A training centre and child-care facility has been built to serve the islands of Uist and Barra. This will combine child-care along with training in modern skills and job-placement, especially for women, in this very remote region. • Tourism and craft industries are also supported.
Shetland	• £5 million has been invested to help the main industries associated with oil, fish-catching and processing, aquaculture, tourism, knitwear, agriculture, and the new skills being developed in these areas.

improvements in transport mentioned above. The HIE website includes the following facts as evidence of their success for the year 2006–2007:

1. Fifty-nine businesses started by young people.
2. Sixty small-scale renewable energy projects are in operation.
3. The skills programme managed by HIE supported 3,000 people into employment.

Exercise

1. With reference to a case study of a local/regional development agency:

(a) Describe the economic problems that occur in this region.

(b) Explain how the regional development agency has attempted to solve these problems.

2. Research one other rural development programme and evaluate the success of the strategies employed.

References

The Highlands and Islands Enterprise Network – www.hie.co.uk
DEFRA website – www.defra.gov.uk/rural/default.htm
England's Regional Development Agencies – www.englandsrdas.com
Department of Agriculture and Rural Development in Northern Ireland – www.dardni.gov.uk

2B PLANNING ISSUES IN RURAL ENVIRONMENTS

The management of the countryside for conservation, recreation and tourism

The countryside has always been viewed as an area of attractive scenery and healthy lifestyle and as such has acted as a magnet to city and town-dwellers who want somewhere to relax either on a day trip or longer holiday. The increase in car ownership and general improvements in transport since the 1950s has meant that the more popular sites can become overused and spoilt. At the same time, many in the rural areas are keen to reap the financial rewards that comes with increases in tourist numbers. With declining incomes in agriculture, many farmers have sought to increase their income by providing bed and breakfast accommodation, campsites and selling local farm and craft produce to city tourists. In some cases farmland has been sold as sites for second homes. Agriculture, too, has changed dramatically – greater intensification of farming, the removal of hedgerows and the increasing industrialization of farming have all threatened to change the character of rural areas. Even in the 1950s, some suggested that control or planning was necessary to protect the scenic attractiveness and environmental quality of many rural areas from irreversible damage.

Resource 35 *Map of Britain's National Parks*

A number of public bodies have had responsibility for the management of the countryside. Currently 'Natural England' is charged with this responsibility. This was formed in October 2006 from an amalgamation of a number of groups that had individual responsibilities for rural communities, conservation and recreation. By integrating all of these responsibilities into one organisation, a more efficient service should be provided. One of the main responsibilities of Natural England is protected land. Protected land refers to land that has some intrinsic scenic, historic or ecological value that requires management to conserve it for future use. Much work on protected land has been ongoing since the 1940s, and Natural England will monitor and evaluate schemes already established and, with funding from central government, will rule on future developments. There is a number of categories of protection for specific types of environment. The most common are Sites of Special Scientific Interest (SSSIs), National Parks and Areas of Outstanding Natural Beauty (AONBs).

Sites of Special Scientific Interest (SSSIs (and ASSIs in Northern Ireland))

These are areas that have special wildlife or rare flora. There are over 4,000 SSSIs in England, covering around seven per cent of the country's land area. They include wetlands, remote moorland and peat bogs. SSSIs support rare plants and animals that now find it difficult to survive in the wider countryside. There is a list of restrictions in force in SSSIs and applications for development within an SSSI must be passed by the SSSI regulators. Most are in private ownership.

National Parks

These are the largest areas that are protected. The first were set up during the 1950s to manage conservation and recreation. There are currently 14 National Parks in the UK (***Resource 35***). In Northern Ireland there are plans to have the Mournes designated as a National Park. About 110 million people visit National Parks annually. Much of the land in National Parks is privately owned and they are living and working landscapes with farming, quarrying and forestry as the main economic activities in addition to tourism. Each Park has a National Park Authority whose responsibilities include conservation of the natural beauty, wildlife and cultural heritage of the Park while at the same time improving opportunities for public understanding and enjoyment of the Park. Money and resources are available to promote and manage tourism within the area but there are restrictions on many types of development. The greatest challenge for the National Park Authorities is attempting to reach a compromise between the interests of the various groups which use the Parks. At times the differences are very great, for example the interests of a quarry owner would seem to be irreconcilable with those of a conservationist or a tourist. If there is a conflict between conservation or tourism, greater weight is given to

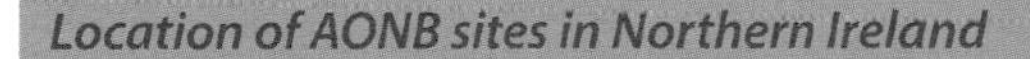

Location of AONB sites in Northern Ireland **Resource 36**

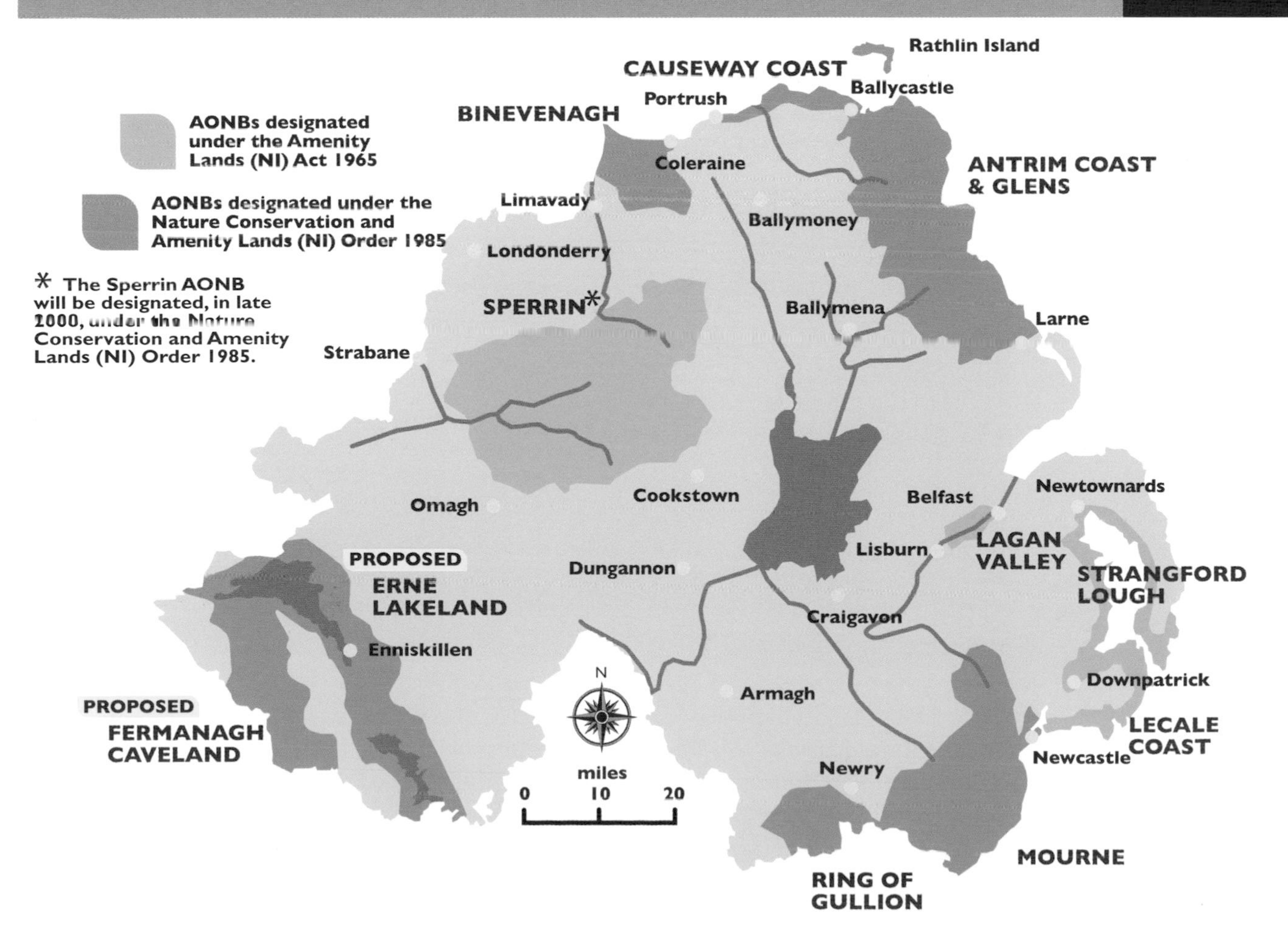

Source: © Environment and Heritage Service (EHS)

conservation. One major issue for all National Parks is access, especially on privately-owned land. In recent times, the rights of access have been increased, meaning increased potential for conflict in the future.

Areas of Outstanding Natural Beauty (AONBs)

These are usually smaller in area than National Parks. There are 36 in all, covering about 15 per cent of England. The smallest is the Isles of Scilly, which is only 16 square kilometres, and the largest is the Cotswolds, totalling 2,038 square kilometres. Natural England is responsible for designating any new AONBs and advising government and others on how they should be protected and managed. In contrast with National Parks where conservation and recreation are joint aims, AONBs are primarily concerned with conservation and enhancing the natural beauty of the landscape, although tourism is an important secondary issue. To achieve these aims, AONBs rely on planning controls and practical countryside management. They also differ from National Parks in that they do not have the equivalent of the National Park Authority, and their management is the responsibility of the local council. Recent legislation aims at strengthening the management and funding of the AONBs. There are nine AONBs in Northern Ireland and two proposed (***Resource 36***).

References

Information on protected land in Northern Ireland, Scotland and Wales is available from the following websites:

www.ehsni.gov.uk/
www.snh.org.uk/
www.ccw.gov.uk/
www.aonb.org.uk
www.naturalengland.org.uk/
www.nationalparks.gov.uk/

Exercise

A scene from the Mournes

Community split over national park

Stretching from Carlingford Lough to Newcastle and Slieve Croob, Northern Ireland's first national park would be a magnet for visitors, creating a tourism boom and up to 2,000 jobs. But local people fear it may be at their expense, pricing young people out of houses, strangling farms and industry, and destroying local communities. Local people have long accepted they must share Mourne with the many visitors who come to enjoy this magnificent landscape but establishing a National Park is proving controversial and has left many people worried about the

Exercise continued

impact on house prices, farms and traditional industries like sand, gravel and granite production.

Those supporting the plan point to the potential boost to the local economy, predicting that tourism revenue could treble from its current level of around £70m per year. The National Park supporters also point to opportunities to brand local produce and claim that the area is likely to benefit from increased investment. But many Mourne people remain unconvinced and are worried about the impact on their lives. One hundred and fifty thousand visitors arrived here in 2006 and many farmers are worried that the creation of a National Park would result in more visitors crossing their farmland and adding to traffic on narrow roads.

Poster campaign against the proposed national park

The recently established Mourne and Slieve Croob Residents' Action Group voiced concern about house prices and the danger of younger people being forced to leave the area. Valerie Hanna of the residents' group says the fear is that wealthy retired people will move in and buy up houses, making it impossible for young couples from the area to set up home. Many people are also concerned about stiffer planning restrictions and fear a national park designation would make it even more difficult to start a new business or even extend their home.

National Park supporters though, paint a very different picture and see a park authority as helping to protect a landscape which has suffered its share of 'bungalow blight'.

Quarry and sand pit owners, who employ about 200 workers, have also set up an action group to fight the national park plan. They fear that a National Park Authority would halt their operations.

Adapted from BBC News
http://news.bbc.co.uk/go/pr/fr/-/1/hi/northern_ireland/6388157.stm
23/02/2007

Question

Read the extract above. Research this topic further and present an argument for or against granting National Park status to the Mournes.

CASE STUDY: Tourism in the Peak District National Park

The Peak District is situated at the southern end of the Pennines and covers most of northern Derbyshire as well as parts of Yorkshire, Cheshire and Staffordshire. In 1951, the Peak District became the first National Park in Britain and since then large numbers of people have been guaranteed access to it by law. It covers 1,438 square kilometres of scenic countryside from moorland in the north to green farmland and limestone plateau in the south and has a wide diversity of habitats and wildlife. The Peak District is also valued for its cultural heritage, including stone circles, ancient hill forts, medieval castles, mills, lead mines and dry stone walls. Some areas within are also designated as SSSIs. Unlike National Parks in other parts of the world, Britain's National Parks are not wilderness areas but have moderate densities of population and a variety of economic activities. The Peak District National Park has a resident population of 38,000. Apart from tourism, the main types of economic activity include manufacturing (cement), quarrying (limestone), and agriculture. There are several large reservoirs to supply the local urban areas. Some of the forested areas are managed for timber supply.

The Peak District National Park is surrounded by many large cities and is easily accessible to the 15.7 million people who live within 80 kms of the boundary. Most visitors come from Derbyshire (14%), South Yorkshire (13%), Cheshire (12%) and the other counties that are

Resource 37 *Map of the Peak District National Park*

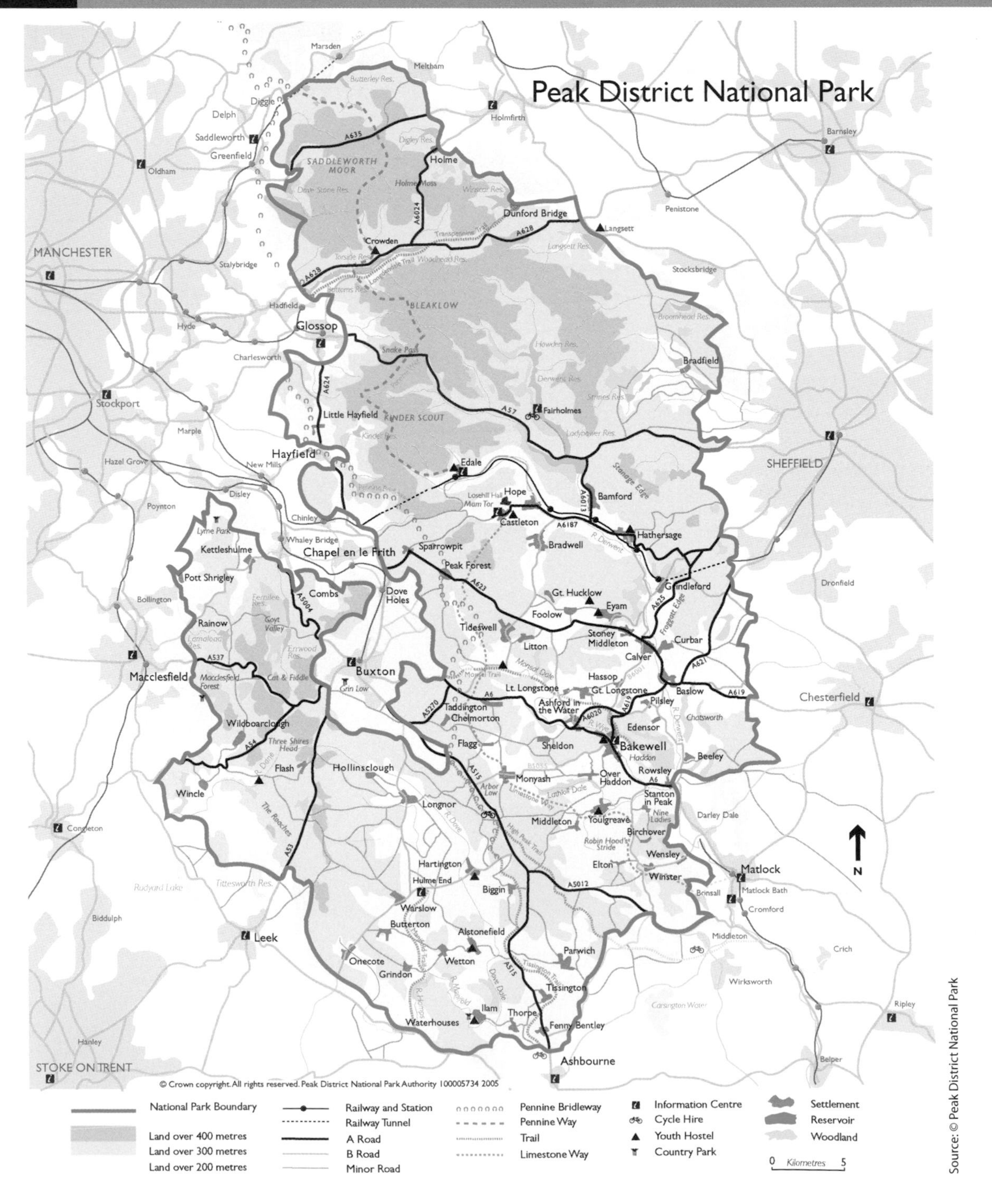

Source: © Peak District National Park

partly within the National Park. Since its designation as a National Park the variety of tourist activities has increased. In the 1950s traditional outdoor activities were the most important but it was the increase in car ownership from the 1960s onwards that brought large numbers of tourists to the most attractive locations. Such sites are referred to as honeypot sites. Among the most popular are:

- Chatsworth, home of the Duke of Devonshire
- Dovedale, a spectacular limestone dale
- Hartington village
- Hope Valley and the village of Castleton
- Upper Derwent and the Ladybower and Derwent Reservoirs

View of Edale and the Hope Valley, Derbyshire, Peak District

In the 1980s there was a considerable growth in action sports such as hang-gliding, water sports and mountain-biking. In recent times, private commercial firms have applied to the National Park Authority for planning permission to build hotel complexes and sporting facilities.

Tourism is by far the most important source of income in the National Park providing around 500 full-time jobs, 350 part-time jobs and 100 seasonal jobs. Tourism also provides the income to keep many historic buildings in good repair. Chatsworth, home of the Duke of Devonshire, is one example. Local people also benefit by providing caravan and camping sites in their fields or offering bed and breakfast accommodation in their homes. This is a very valuable source of added income, in the upland areas especially; without it many would have been forced out of farming altogether. The number of households offering farm-based holiday accommodation increased by 45% between 1991 and 2000. Local shops also benefit from tourist trade. The popularity of 'honeypot' villages, such as Castleton, means there is a greater level of local employment than is usual in a village of this size. Continuation of traditional crafts is also encouraged. In 2000 there were over 300 local and traditional events, reflecting the importance of local history and customs.

Conflicts in the Peak District National Park

Between 16.2 and 20 million visits are made by car to the Peak District National Park every year and another 1.5 million visits made by public transport. Such large numbers undoubtedly present a considerable challenge to the Park Authorities. In particular, some of the most popular honeypot areas attract large numbers of visitors resulting in overcrowded car parks, blocked roads, and overstretched local facilities especially in summer. Over 60% of all recreational visits to the National Park are made during the months of May to September. In a typical summer week over 500 000 visits are made.

There are 3 000 km of public rights of way in the Peak District National Park. Some routes, especially those over moorland, have been seriously eroded by the heavy use of off-road vehicles and mountain bikes.

Wildlife may be disturbed by such activities as orienteering, mountain-biking and hang-gliding. Climbing can result in the decline of some species such of mosses, lichens and cliff-nesting birds. Meanwhile, litter of all kinds is both unsightly and can cause pollution, damage to livestock and wild animals. Broken glass is a particular danger to people and a possible cause of fire. Farmers are also concerned by loss or injury to livestock and by the actions of tourists; trampling crops of grass reduces the amount of winter feed for farm animals. Walkers who stray from footpaths cause damage to dry stone walls.

Gift shops and cafes that cater for the needs of tourists are often more profitable than shops selling everyday goods for local people (eg butchers or bakers). In some villages where tourist

Resource 38 *Map of zoning in the Peak District National Park*

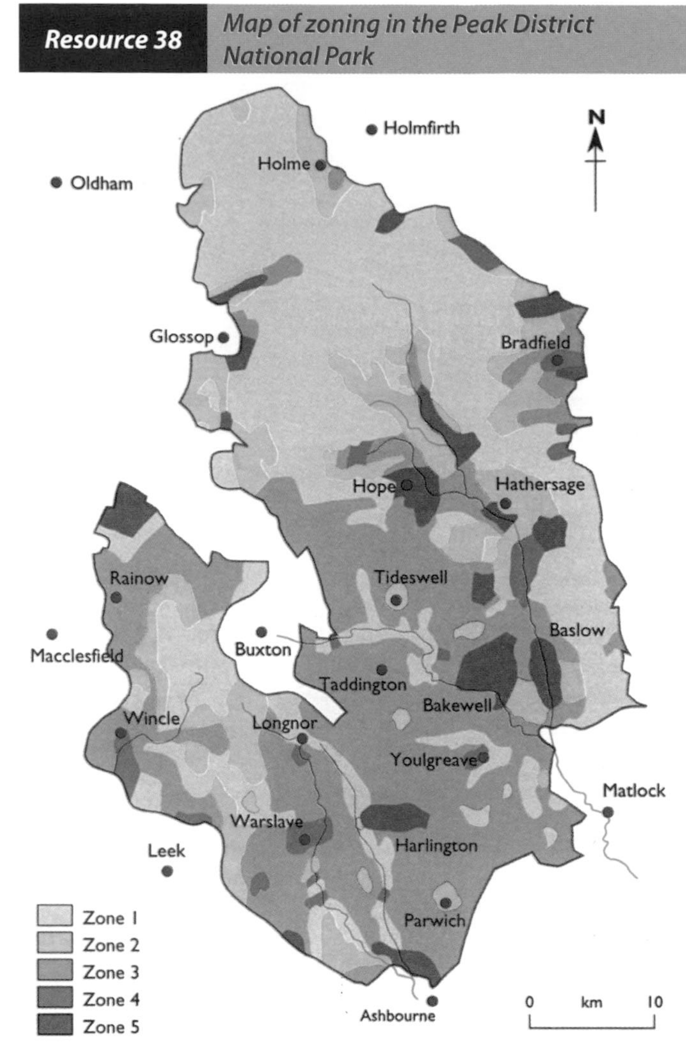

Explanation of zones

Zone 1: Wild areas with a general absence of human influence; low density, quiet recreation

Zone 2: Remoter areas of farmland and woodland with poor access; low density recreation opportunities

Zone 3: More generally accessible areas; moderate visitor densities

Zone 4: Specific localities suited to moderate levels of recreational use

Zone 5: Accessible areas of 'robust' landscape, able to take high visitor densities and facilities

shops are in the majority and few shops cater for the local people, the local community may feel 'pushed out' by tourists.

The Peak District Sustainable Tourism Strategy

The National Park Authority (NPA) is responsible for drawing up policies for planning and management of the National Park and it is their task to seek a compromise between the interests of the various groups which claim access to the Park such as the Ramblers Association, local communities, tourists and conservation organisations. The diversity of interests among such groups is such that this NPA, like many others in Britain, has opted for a zoning of the park. In this way special zones are set aside for particular types of activity in an effort to minimise potential conflict (***Resource 38***).

In 2000, the Peak District Sustainable Tourism Strategy was developed. The intention was to further promote the park as a major tourist attraction, which would help the economy of the area but at the same time protect the environment in the long term. A number of partnerships with local communities has been set up emphasising that the park is an integral part of the surrounding countryside and that its maintenance is not simply the responsibility of the Park Authorities. This is very important considering that approximately 80% of the land in the park is privately owned. The Authorities try to meet the needs of different users of the park by providing targeted facilities including:

Strategies relating to honeypot sites

- An increase in the number or capacity of car parks in popular villages and at beauty spots (honeypot sites), often with public toilets and information boards. Picnic areas are often sited close to car parks. Facilities like these are carefully designed to minimise their impact on the landscape.
- Increased provision for disabled tourists; routes suitable for wheelchairs have been developed; adapted fishing platforms have been provided and buses have been adapted to facilitate disabled passengers.
- Grants are available to assist residents set up small businesses to market local products such as crafts, cheeses and honey, to sell to visitors in order to enhance the local economy.
- There are some 13 Information Centres in the Peak District. Centres are open all year at Bakewell, Castleton, and Fairholmes in the Upper Derwent. Touch Screen information points are also open in some villages. Bakewell Centre has an average of 500 visits per day. With more information available, tourists may make use of a wider area within the park.

Strategies relating to active recreation

- A cycle-hire scheme and special cycle routes along disused railway lines or quiet roads have been constructed.

- Fishing and sailing are permitted only in a small number of the many reservoirs.
- Activities such as hang-gliding are permitted only in certain designated areas.
- Motor sports are very strictly controlled as they can cause damage to the landscape, intrusive noise and pollution to air or water.

Strategies dealing with traffic problems

- The South Pennines Integrated Transport Strategy (SPIT) includes the use of traffic-calming measures and restrictions on freight traffic using the most scenic routes across the Park.
- In the Upper Derwent valley, where congestion was particularly severe at peak times, a special Joint Management Scheme has been developed by the National Park Authority, local landowners and other bodies to restrict car access and encourage greater use of public transport. Increased parking has been provided.

Strategies dealing with environmental issues

- The National Park Authority assists the Highway Authorities with the management and maintenance of the public paths network. The Pennine Way, the most popular of the footpaths, has been paved to prevent further erosion.
- A Local Countryside Access Forum has been set up to discuss how new legislation allowing greater access to the park can be implemented without causing long-term damage.
- Planning permission to build new housing will normally only be given to local people already resident in the area and there are strict controls on the nature of new buildings or alterations to existing buildings.
- Quarry owners must landscape any unused works according to agreed regulations.

Strategies dealing with conservation issues

- Thirty-eight percent of the landscape in the Peak District National Park has already been given additional protection either as SSSIs or AONBs and local authorities are to be given increased powers to safeguard these areas from unsuitable developments.
- Biodiversity will be protected through a range of measures involving local partnerships with organisations such as RSPB.

The National Park Authority regularly reviews its management policy taking into consideration changes in government policy towards funding and recently changes in legislation regarding access to the countryside. The most recent plan sets out the NPA's targets for 2006–2011.

Conservation issues in the Peak District National Park

Climbers help protect rare chicks in Peak District National Park

Climbers voluntarily avoided some of the best climbing cliffs in the Peak District this spring to give rare chicks a fighting chance of survival. As a result, ten ring ouzel chicks were successfully raised in four nests on the highly-popular Stanage Edge, near Sheffield. This figure is up from six chicks in two nests last year. Ring ouzels are officially listed as of major conservation concern as they are in national decline, and Stanage is the south eastern outpost of their rapidly-retreating range.

Owned by the Peak District National Park Authority, Stanage is visited by a quarter of a million people each year and is managed in liaison with the Stanage Forum – a partnership of user groups, local people, land managers and conservation groups.

Exercise continued

Authority seeks public's views on quarry extension plan

The Peak District National Park Authority is inviting people's views on a planning application to extend Dale View quarry on Stanton Moor. Quarry operator Stancliffe Stone has applied for a 3.18 hectare extension to the current 9.09 hectare quarry, and proposes to give up its rights to quarry the nearby dormant Lees Cross and Endcliffe sites, should it be given permission.

The proposed extension-land is on fields off Lees Road, half-a-mile to the east of Stanton-in-Peak village, and would produce up to 62,500 tonnes of gritstone per year, mainly for building work. The life of the extension would be 21 years, and would involve a maximum 25 lorry movements in, and 25 out per day, using existing roads.

Stancliffe Stone mounted an exhibition of its plans in Stanton-in-Peak on Friday and Saturday (June 16-17), prior to submitting the scheme to the National Park Authority.

Source: www.peakdistrict.org

Examples of planning restrictions imposed by the Peak District Park Authority

Satellite dishes

You will need planning permission to install, alter or replace a satellite dish within the National Park if it is on a house, or in its grounds, or on any other building which is less than 15 metres high in the following cases:

- If the dish is facing a road or a footpath
- If the satellite dish is larger than 70 cm when it is measured in any direction
- If it is positioned above the highest part of the roof
- If it is on a chimney
- If there is already a dish on the building or in its grounds
- If the building upon which it is placed is taller than 15 metres

Extensions to buildings

If you are considering building an extension to your property you may need planning permission.

When considering extensions and alterations you should try and ensure they are in scale and harmony with your house, paying attention to such details as window openings and matching materials. The Authority encourages people to consider employing a skilled designer when preparing plans for extensions and alterations as their knowledge and experience can be useful in preparing planning applications. The Authority's Planning Officers are able to offer general design guidance prior to the submission of your scheme.

Source: www.peakdistrict.org

Questions

1. In relation to National Parks, what is meant by sustainable development?
2. To what extent do the strategies outlined above conform to the aims of sustainable development?
3. Describe and explain how this National Park has been managed for conservation, recreation and tourism.
4. Research one other National Park in Britain and contrast its planning procedure with that of the Peak District.

2C CHALLENGES FOR THE URBAN ENVIRONMENT

Issues in Inner Cities in MEDCs

Inner city areas were once at the centre of economic activity when they accommodated the workers and the factories that helped make the UK a major industrial power in the nineteenth and early twentieth centuries. In those days workers needed to live close to their place of work and so the building of large numbers of terraced houses at high density seemed the most economical way of providing homes for the growing number of factory workers. That these high density terraced dwellings were overcrowded and unhealthy went largely unnoticed by the local authorities.

Things began to change significantly from the 1960s onwards.

1. Housing had become an issue and local authorities began taking steps to improve living conditions in the inner city areas. The first phase of redevelopment involved removing many of the terraced houses and replacing them with blocks of high-rise flats. These could not accommodate the same numbers as the old terraced houses and the overspill was housed at the edge of the city in the rapidly growing suburbs.
2. Heavy industry, which offered employment to large numbers living in the inner city, declined in the face of competition from cheaper manufacturers abroad, and so unemployment increased, bringing with it financial hardship and a decrease in the amount of disposable income. New industry preferred sites in the suburban areas where the workforce was younger and had more appropriate skills. This marked the beginning of the downward decline that typified many of the UK's major cities.
3. Under such conditions, those who could, left the inner city to find a better quality of life elsewhere and eventually the inner city areas became populated by the least well off members of society – the elderly, the unskilled and new immigrants.
4. In Britain, many of these declining inner-city areas were occupied by the growing number of immigrants looking for affordable accommodation close to their place of work in the service sectors in the city. Often a chain of migration followed and many cities had clusters of immigrants from similar backgrounds living in close proximity. There are many socio-economic reasons for this clustering, such as a feeling of safety, friendship and access to ethnic services and affordable housing, but it has led to a ghetto mentality especially if the immigrants maintain their separateness. This was often emphasised by the out-migration of the white population from the streets close to an immigrant cluster.

Social and Economic Deprivation

It is easy to see that such areas today will not attract much investment and the jobs that are available will be lowly paid. With little money and often poor educational attainment, many of the younger people feel alienated and see little opportunity of escaping this cycle of deprivation. Many such areas are associated with anti-social behaviour involving drug-taking, crime and gang culture. Individuals become part of a self-perpetuating cycle of poverty (***Resource 39***). In such circumstances, the cycle of poverty is passed on from one generation to another. Children from disadvantaged homes are more likely to do poorly at school and have few qualifications,

Resource 39 *The cycle of poverty*

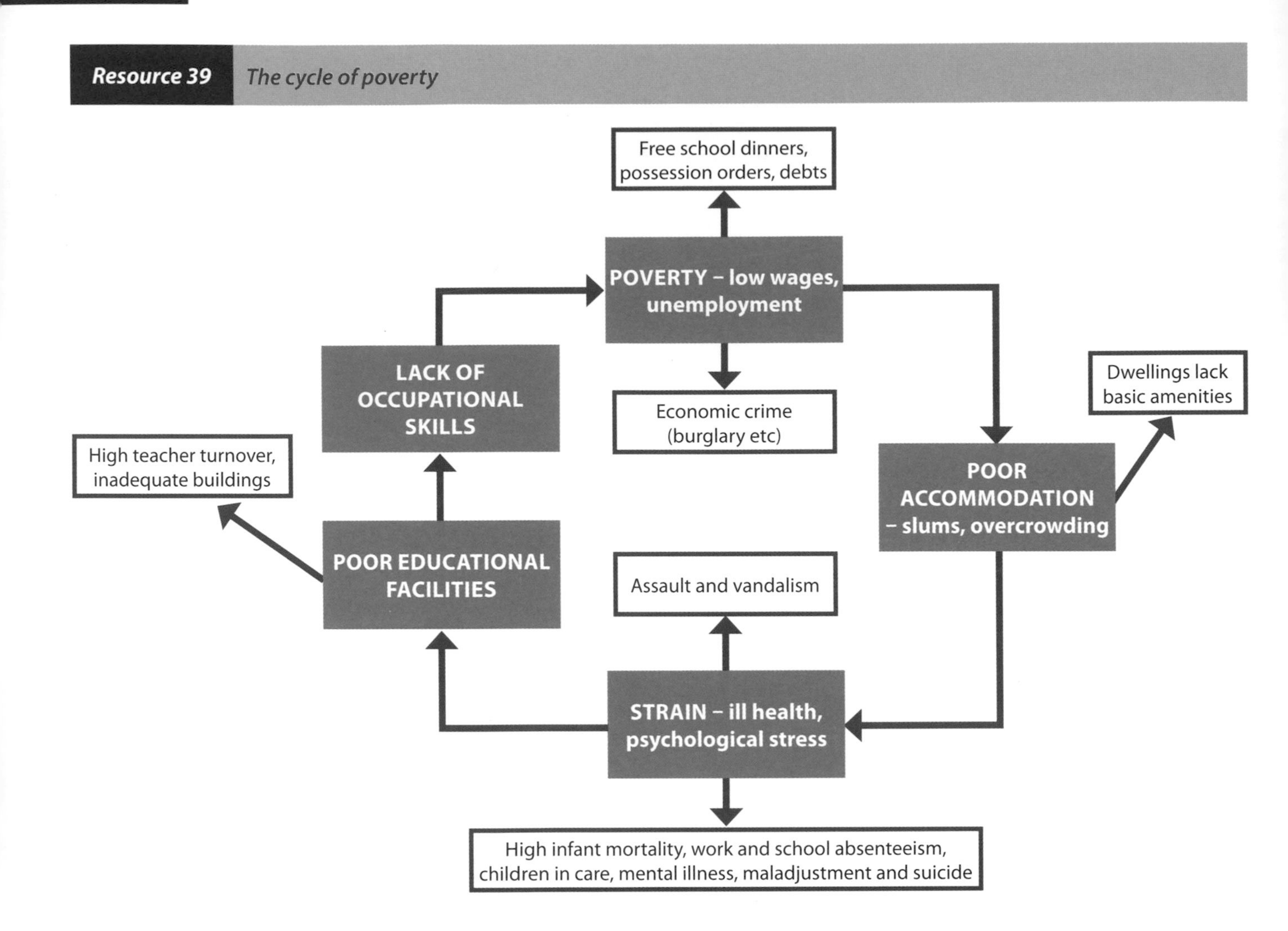

which reduces their opportunities for economic improvement. They are unable to secure a well-paid job and so they become victims of multiple deprivation. Several British cities, including London, Liverpool and Manchester experienced riots in the early 1980s. Whilst the riots were for the most part clashes between black youths and the police, they highlighted the social and economic deprivation of these areas. Belfast had its own particular problems, but much of the paramilitary activity emanated from the inner city areas.

Alongside the problems of social and economic deprivation, inner cities were also affected by visual and land pollution. Often factory buildings were left derelict adding further to the sense of decline that pervaded much of these areas. Some of the land was also polluted from earlier industrial usage and required expensive decontamination before it could be reused (the gasworks in Belfast is an example). Such derelict sites (brownfield sites) had little attraction for modern industry.

Redevelopment

It was clear that the inner cities needed modernisation and that this would have to come from the local authorities. From the 1960s governments had made attempts to improve living conditions in these areas but it has been shown that these early measures had little success. In the 1970s there was a change in the direction of redevelopment. Many of the large blocks of flats were removed and replaced by small two or three storey houses. Today, more attention is paid to landscaping housing areas with more open space and traffic calming measures, but these inner areas still remain socially and economically disadvantaged. **Resource 40** compares an inner ward of North Belfast with the rest of Belfast and with Northern Ireland using figures from the 2001 census.

Social and economic comparisons of an inner city ward (North Belfast) with Belfast and Northern Ireland (2001 census)

Resource 40

Percentage	N. Belfast	Belfast	N. Ireland
Single parent families	17.4	11.0	8.1
Car ownership	30.2	56.2	73.7
General health – not good	21.4	14.4	10.7
Age 20-29 with no qualifications	31.4	12.8	13.8
University degree	5.4	19.2	15.8
Unemployed	8.8	5.4	4.1
Unemployed, who were long-term unemployed	47.7	42.6	40.4

Source: ninis.nisra.gov.uk

Exercise

Describe and suggest reasons for the patterns shown in this table.

Urban Development Corporations

Following the riots in some British cities in the 1980s, the Conservative Government formulated plans to regenerate the inner cities. This was largely done through Urban Development Corporations (UDC). These corporations were given government grants to improve the basic infrastructure of the inner cities to enable them to attract private investment. The first of these was the London Docklands Corporation, which operated between 1981-1990 to modernise this deprived area. The Dockland economy has been transformed and has attracted more than 600 new firms associated with finance, publishing and high technology computer industries as well as new residential areas. It includes the prestigious Canary Wharf Tower, an office block built to attract financial companies from the city of London. Similar UDCs were established in other cities, including Belfast (Laganside) and Liverpool.

A decrease in population was one of the most obvious signs of decay in the inner cities. If the UDCs were to be regarded as successful there would have to be at least a slowing down of that trend. And now because there are more jobs available and many workers are choosing to live close to their place of work, people are moving back into the inner cities, but the social class has changed. The new apartment blocks offer modern but expensive accommodation, well beyond the budget of the typical inner-city dweller. This movement of high income groups into high status accommodation in the inner cities is referred to as **reurbanisation.** Closely associated with this is the process of **gentrification**. This refers to areas of old housing in the inner city that have been bought by middle class buyers or developers who have sufficient funds to modernise them. As a result, house values increase and the area becomes more fashionable. The UDCs have now completed their remit and recent government policy towards the regeneration of the inner cities involves partnerships with local authorities, and is generally more neighbourhood-based.

References

www.ninis.nisra.gov.uk/nra/ – Information on inner city wards in Belfast

www.dsdni.gov.uk/index/urcdg-urban_regeneration/belfast_regen_office.htm – Urban regeneration in Belfast

www.laganside.com/ – Laganside

Geoactive online no. 320. Urban Regeneration in Paris

Geoactive online no. 324. Inner City Redevelopment in Birmingham

Geoactive online no. 373. London Docklands: Post LDDC Developments

CASE STUDY: Issues in the Inner City of Liverpool

The city of Liverpool situated on the west coast of Britain has had a long tradition as a major port. The port was in an excellent position to take advantage of trade with America and West Africa in colonial times and the city's population grew rapidly throughout the nineteenth century. A number of industries set up in the port are associated with this trade. The port was also the exit point for large numbers of emigrants who left Britain for America. Liverpool was the second most important city outside of London in the British Empire for most of the nineteenth century. Its population peaked in 1937 when it was 850 000. From then until relatively recently, the city had gone into decline and the areas most affected were the inner wards.

There are many reasons for the decline of Liverpool, including:

- The docks became outdated and the water was too shallow to deal with the larger container ships.
- Much of the off-loading of containers relied on mechanised lifting gear resulting in unemployment for dock labourers.
- In the twentieth century, trade with Europe became more important than with America and ports on the east of Britain offered better locations.
- There was a general decline in manufacturing industries associated with ports.
- The new industries favoured locations in the south and south east of Britain.
- Abandoned docks, derelict buildings, high unemployment and falling populations created a negative image for investors.
- The high crime rates and anti-social behaviour in the high-rise flats that replaced the old terraced housing in the 1960s, contributed to the decline.

The social and economic situation in Liverpool continued to decline and by the 1980s it was one of the poorest cities in Europe.

Resource 41 *Social and economic characteristics of Kirkdale Council Ward (inner city) and Liverpool*

Characteristic	Kirkdale	Liverpool
Persons not in good health (%)	18.7	13.3
Unemployed (%)	9.7	6.7
Long term unemployed (% of unemployed)	21.2	12.9
No qualifications (%)	53.8	37.8
Single parent families (% of all households)	21	16.3

Source: UK Census 2001 figures published on Liverpool City Council's website

Exercise Explain why this table shows that Kirkdale is a deprived region within Liverpool.

Much of the recent regeneration of Liverpool occurred under the guidance of the Merseyside Development Corporation, which operated between 1981 and 1998 with funding from central government. The aim was to create an environment that would be attractive to outside investors and in that way the decline in population numbers could be halted. One of the stumbling blocks to development in Liverpool was its backward and outdated image. MDC directed its efforts towards the dockland area and initiated a series of projects aimed at modernising and advertising the potential that this area possessed. Among the development projects undertaken were:

A view of the docks in Liverpool

1. **The International Garden Festival:** This involved clearing contaminated land in the dockland area and building an exhibition parkland. This parkland displayed elements of Liverpool's past and present, such as its maritime history, along with modern elements such as the Beatles. It was a prestigious development, which improved the environmental appearance of the docks, but it brought little long-term economic advantage.
2. **The Albert Dock:** This was the most ambitious project, covering large areas of the old port of Liverpool. The historic architecture of old dockside buildings was restored. The upper floors were converted into expensive apartments while the ground floors were used for a variety of activities including offices, recording studios and shops. The historic setting was used to develop tourism through the building of an art gallery, maritime museum and the 'Beatles experience'. About six million tourists visit this area annually but local people have not benefited greatly from either the accommodation or the new jobs that were created. These new apartments, which include parts of Toxteth, had a starting price of £260,000, far beyond the budget of the original residents.
3. **The Marina:** A middle-class housing development and yacht club was built to attract middle class residents back into this inner city location.
4. **The Queen's Dock:** This was an area of modern office development which provided a range of additional office jobs, few of which were available to the local unskilled workers.
5. **Brunswick Dock:** A modern business park was built here and this did provide employment opportunities for local people.

Towards the end of the 1980s a new approach was adopted towards regeneration of inner Liverpool. These schemes were aimed less at the middle classes and large scale developments such as the Albert Dock, but focussed on local communities. There was less input from central government and a greater emphasis on forging partnerships with various groups in the city and the city council.

Examples of this new approach include:

- **The Eldonian Initiative 1987**: This was a partnership between Liverpool City Council and the local community in an area of terraced housing to the north of the city centre. There had been a strong community spirit in the area, but high unemployment and limited facilities. Local residents were consulted about their housing needs and possible locations for the new development. The final decision was to replace the existing terraced houses with new bungalows and to provide a number of community-based facilities for people of various age groups. This has been very successful and has attracted external investment. There are also schemes to increase the skills level of the long term unemployed.

Map of Liverpool city centre **Resource 42**

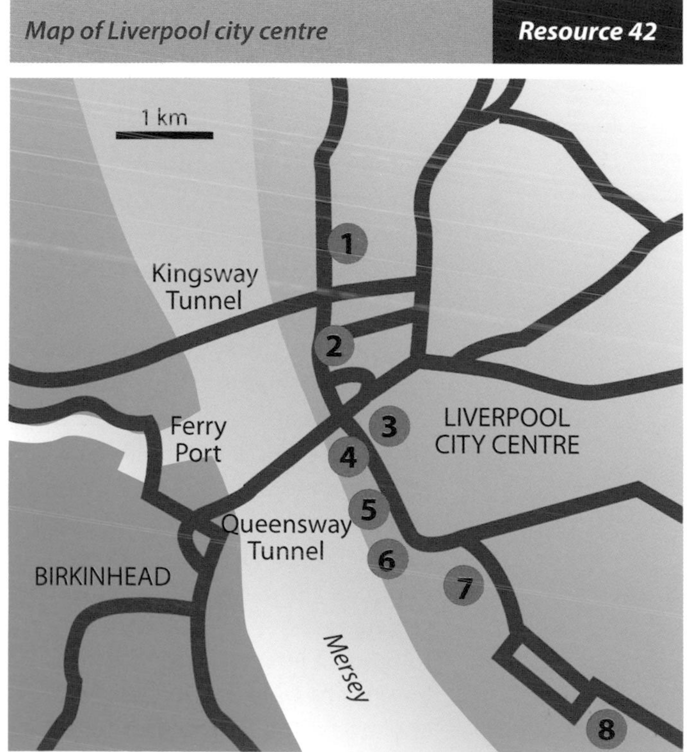

1. Eldonian Village
2. 100 Old Hall Street – Littlewoods HQ
3. Paradise Street
4. Albert Dock
5. Kings Dock
6. Liverpool Marina
7. Toxteth
8. Site of the Liverpool Garden Festival 1984

- **Priority Neighbourhood Action Plans 2001**: The Labour government launched these plans nationwide to tackle the underlying causes of social and economic deprivation in inner cities. Twenty-one neighbourhoods in inner Liverpool were identified for intensive action. Five of these neighbourhoods were in the areas affected by the Toxteth riots in the 1980s. Funding for the project comes from a number of sources including central government, Objective 1 funding from the EU, and local councils. The Action Plan has five main targets:
  1. improve training and skills of the working age groups
  2. improve local service provision
  3. improve local housing and the environment
  4. improve health awareness and provide greater access to health services.
  5. reduce local crime

Resource 43 *Indicators of deprivation in Granby/Lodge Lane (Toxteth) neighbourhood and Liverpool*

Characteristic	Granby/Lodge Lane	Liverpool
Population change 1999–2000	-2.5	-0.7
Unemployment (%)	14.1	6.7
5+ GCSEs grades A*–C	19	33
Teenage pregnancies (per thousand)	88.3	47.4

Source: UK Census 2001 figures published on Liverpool City Council's website

Impact of regeneration on Liverpool

There is no doubt that the many development projects undertaken in Liverpool have greatly improved the physical environment and the general infrastructure. Many of the derelict sites in the docks in particular have been cleared and now have a vibrant economy again. Population has increased in the inner city. Between 1971 and 1991 the population in the city centre of Liverpool decreased from 3,600 to 2,340 but now the latest figures are in excess of 12,000. The improvements brought by the MDC include 17,000 new jobs, 80 kms of new or improved road and almost 4,000 new housing units. Many of these residents are high-salaried workers employed in the new offices and high technology industries. In other words, the inner city is now populated by the middle classes, indicating that Liverpool has been reurbanised.

There are also examples of gentrification where former working-class housing has been replaced by high quality private residential developments, especially close to the King's Dock at Paradise Street (***Resource 42***). Liverpool has an improved physical environment and an attractive city centre. The higher income groups have more disposable income and that will attract new services as well providing more tax revenues for local authorities. Liverpool has had a variety of new entertainment and leisure facilities and these have attracted a growing number of tourists. It is estimated that about six million tourists have visited the Albert Dock complex annually. Liverpool will be the European Capital of Culture in 2008 and this is likely to generate over 14,000 new jobs as well as considerable investment and a further increase in tourism. Liverpool City Council and its government-sponsored partner, Liverpool Vision, have plans to bring further development to the central business district of Liverpool. They are largely concerned with offices and retail outlets as well as entertainment facilities. Much of this will be centred on the Albert Dock area. These new developments will provide some employment opportunities for local people.

In spite of all the positive outcomes of regeneration, Liverpool, like many other large cities, still has areas where poverty is very much in evidence. There does not appear to be an easy method of tackling this but currently the government is looking at partnerships with local communities to improve neighbourhoods (see Priority Neighbourhood Action Plans, opposite) and trying to raise the skills of local people. In that way, it is hoped to create a workforce with the skills required for new industry.

References

Regeneration and Development in Liverpool City Centre 1995–2000 – www.liverpoolvision.co.uk

The Eldonian Initiative – www.eldoniangroup.com

Liverpool City Council website – www.liverpool.gov.uk

Re-development of an inner city area – Liverpool Geofile September 2005

Toxteth's long road to recovery – http://news.bbc.co.uk/1/low/uk/1416198.stm

The following Geofile articles have information on inner-city renewal: January 2005, January 2006, April 2006

Exercise

1. Research the theme of regeneration further, using some of the references above, and then present your views on the following statement:

   'The developments outlined have greatly improved the quality of life for the residents of inner Liverpool'.

2. Contrast the approach towards development adopted by the MDC with the later neighbourhood schemes.

Issues of rapid urbanisation in LEDCs

According to the United Nations report on the State of World Population 2007, the world's urban population will rise to five billion by 2030 and most of this growth will occur in the LEDCs. The urban population of Africa and Asia will double during this time while in the MEDCs the urban population will grow very slowly from 870 million to just over one billion. Rapid urbanisation has been one of the most pressing issues facing practically all LEDCs in the latter half of the twentieth century and there is little sign that the problems are diminishing as we enter the twenty-first century. The problems of rapid urbanisation are exacerbated by the growth of very large cities.

Urbanisation in LEDCs

The current situation in LEDCs shows marked contrasts to the urbanisation phase that occurred in MEDCs in the nineteenth and early twentieth centuries. At that stage large numbers migrated to the towns and cities to take advantage of the new jobs in the factories and associated service industries. In other words, urbanisation was both a result and a cause of economic development. It took place over a period of about 150 years during which time society underwent tremendous social and economic change. In LEDCs, urbanisation is a relatively recent occurrence. More importantly, the underlying causes are as much to do with rural poverty as urban prosperity. Medical advances have resulted in a falling death rate in many regions but the fall in fertility that characterised the MEDCs has not happened to the same extent in the LEDCs. As a result, population numbers are increasing, putting greater pressure on already scarce rural resources.

Population of selected LEDC cities 2000 **Resource 44**

City	Population (m)	% of total population
Mexico City	18.1	24.7
Mumbai	18.0	6.3
Lagos	13.4	27.4
São Paulo	17.8	12.8
Calcutta	12.9	4.5

Source: Geofile January 2006

For many, life in the countryside holds no future and they feel they have little choice but to flee from rural poverty to the cities. In parts of sub-Saharan Africa natural disasters such as droughts, famine and war are primary causes of rural to urban migration. High rates of natural increase contribute to the growth in the size of LEDC cities. A recent UN report states that natural increase accounts for 60% of urban growth in LEDCs (whereas in MEDCs rural to urban migration was the main cause). This means that these cities have a substantial youth dependency and significant potential for future growth.

Economic Activity

Migrants arrive in the cities with few of the necessary skills required for the limited number of jobs available and are very often unsuccessful in finding full-time employment. The readily-available and low-cost labour force has been an attraction for multi-national companies in some cities, but the pace of urbanisation is such that the demand for jobs rapidly outstrips the supply. The lack of employment opportunities in regulated or formal sector jobs has driven many into eking out an existence in the informal sector such as street-sellers, shoe-shiners, beggars and prostitution.

Service provision

Unemployment is only one problem caused by the pace of urbanisation in LEDCs. The large number of people moving into the cities puts added demands on essential services such as clean water supply, sewerage, waste disposal, health care and education. These services require money and expertise and these are often not readily available. One striking characteristic of LEDC cities is the growth of very large or mega-cities with populations in excess of ten million. A study of city size in many LEDCs shows one or two very large cities while the remainder are very much smaller. The reason for this lies partly in the uneven nature of development in the LEDC, which results in urban resources being concentrated in the larger cities. Once this process begins these larger cities exert an even greater pull or attraction for migrants. The end result is to further increase the demand for jobs and other services in just a few cities.

Growth of informal settlements

Although the rate of increase in some of these mega-cities is slowing down, the absolute numbers involved are still a major concern. Mexico City receives approximately 5,000 extra people daily. Consequently, many of the poorest people do not have access to the most basic requirements of life – shelter, clean water etc and have little alternative to joining earlier migrants in the growing number of slums or informal settlements. These settlements are built using whatever materials are available – corrugated iron, timber, even plastic sheeting. They are not served by services, although at times a few basic services may be provided. The people living in these informal settlements have no legal right to occupy the land and local authorities do forcibly remove them on occasions.

Informal settlements beside Mumbai Airport, India

Above all else it is the absolute increase in the numbers of people living in such settlements that poses the greatest challenge to urban authorities. One third of all city dwellers in LEDCs live in an informal settlement. In total, they number close to one billion or one sixth of the world's total population. There are distinct regional variations in the proportions of informal settlements. In sub-Saharan Africa 72% of the urban population lives in such settlements compared to 56% in south Asia. Finding a solution to this growing problem will not be easy and it is clear that no single solution is possible. In 2000, the

United Nations Millennium Declaration detailed targets to fund projects to deal with some of the poverty in these settlements.

Advantages of informal settlements

Informal settlements do have some advantages. For one thing they offer the opportunity for rural migrants to access the opportunities of work and services which, although inadequate, are still better than in most rural areas. Secondly, they provide a pool of labour for new industries. Thirdly, for some migrants life in an informal settlement may only be a temporary hardship. **Resource 45** shows three possible stages that new migrants may experience:

1. **Bridgeheaders**: These are the newest migrants to the city. Their priorities are to be close to opportunity, possibly in the centre of the city, and have some basic shelter.
2. **Consolidators**: Over time the new migrant earns some money and moves to an established informal settlement. The priority now is to have some security of residence and still be close to work to avoid transport costs.
3. **Status Seeker**: The migrant now wants to be fully integrated into the urban way of life. He has a well-paid job and the main priority is to obtain better quality housing.

It should be noted that not all residents of informal settlements will pass easily into the third stage.

Changing priorities by residents of informal settlements — ***Resource 45***

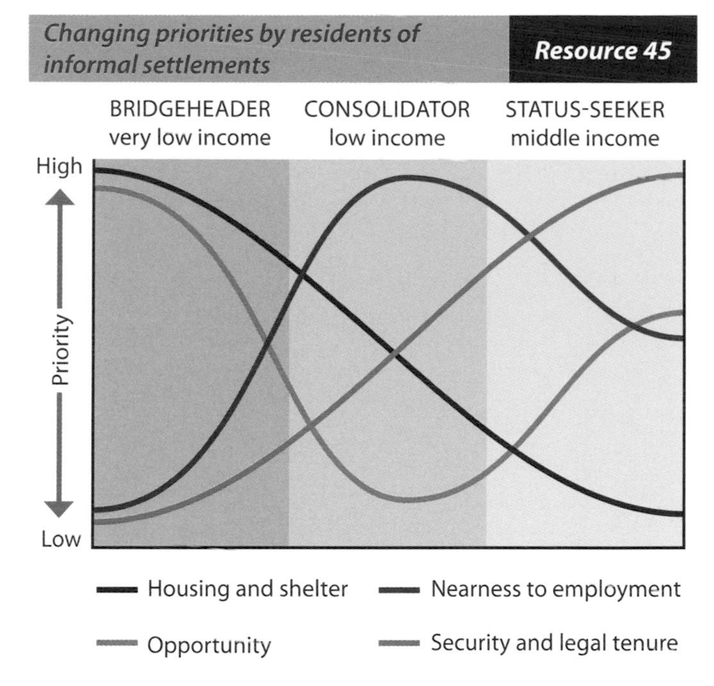

Urbanisation in MEDCs

Year	% urban population	Numbers in cities
1750	10	15 million
1950	52	423 million

Urbanisation in LEDCs

Year	% urban population	Numbers in cities
1950	18	309 million
2030	56	3.9 billion

Study the tables above.

(a) Contrast the urbanisation process in the LEDCs with that in the MEDCs.

(b) Explain why the situation in the LEDCs presents a challenge to the authorities in terms of service provision and economic activity.

References

Rural Urban Migration in the Developing World Geofile January 2001

Issues in an LEDC City: Beijing Geoactive No.310

Housing Solutions in LEDC Cities Geofile January 2006

CASE STUDY: Cairo – Issues of rapid urbanisation in LEDCs

Cairo, the capital of Egypt, is the largest city in Africa and the Middle East. Population totals for the city of Cairo are about 11 million while the total for the greater Cairo region is estimated between 16 and 17 million, a figure that represents just over a quarter of Egypt's total population and almost half of the country's urban population. Cairo is three times larger than the second largest city, Alexandria and is a classic example of a primate city. Cairo is situated on the Nile at the head of the delta. The physical growth of the city is constrained by hills to the west and the east and by the river itself. During European rule a number of bridges were built that allowed development on both sides of the river. The city shows three distinct phases of growth.

1. The Medieval City

This is a walled city with narrow streets, dominated by a citadel, and beyond the walls by the cemetery cities. These cemeteries made up about a quarter of the land area of the old city and had a restricting influence on the future growth of Cairo.

2. The European City

Cairo remained relatively unchanged until France and Britain took control of Egypt. The course of the Nile was controlled and bridges were constructed, so the city expanded in area and was modernised. The old city remained the home of the Egyptian population while the new developments provided middle and high-class residential areas for the European settlers.

3. Modern Suburbs

Most suburban development occurred in the north of the city along the main roads, but as this was threatening fertile agricultural land some suburban developments have been directed into the hills to the east and towards Gisa on the west.

Rapid urbanisation has affected Cairo in the latter part of the twentieth century. The population has almost doubled in the last thirty years, from 6.4 million in 1975 to 11 million in 2005. Migration of poor people from the rural areas partly explains this rapid growth, but a high rate of natural increase is the main cause. Over 33% of the population of Greater Cairo is under the age of 15 creating huge potential for future growth. It is estimated that the population of Greater Cairo is growing at a rate of 2% per annum but the labour force is growing at 3% per annum due to the large number of young people moving into the working age group. As is so often the case in LEDCs this rapid increase in population has not been matched by a similar increase in job creation or service provision. The problems facing Cairo include:

Resource 46 *The growth of Cairo*

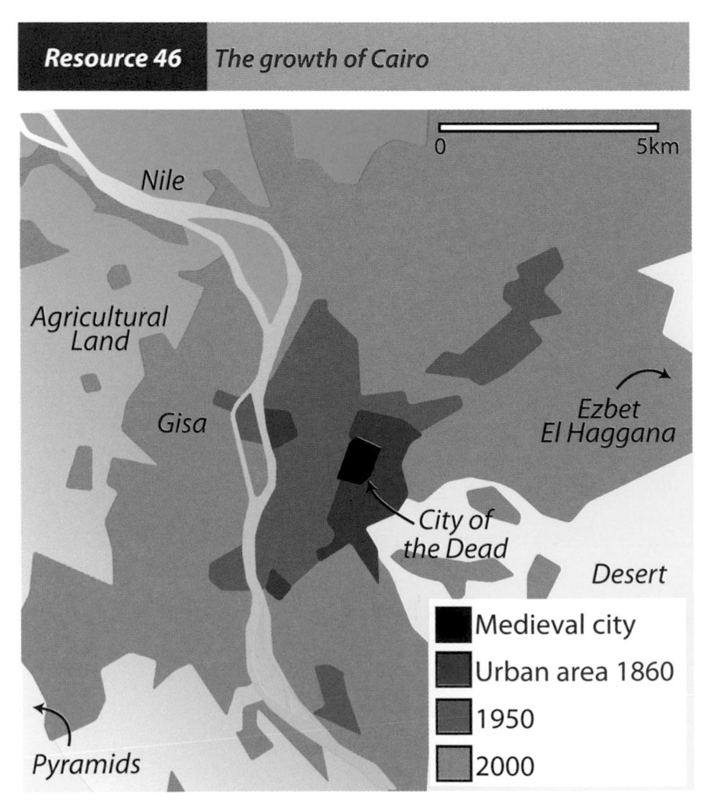

1. Lack of adequate housing

The rapid growth of the city and the inability of the Authorities to provide housing have resulted in many people living in one of 67 informal settlements in the Greater Cairo Region. In these settlements people live in poverty, lacking even the most basic services of sanitation, clean water supply, health and education provision. Informal settlements occur in three types of locations in Greater Cairo:

- On agricultural land surrounding the city
- In run down parts of the old medieval city including the cemetery City of the Dead
- Refuse-tip dwellers – the Zabaleen

The residential areas of Cairo **Resource 47**

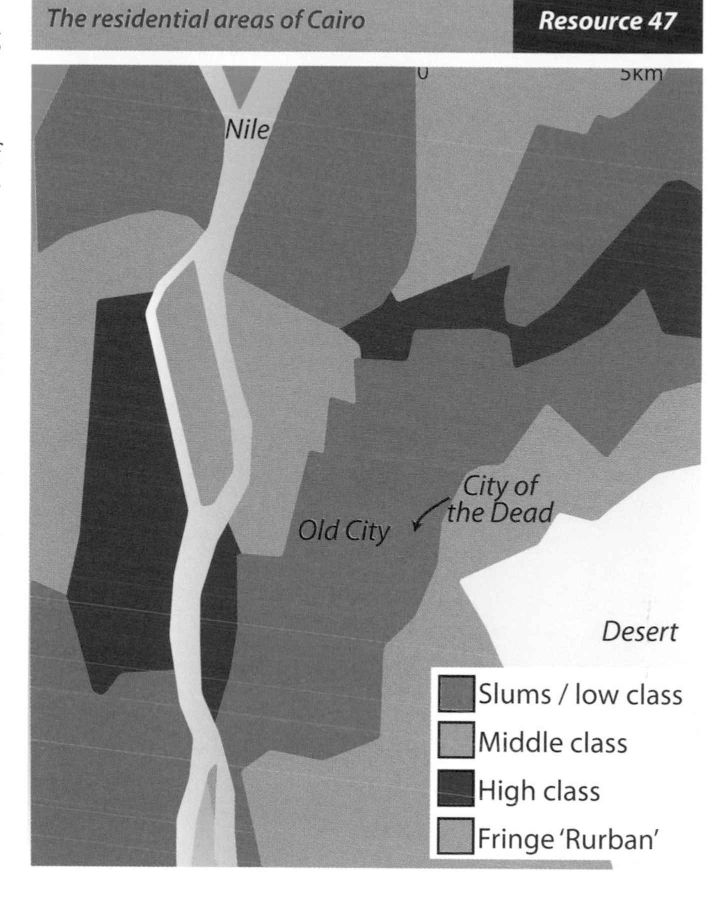

Informal settlements in Cairo consist mostly of permanent constructions and in that way they differ from informal settlements in other parts of the world. The newest settlements – those built on the city's edge – are crudely constructed blocks of flats. Over two thirds of the new dwellings built here since the 1960s have been built illegally and without any form of building control; most are unsafe. Furthermore, the increase in the numbers of these new dwellings is using up valuable agricultural land, which is also in short supply. In the old city, many are dilapidated buildings that the new settlers have partitioned or extended upwards; some originally built as five storeys now exceed ten storeys, again with no planning permission or building regulation. With such haphazard building and lack of regulation it is not surprising that many dwellings collapsed in the Cairo earthquake in 1992. Streets are narrow, often less than three metres and there is obvious overcrowding. Population density exceeds 109 000 per km^2 in some parts. Add to this the fact that the buildings are inadequately maintained and suffer from dampness due to leaks of water and sewage from poor drains and it is easy to understand that over 60% of dwellings in Cairo are regarded as substandard. However, some people have been unable to find any accommodation and are forced to live on roof tops in this old part of Cairo. Also, up to three million live in the City of the Dead. These are traditional burial places that included one or two adjacent rooms where relatives of the deceased could pay extended visits. Initially they had no services but in recent times they have been supplied with essential services. The third type of informal settlement – the refuse tip dwellers – is dealt with in the section on waste disposal below.

2. Waste disposal

Cairo generates over 7,000 tonnes of waste each day and this will obviously increase as the population grows. Traditionally, much of Cairo's waste was collected by the Zabaleen. This group is an ethnic minority within Egypt. Their lifestyle, which involves rearing pigs, as well as their Christian religion is at odds with the majority Muslim population and they have lived apart from the rest of Egyptian society since they moved into the country in 1952. Initially there were approximately 120,000 Zabaleen living in communities of about 17,000. Their main occupation is recycling. The men collect waste material from Cairo daily, bring it to the refuse tips where the waste is sorted by women and children, either to be reused by the Zabaleen themselves or sold. About 80% of the rubbish collected is recycled. These settlements are among the most unsanitary in the Cairo region and are rat and snake-infested. With no water or sanitation, there used to be frequent fires on the waste tips. However, since 1981 there have been some moves to improve the living conditions of these people (see box overleaf).

Zabaleen sorting rubbish. Available scraps of food are fed to pigs.

Zabaleen Environment and Development Programme 1981

In 1981 the Egyptian government partly funded a development programme in some of the Zabaleen communities. Some of the schemes included:

- the provision of piped water, a sewerage network, roads, primary schools and health centre;
- basic permanent housing;
- improved recycling facilities, including a composting plant that would produce fertilizer from organic waste;
- a female education programme.

3. Transportation problems

As the population increased so also did car ownership. There are 1.2 million cars in Cairo and together with public transport there are 14 million journeys across Cairo per day. The road network in the city does not cope well with these numbers. The worst gridlock occurs on the city's six bridges. Public transport is overloaded and unsafe. A controversial new ring road, which passes within 2.5 kms of the Pyramids was completed in 2000 and should relieve some of the congestion. There are also plans to extend the metro system.

4. Water supply

The Cairo water authority does not have the capability to provide enough clean water for all of the city's inhabitants. Many of the recent dwellings are not connected to the main water supply lines and the existing network is old and badly in need of repair, resulting in more than 25% of the water supply being wasted. There is a similar problem with the sewerage network. The original sewerage system was designed to serve a compact city with a population of one million, not a sprawling city with over 16 million people. There are frequent overflows with obvious health risks. The Cairo Waste Project, started in 1983 and funded by international aid, is the largest water treatment project in the world and should ease the problems considerably.

5. Air and Water Pollution

Air pollution is a serious problem in Cairo. The main sources of this pollution are the 11 oil-fired power stations, the growing number of manufacturing industries and the increase in vehicle emissions. Cairo contains just over a quarter of Egypt's population, 60% of the country's industry, 50% of the total cars and uses almost half of the total electricity. The air pollution problem is made worse by the anticyclonic conditions that exist for much of the year. This results in light winds and temperature inversions at night, which trap the polluted air over the city. A study in 1994 identified air pollution as the highest environmental risk in Cairo. Lead levels are among the highest in the world and each year 20,000 people die from the effects of lead pollution alone. Water pollution is mainly caused by the sewage problems mentioned earlier but untreated industrial waste disposal adds to this problem.

Informal housing (mostly roof top dwellers) in eastern Cairo, viewed from the Muqattam Hills. The Citadel of the Medieval City can be seen on the horizon (top left).

6. Unemployment

Cairo as the capital city offers the best opportunity for employment. The city is the main manufacturing centre in the country and is also the focal point for administration, finance and tourism. In addition, the government has attempted to attract industry to the new towns that surround Greater Cairo. However, there is still considerable unemployment and underemployment. Many people, perhaps over 50% in the poorest areas, work in the informal sector. In this type of employment wages are low and unreliable and work is not guaranteed, leaving this group of people in a vulnerable situation.

Exercise

Ezbet El Haganna- an informal settlement on the periphery of Cairo

Ezbet El Haganna is one of the informal settlements on the north eastern edge of Cairo, with a population of one million people. The following points illustrate the poverty of the inhabitants:

- 80% of households have no running water in their house
- 60% are not connected to a sewerage system and raw sewage flows in the streets
- 52% of those aged nine and over are illiterate (more than twice the average figure for Cairo as a whole)
- 40% of the population is under 15
- Average monthly household income is the equivalent of 44 US $ (the equivalent figure for Cairo is 206 US $)
- Population density exceeds 220,000 per km^2
- 30% of families live in one single room, sharing toilets
- 65% of those employed were working in the informal sector
- Many residents regularly suffer frequent health problems (especially diarrhoea, kidney, liver, eye and skin ailments). Infant and general mortality rates are thought to be high although no accurate figure is available, as births and deaths are often not recorded
- Many dwellings were in need of structural repair. Between 3,000–4,000 dwellings had no roofs.
- Only 1,200 of the residents have voting cards – a requirement to vote at elections
- Most residents, especially women, have no legal documents, such as birth certificates or identity cards.

Source: State of the World Population – report 2007

Questions

1. Study the information above. Describe the social and economic problems of the inhabitants of this informal settlement.
2. Describe the location and characteristics of the three types of informal settlements in Cairo.
3. Describe and explain the problems caused by rapid urbanization in a LEDC city.
4. How do the issues in a MEDC city differ to those in a LEDC city?

References

City of the Dead: http://news.bbc.co.uk/1/hi/programmes/crossing_continents/africs/1858022.stm

Informal settlements in Cairo: www.ucl.ac.uk/dpu-projects/global_report/pdfs/Cairo.pdf

State of the world population: www.unfpa.org/swp/2007/english/chapter_1/index.html

3A THE NATURE AND MEASUREMENT OF DEVELOPMENT

The Problems of Defining Development

In the first United Nations Development Report 1990, development is described as a process "that creates an environment in which people can enjoy long, healthy and creative lives". This was seen as a new departure in development studies which had, until the 1990s, focused on the creation of wealth as the main objective of development. The 'north south divide' map published in the 1980s simply grouped countries according to their wealth. The current emphasis on people and their quality of life does not mean that wealth creation is not important, but rather shifts the emphasis towards the benefits that wealth can bring to people. People are placed at the centre of development, and wealth is simply the means to enable improvements in our quality of life. Modern development studies see development as made up of two main components:

Economic development

Countries regarded as being developed (or MEDCs) have all experienced restructuring of their economies, and their employment structures show a move away from predominantly primary occupations into secondary and now into tertiary and quaternary occupations. This restructuring seems to be a vital requirement in the development process. If wealth is to be created, a country must have a manufacturing industrial economy so that value-added manufacturing goods are available for export. With time, routine manufacturing jobs are often relocated to LEDCs, leaving only the skilled and Research and Development in the MEDCs. This is the reason why many LEDCs strive to establish a manufacturing industrial base to their economy. However, the wealth needs to be distributed fairly among the population as a whole if a country is be regarded as developed. If wealth alone created development, then the oil-rich countries in the Middle East would be regarded as developed. However, in these countries the wealth remains in the hands of the minority and their record of human rights abuse is very poor. In other words, the creation of wealth has not facilitated social development.

Social development

This relates to the use of the wealth created to fund improvements in health care, education and legal provision for all members of society, irrespective of gender, race or religion. These services should be accessible to everyone as a right and not only to those who can afford to pay. Development requires the legal system to protect the rights of all citizens, including the right to vote in elections and to freedom of speech. All MEDCs operate a democratic style government chosen by the electorate. In that way, the population has a say in the policies affecting their lives. Fundamental to the development process is social mobility, where progress results from merit and not from privilege or from membership of a particular religious, racial group or caste. It is these aspects of development that are the most difficult to achieve. Whilst economic development can occur in most countries, a fundamental change in attitude and outlook is required to bring about a developed society. Some critics of this approach to defining development claim it is Eurocentric and does not adequately reflect different cultures. It is important therefore to view a country's progress on the development scale solely with the UN definition in mind.

Measures of Development

The problems in defining development are reflected in the measures used to gauge a country's position on the development continuum.

Economic measures

Economic measures are still very much in use but they are only used to make a very broad classification of countries into major income bands. They help to provide an indication of potential improvements in quality of life in the future. The most common economic measure is **Gross National Income per capita (GNIpc)**. GNIpc, formerly known as GNPpc, is the total value of goods and services produced in a country plus taxes and income from abroad in one year, divided by the total population. In order to allow international comparisons to be made, it is always expressed in US dollars. This is a straightforward measure and it does give an indication of the potential for social development, but it has many weaknesses. It is not a measure of individual income as it apportions equal wealth to all members of the country – a situation that never occurs. In addition, GNIpc gives no indication of wealth distribution. In most countries, and in LEDCs in particular, wealth distribution is skewed, with wealth being concentrated amongst a privileged elite. The GNI of a country is affected by fluctuations in the currency exchange rate. The real purchasing power of a given amount of money varies from one country to another but this fact is lost if only GNIpc is used. In many LEDCs a considerable amount of internal trade is non-marketed products from subsistence living and these are not counted in GNI calculations. Currently, development programmes are often judged on the merits of their impacts on the environment. GNIpc gives no indication of environmental impacts. The World Bank divides countries into four categories based on their GNIpc.

Classification of countries using GNIpc – figures for 2007 **Resource 48**

Income group	GNIpc US$	Number of countries	Examples
Low Income	Less than 935	49	Bangladesh, Benin
Lower middle income	936–3705	54	Ukraine, Bolivia
Upper middle income	3706–11455	41	Libya, Brazil
High income	11456 +	60	USA, France, UK

Source: drawn from data published by The World Bank

Attempts have been made to correct some of the inadequacies of this measure. One of these is **Purchasing Power Parity (PPP).** This takes account of the real purchasing power of a given amount of money in different countries and as such is a more equitable measure than GNIpc for LEDCs. However, apart from this, it still has all of the weaknesses of GNIpc. **Resource 49** shows how these two measures affect the global patterns of income.

Comparisons between GNI per capita and GNI(PPP) per capita **Resource 49**

Country	GNI per capita	Rank	GNI (PPP) per capita	Rank
Luxembourg	65 630	1	65 340	1
Denmark	47 390	5	33 570	11
Iceland	46 320	6	34 760	8
UK	37 600	12	32 690	13
Czech Republic	10 710	57	20 140	49
India	720	159	3460	144
Bangladesh	470	175	2090	166
The Gambia	290	192	1920	172
Uganda	280	194	1500	179
Dem. Rep. Of Congo	120	207	810	200

Source: The World Bank Database 2006

Resource 50 *GDP per capita (2004)*

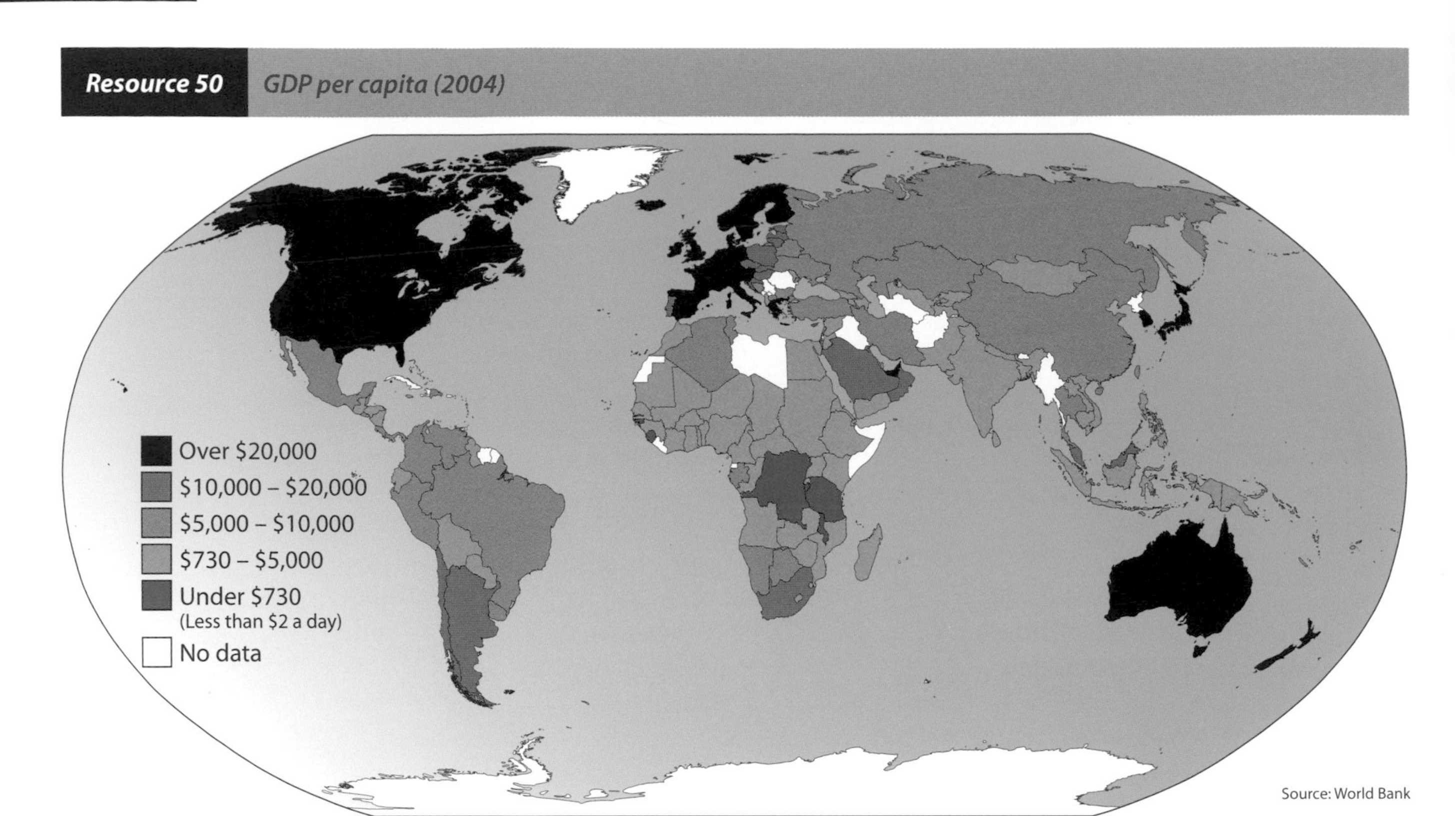

Gross Domestic Product per capita (GDP pc) is a similar measure but it excludes overseas earnings or profits that go to overseas investors.

Social Measures

Social Measures are increasingly used as indicators of development because they show the impact of development on society. **Life expectancy at birth** is one of the most common social measures used. Life expectancy at birth refers to the number of years a person is

Resource 51 *Life expectancy at birth (2007)*

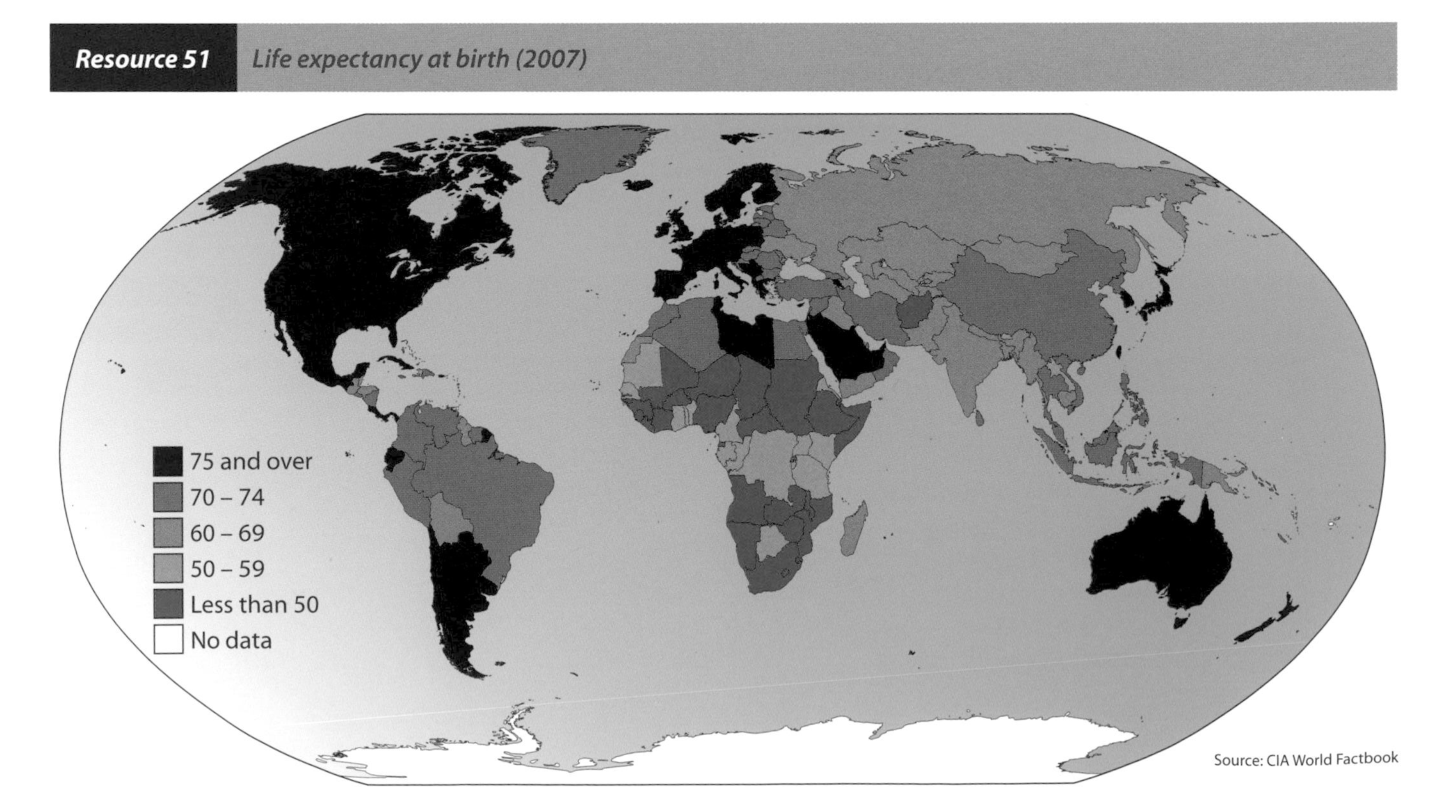

expected to live, and it is a clear reflection of the social conditions pertaining at a given time. The figures for life expectancy are traditionally higher in MEDCs than in LEDCs. Over the last thirty years there has been an increase in life expectancy worldwide, with some of the largest increases recorded in some LEDCs. The exception to this is Sub-Saharan Africa. For this region as a whole, life expectancy is lower today than it was three decades ago. Several countries in this region show very dramatic decreases in life expectancy: 20 years in Botswana and 16 years in Swaziland. The main reason for this is the spread of AIDS. The situation is even worse for women. Women account for 57% of HIV infections and young African women are three times more likely to be infected than men.

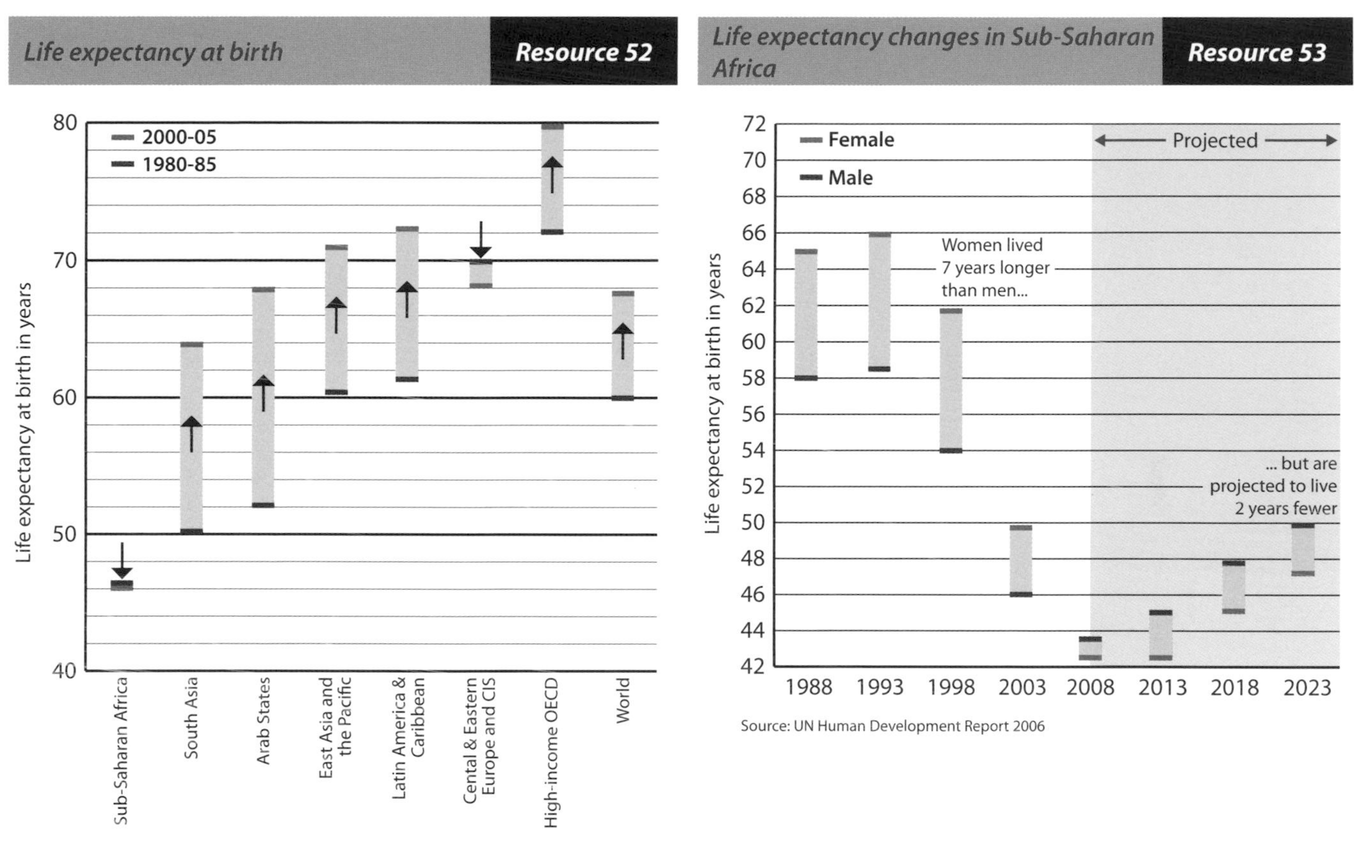

Life expectancy is a very reliable measure, especially for LEDCs, where changing trends are examined. However, it is still an average figure and may well hide variations in society as the case relating to women in Southern Africa showed. Since it is calculated at birth there is no accounting for new developments in medicine that may occur. As conditions in LEDCs improve, a vaccination programme could eliminate some diseases or have a dramatic impact on survival rates from endemic disease. Life expectancy figures obviously cannot take account of this.

Progress in education is seen as critical for human development. There are various measures used to chart a country's progress in this field such as **adult literacy rates, enrolment in primary, secondary and tertiary level education**. Once again, progress appears to be made in most countries whichever indicator is used, but the improvements are much less noticeable for women. In 2006, women still account for two thirds of adult illiteracy – the same proportion as in 1990. In terms of enrolment in primary education, out of the 115 million children not attending school, 62 million are girls. As noted above, these measures go some way towards describing a country's level of development but they are more effective if female literacy rates or educational attainments are considered.

Resource 54 *Percentage of population aged 15 and above who are illiterate (2007)*

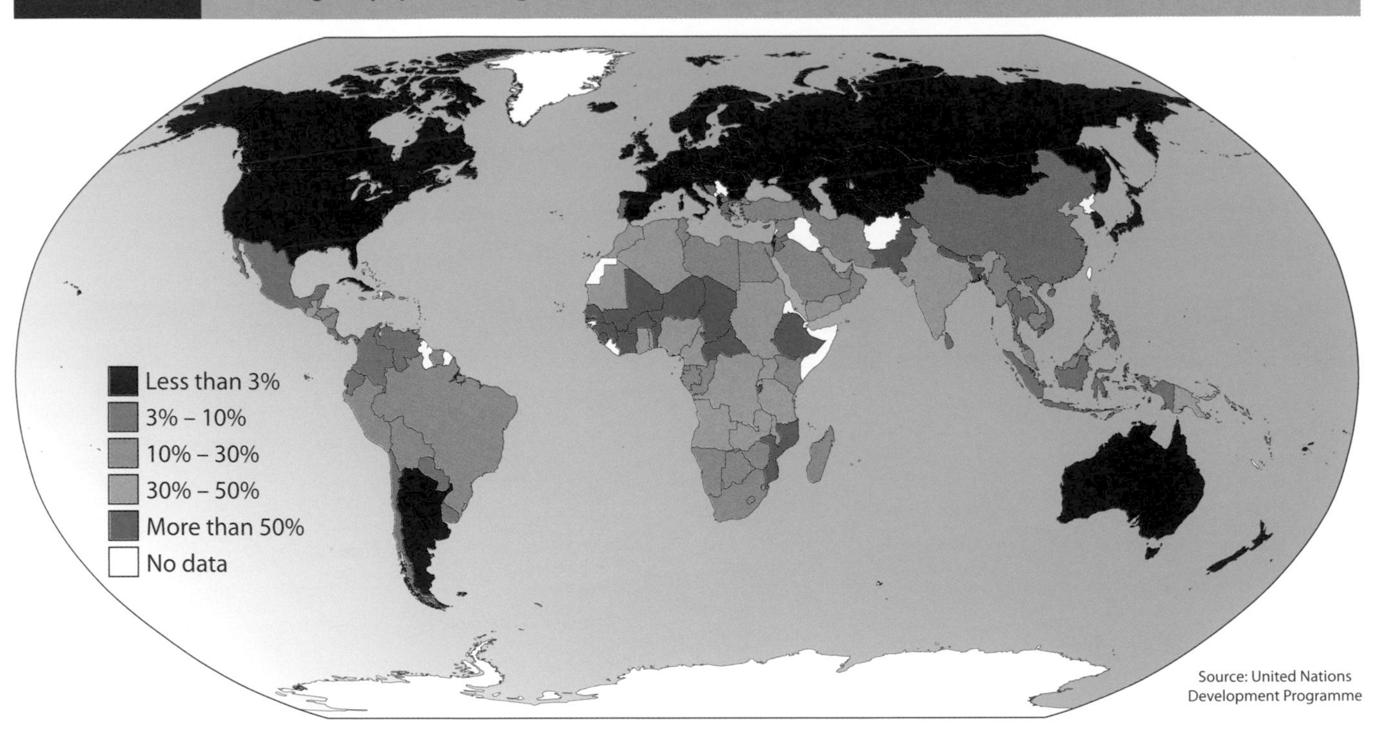

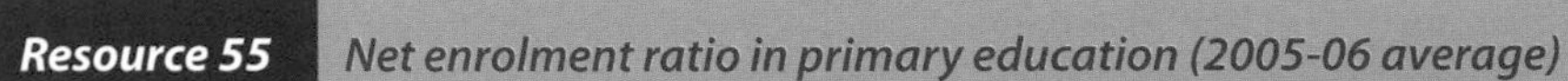

Resource 55 *Net enrolment ratio in primary education (2005-06 average)*

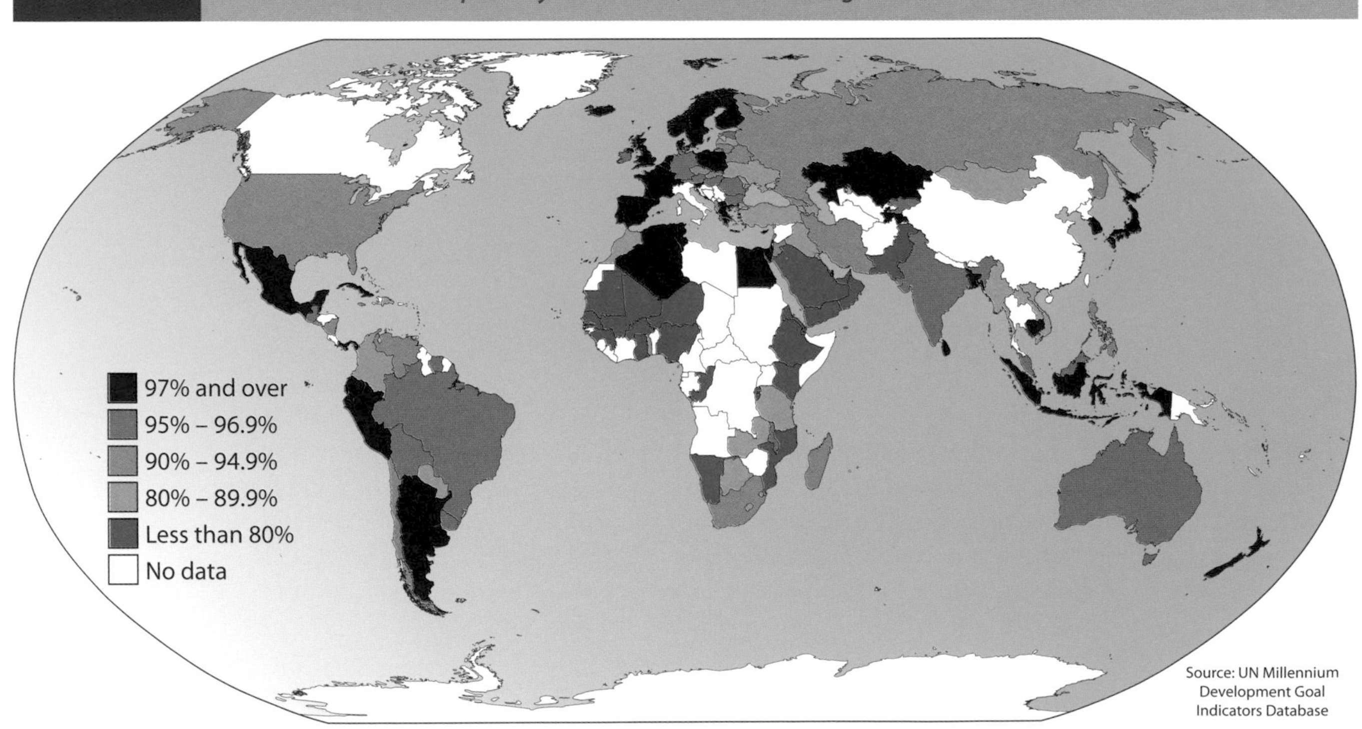

Infant Mortality Rate which records the number of live children per thousand who die within the first year of life is another effective measure of the quality of life. In recent times greater emphasis has been placed on the environmental impacts of development, and there is an increasing urgency to foster a more sustainable use of resources, particularly in the light of the latest warnings concerning climate change. **Resource 57** includes details of various measures of development for a selection of countries.

Literacy rates for major world regions (1970–2004) **Resource 56**

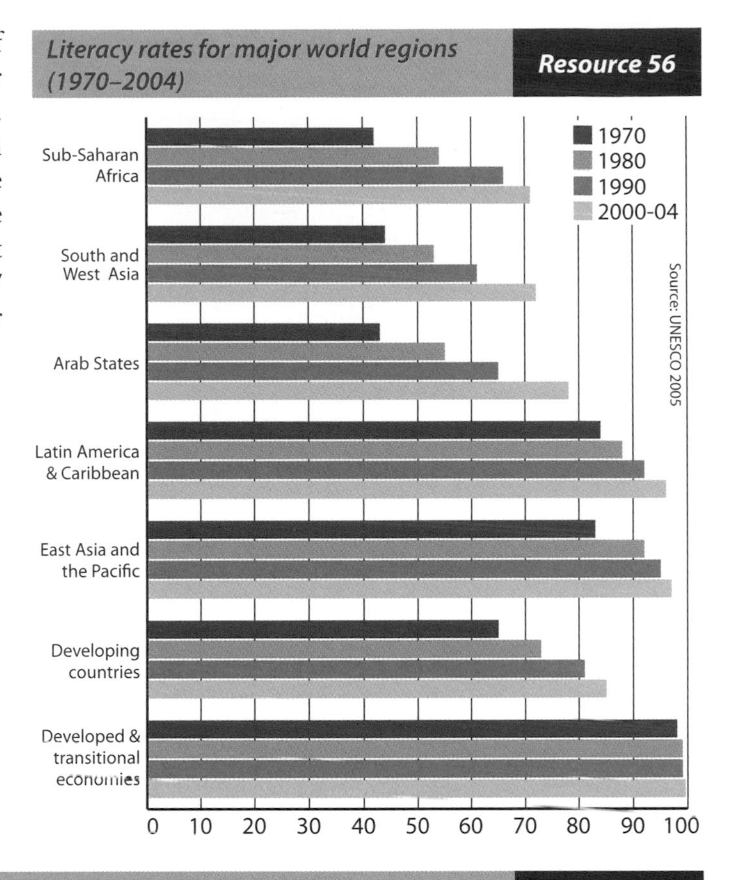

Development indicators for selected countries **Resource 57**

Country	Life Expectancy in years at birth	Infant Mortality Rate per 1000	% children under 5 underweight	GNI per capita PPP in US$ 2006	CO_2 tonnes per capita 2002
Egypt	71	33	5	4680	2.0
Zimbabwe	37	60	17	1940	0.9
Somalia	48	117	23	-	-
Chad	51	102	34	1230	0
Argentina	75	14.4	2	15 390	3.3
Brazil	72	27	4	8800	1.9
Kuwait	78	8	2	29 200	25.3
UAE	79	9	-	23 990	23.6
India	64	58	46	3800	1.1
Sri Lanka	74	65	23	5000	0.6
Norway	80	3.2	0	43 920	8.3
UK	79	4.9	0	35 690	9.2
Poland	75	6	-	14 530	7.7
Russia	65	10	-	11 620	10.6
China	72	27	6	7730	2.9
Australia	81	5	0	31 860	17.3
Japan	82	2.8	0	33 730	9.5

Source: Extracted from world population data sheet published by The Population Reference Bureau July 2007

Once development came to be associated with quality of life it was obvious that the measures used would have to change if a more realistic picture of a country's position on the development scale was to be achieved. It was soon realised that no single indicator would give a true picture of quality of life. In 1977 the first composite measure was developed. A composite measure uses several indicators to provide an index which can then be used to rank countries. The first composite measure was the **Physical Quality of Life Index (PQLI)**. This used three indicators: life expectancy at age one, infant mortality, and literacy – each having equal weighting. Values of PQLI fall within the range of 0–100 and countries are ranked according to their calculated value. PQLI has been criticised because the indicators are all likely to illustrate similar trends. In other words, a country with a low life expectancy will most likely have high infant mortality and low literacy rates. In light of these criticisms the United Nations has developed a more comprehensive index called the **Human Development Index (HDI)**. This index measures the average figures for life expectancy, enrolment in primary, secondary and tertiary level education and Purchasing Power Parity (PPP). HDI attempts to overcome some of the weaknesses of PQLI by broadening the measures used, particularly in respect to education and wealth. The values of HDI fall in the range 0–1 and countries are ranked according to their calculated value.

Comparison of a country's ranking according to HDI and GNIpc reveals some interesting patterns, as shown in **Resource 58**. Bahrain has an average income almost twice the level of Chile but a lower HDI rank because it underperforms in education and literacy. Angola and Tanzania have similar HDI rankings even though Angola has a much larger GNIpc; however, Angola has had a costly civil war which has had a clear impact on development within the country.

Try to find explanations for the other differences in HDI and GNIpc rankings shown in **Resource 58**.

Resource 58 *Selected HDI rankings*

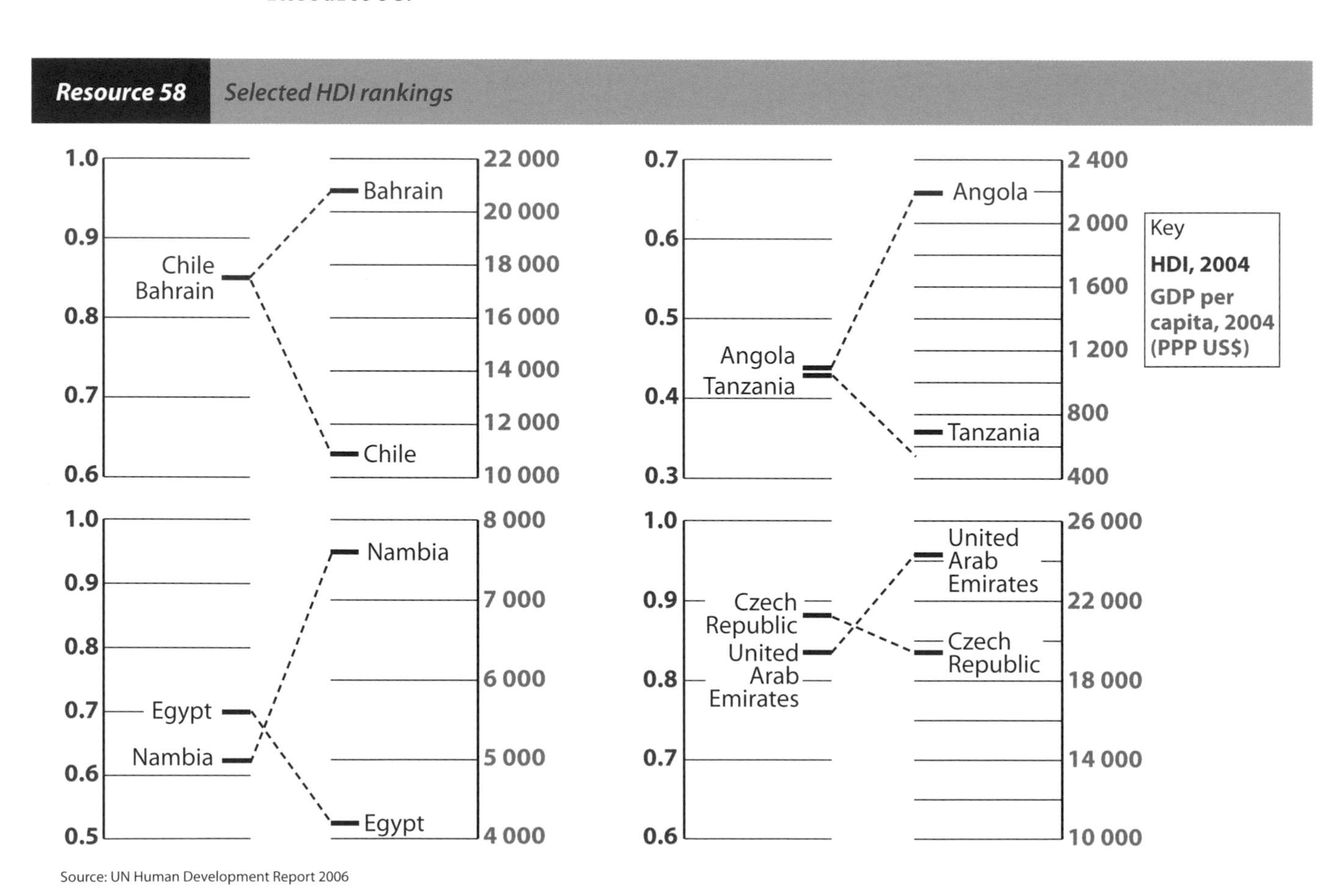

Source: UN Human Development Report 2006

HDI can also be used to see how a country has progressed through time. **Resource 59** shows almost all major world regions have progressed since the mid 1970s. The only exception is Sub Saharan Africa which has 28 of the 31 lowest HDI ranks.

HDI has been further extended to measure inequalities between males and females to give a gender development index (GDI). This was seen as a major improvement as it is often females who are considerably underprivileged in LEDCs.

Composite measures do appear to give a more realistic view of variations in quality of life than single measures but they too have weaknesses. HDI does not include any measure of human rights or democracy. HDI is also less effective at showing variations at the top end of the development spectrum because most MEDCs will have high life expectancy and literacy and educational enrolment. Nevertheless, HDI and GDI are widely used because they highlight differences in LEDCs. In 2000, the UN set out a series of development targets or goals (MDGs) to raise the quality of life in those countries at the bottom of the HDI rankings.

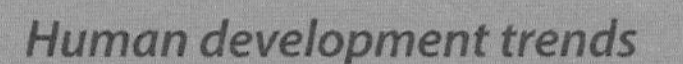

Human development trends **Resource 59**

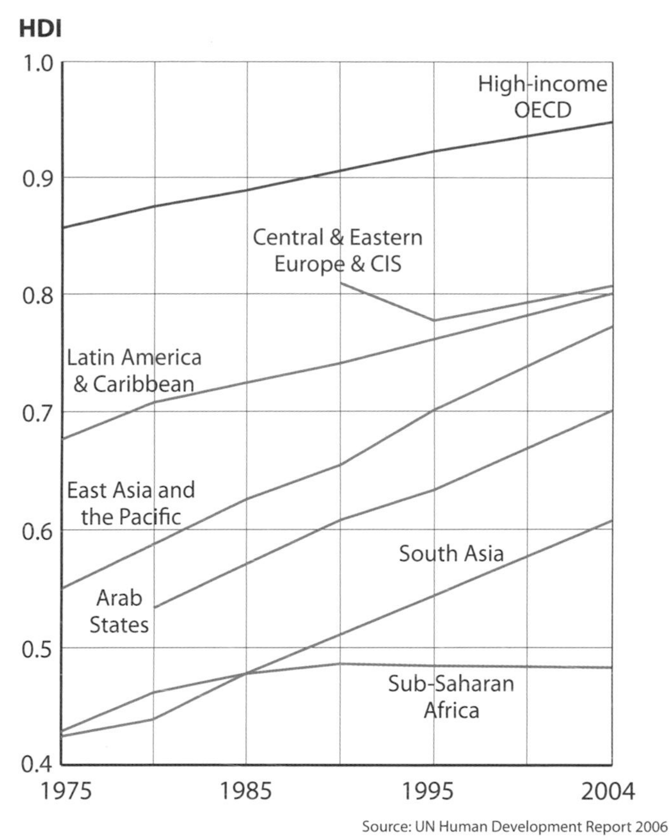

Source: UN Human Development Report 2006

Exercise

1. (a) *Question from CCEA June 2001*
Study **Figure 1**, which shows development indicators for two countries. One country is a MEDC and the other is a LEDC.
(i) Complete a copy of the box below to identify the correct country in each case. (1)

Figure 1

	Gross National Product per capita (US $)	Birth rate per thousand	Death rate per thousand	Adult illiteracy %	Population per doctor
Country A	250	51	19	77	36 225
Country B	24 000	12	9	0	399

MEDC	
LEDC	

(ii) Explain how social and demographic conditions reflect the level of GNP per capita. (6)

(b) Study **Figures 2** and **3** (overleaf) showing information on regional variations in development in England.

Exercise continued

Figure 2

	% unemployed	GDP per capita England average = 100	% 16 year olds gaining 5+ GCSEs	Mean house price £000
North	9.8	86	38	51
North West	6.9	91	43	58
Yorkshire & Humberside	6.9	90	38	56
East Midlands	6.3	98	42	59
West Midlands	6.8	95	41	64
East Anglia	5.9	108	48	73
South East	5.2	111	51	88
South West	5.2	95	55	68

Figure 3

(i) Using the information in **Figures 2** and **3**, describe the pattern of regional inequality in England. (4)

(ii) Which one of the indicators in **Figure 2** do you consider to be the most effective in measuring development? Give a reason for your choice. (4)

2. *Question from CCEA January 2003*
Study **Figure 4** opposite, showing GDP per capita and infant mortality rate for nine selected countries in 2002.

(i) Identify the infant mortality rate for Estonia. (1)

(ii) Describe the relationship between infant mortality rate and GDP per capita. (2)

Exercise continued

Figure 4

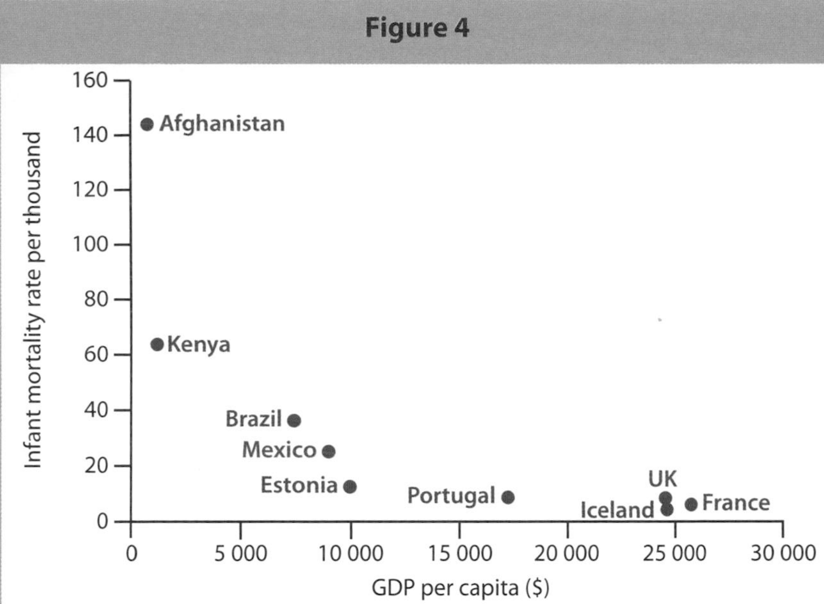

3. *Question from CCEA June 2005*

Study **Figures 5** and **6**, which show wealth measured as Gross Domestic Product (GDP) in 1970 and 2000 for a number of countries and regions. The size of each country or region is drawn in proportion to the size of its GDP.

Figure 5

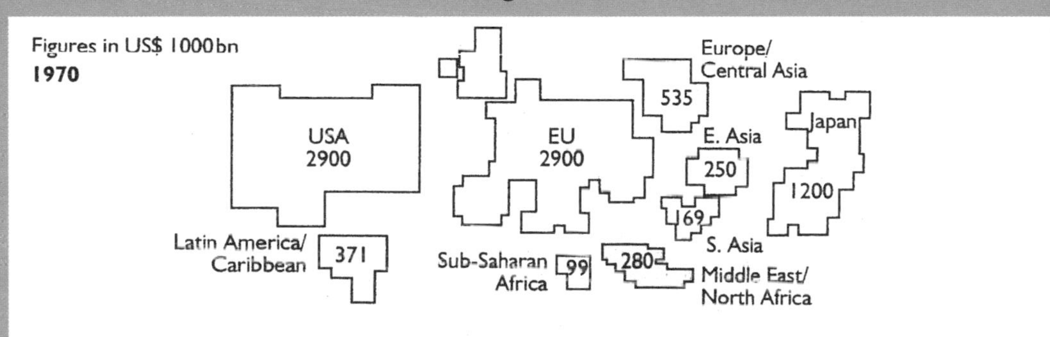

Figure 6

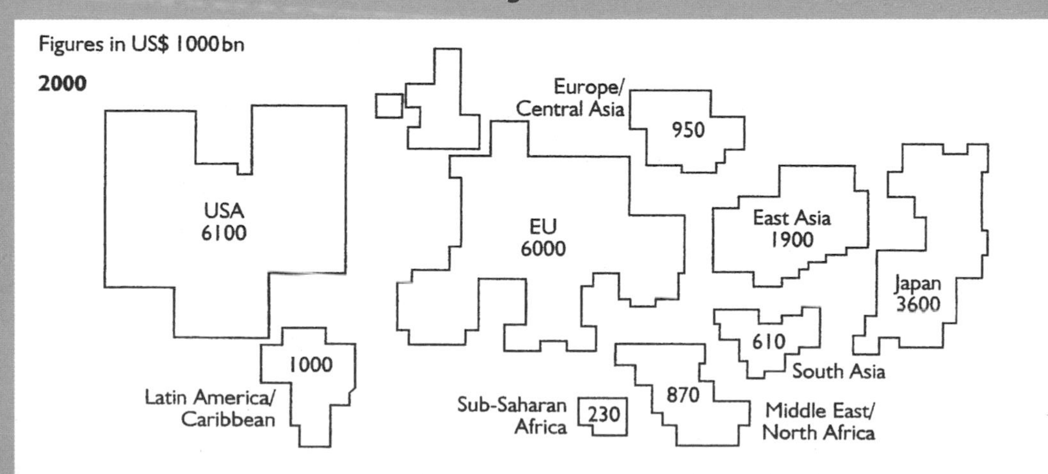

(a) With reference to **Figures 5** and **6**, describe what has happened to the total amount of GDP between 1970 and 2000.

(b) Describe and explain the pattern of global GDP in 2000, as shown in **Figure 6**.

Regional Contrasts in Development

All countries will show variations in development between one region and another. There are many reasons for this, ranging from physical, economic and political, and the pattern may change over time. For development to take place there has to be some economic potential which sets the development process in motion. Once started, the process of development gains momentum and leads to yet more development. In the nineteenth century the discovery of coal and iron ore brought considerable industrialisation to the coalfield regions. Once coal mining and iron ore mining were developed, ship building and other heavy industry followed. New industries and services were established to meet the increased demand, which resulted from a higher disposable income. People moved from the poorer regions to work in the new factories. Two possible outcomes can follow from this situation. Development can spread throughout the country in time or the core region may continue to grow at the expense of the poorer region. **Resource 60** illustrates this.

Resource 60 *Regional inequalities*

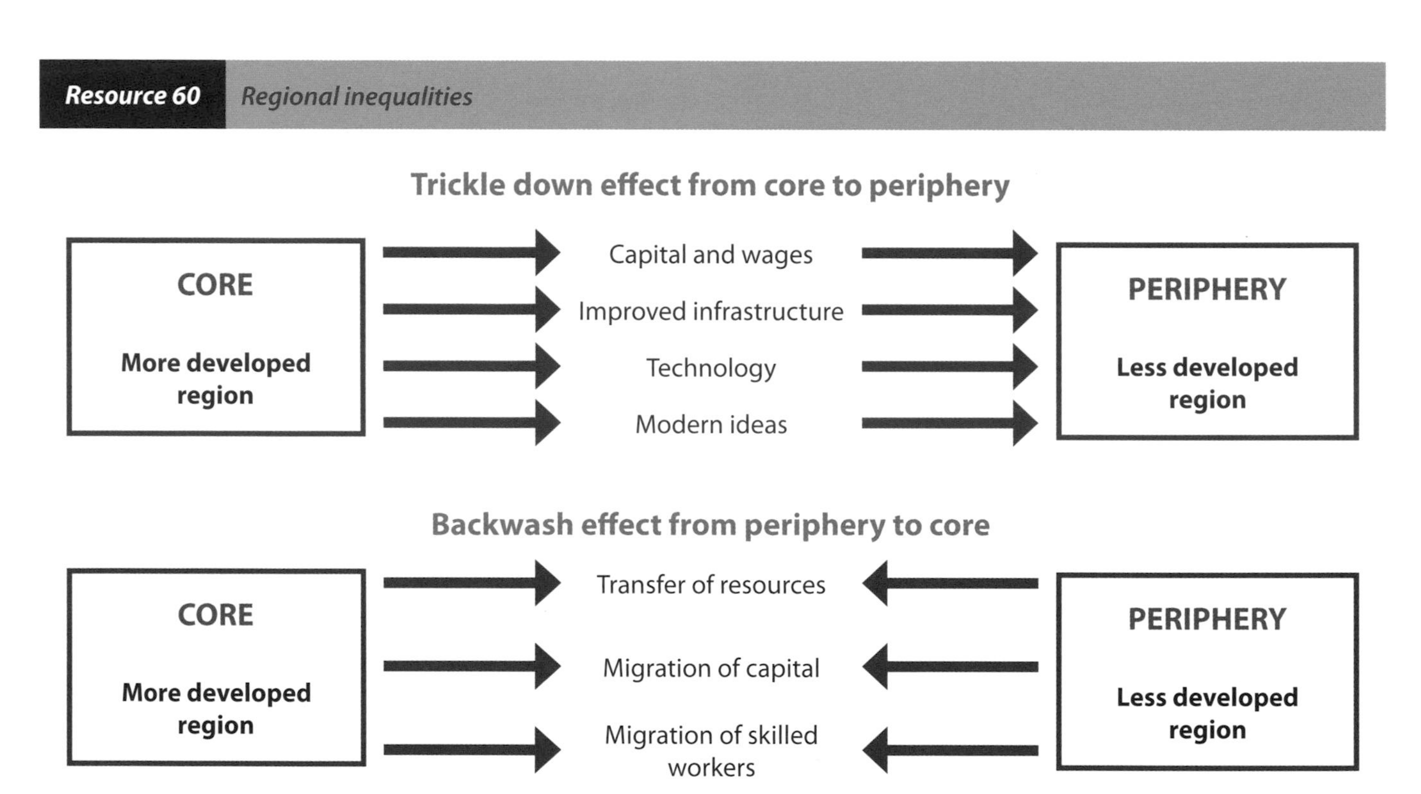

A region may develop rapidly because of the discovery of an important mineral or fuel resource but may face a period of decline when the resource is depleted or of little value. Similarly, industrial regions grow and decline in the face of fluctuations in demand and competition from other manufacturers. The changing fortunes of the coal mining and traditional manufacturing areas in the UK are prime examples of this change over time.

Apart from the situations described above, some regions fail to attract development. Among the many reasons for this are:

1. Physical reasons

Highland areas are remote and inaccessible with a harsh climate. As such, they offer little economic potential either for agriculture or for large scale economic development and for the most part they are areas of primary industry associated with farming or forestry. As urban development is restricted to the more accessible parts of the country, the remote areas are typified by low population densities. The Highlands and Islands of Scotland, the Welsh mountains and the Pennines are examples of such regions in Britain. On the other hand, lowland areas with fertile soils, mild climates and greater opportunities to develop communications present few obstacles to development.

2. Economic reasons

The discovery of an economically viable mineral or fuel resource is one of the main reasons for economic prosperity. As mentioned earlier, the discovery of coal and iron in areas such as Northumberland and Durham, Yorkshire and South Wales in the nineteenth century heralded an economic boom. The discovery of oil and natural gas in the North Sea in the twentieth century has had a similar effect on some towns on the east coast of Scotland. Other economic reasons for development include trading potential, skilled labour force or favourable geographical location. The south east of England is a highly developed region largely due to a combination of favourable physical factors: its proximity to the capital, London, and to Europe.

3. Historical factors

A region may have developed due to some factors that are no longer relevant but because of inertia it remains attractive to investors and industrialists. Capital cities and their neighbouring areas often fall into this category.

4. Political factors

Political factors often help direct development towards economically deprived parts of the country. This can happen in a variety of ways but usually through regional development policies. In the UK these have operated since the 1930s when assistance was given to those areas which had unemployment levels above the national average. It is in the government's interest to try to even out imbalances in economic development. If serious regional imbalances occur there is likely to be out-migration towards the more prosperous regions, leading to a downward spiral of decline in the poorer region and pressure on resources in the richer region. In recent times the task of promoting regional development has been carried out by Regional Development Agencies. There are nine RDAs in England, each one working to attract investment to their region. Wales and Scotland have their own development agencies. The European Union also operates a European Regional policy. Areas where the average income is less than 75% of the EU average are awarded Objective 1 status. An area which is suffering from serious industrial or agricultural decline is awarded Objective 2 status. In each case funding is available from the EU and a similar amount is made available from the government of the region concerned. Within the UK, Northern Ireland, the Scottish Highlands, Merseyside, South Yorkshire and parts of Wales have been awarded Objective 1 status. Northern Ireland and the Scottish Highlands have progressed and since 2000 both regions have ceased to qualify for Objective 1 funding. Rural areas facing difficulties in attracting investment are often awarded Objective 5b status. In these areas grants are available to assist in widening and diversifying the economy outside of agriculture. Within urban areas, development policies have been carried out by Urban Development Corporations such as the Merseyside Development Corporation.

The decline of Harland and Wolff has created economic hardship for East Belfast.

Exercise

1. *Question from CCEA January 2006*
   Study **Figure 1**, which compares the distribution of wealth in the whole of Brazil with that of its cities of São Paulo and Rio de Janeiro.

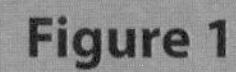

Figure 1

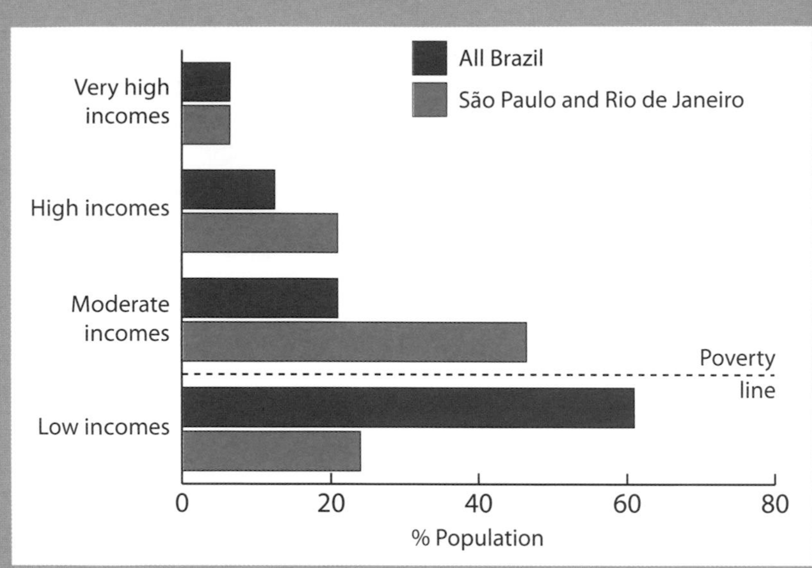

(a) Describe what **Figure 1** shows about inequality in development between Brazil as a whole and São Paulo and Rio de Janeiro. (2)

(b) Suggest **one** reason for the difference shown in the low income category in **Figure 1**. (2)

CASE STUDY: Regional contrasts in development in Italy

Italy is a country of considerable regional contrasts in terms of its physical and cultural landscape as well as its economy and levels of development. As a political unit, Italy has only been in existence since 1861. Prior to that, there were a number of small independent states. The Vatican City and San Marino still remain independent within Italy to the present day. **Resources 61–63** demonstrate the variations in relief and climate that exist between northern and southern Italy.

Northern Italy

In the north there are extensive areas of flat and fertile land drained by the River Po and its tributaries. This area experiences a continental climate with adequate rainfall of 800 mm and plentiful supplies of water for irrigation where necessary. Temperatures range from 0–23°C. The fast flowing streams from the Alps, which form the northern border of Italy, provided early sources of power and also provided trade routes and access to northern and western Europe. Agriculture is modern, highly mechanised and commercial. The main products are wheat and livestock products as well as vines and olives.

A typical town in Central Italy.

There are some mineral resources, including coal, natural gas and oil, but these are in small quantity, and Italy is a net importer of fuel resources. Part of the real potential of this north Italian plain lies in its accessibility to the economic core of Western Europe.

Industrialisation came late to northern Italy, not really developing until the second half of the twentieth century. This late development meant that, when it did occur, industrialisation was modern and able to benefit from developments in other parts of Europe. Much of the industry is associated with privately-owned Italian entrepreneurs who have developed huge industrial empires. Fiat and Pirelli are two of the best known examples. Much of the industrial development is concentrated in the industrial triangle of Milan, Turin and Genoa. Between 1950 and 1963 GDP increased by an average of 6% annually and in 1961 the increase was 8%. The oil crisis in the mid 1970s resulted in a slowing down of the Italian economy but Italy has restructured and still ranks in the top ten industrial nations in the world. For the most part, these industrial facts and figures apply to northern Italy as there is little growth in the remainder of the country.

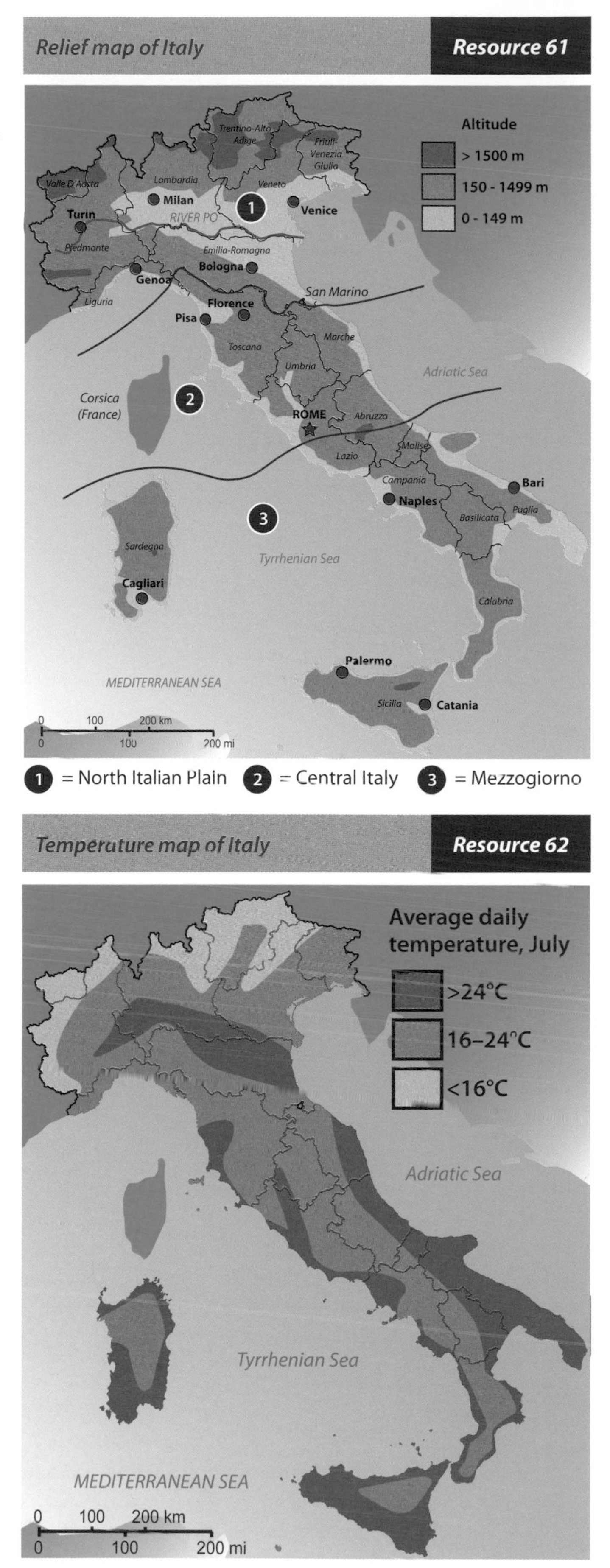

Peninsular Italy

South of the North Italian Plain is peninsular Italy – an area that suffers from increasing remoteness from the core of Italy and Europe. The Apennine Mountains form the backbone of this region. They form a significant barrier to communications especially between north and south. This is because the rivers, which are important natural routeways in highland areas, flow either to the Adriatic Sea to the east or the Tyrrhenian Sea to the west. Areas of lowland are scarce apart from a narrow coastal plain and in the river valleys. With the exception of the volcanic soils close to Naples, most areas have poor and infertile soils. In addition, climate becomes more severe with higher summer temperatures and seasonal rainfall. The peninsula of Italy forms two distinct zones:

1. Central Italy stretches from the southern edge of the Po Valley to Rome. This is a region experiencing considerable economic progress in recent times. This is partly due to its proximity to the highly industrialised area to the north.
2. The Mezzogiorno which lies south of Rome and includes the islands of Sicily and Sardinia. It is this region which presents the most striking contrasts in regional development in Italy.

The Mezzogiorno comprises about 40% of Italy's land area and has 33% of the Italian population. It forms a striking contrast to the affluent North Italian plain. There are many reasons that combine to make this one of the poorest regions in all of Europe. Historically, this region has experienced colonial exploitation by Spain and North Africa, which resulted in economic deprivation. In more recent times the influence of the Mafia has had a negative influence as far as attracting outside investment is concerned.

Resource 63 *Rainfall map of Italy*

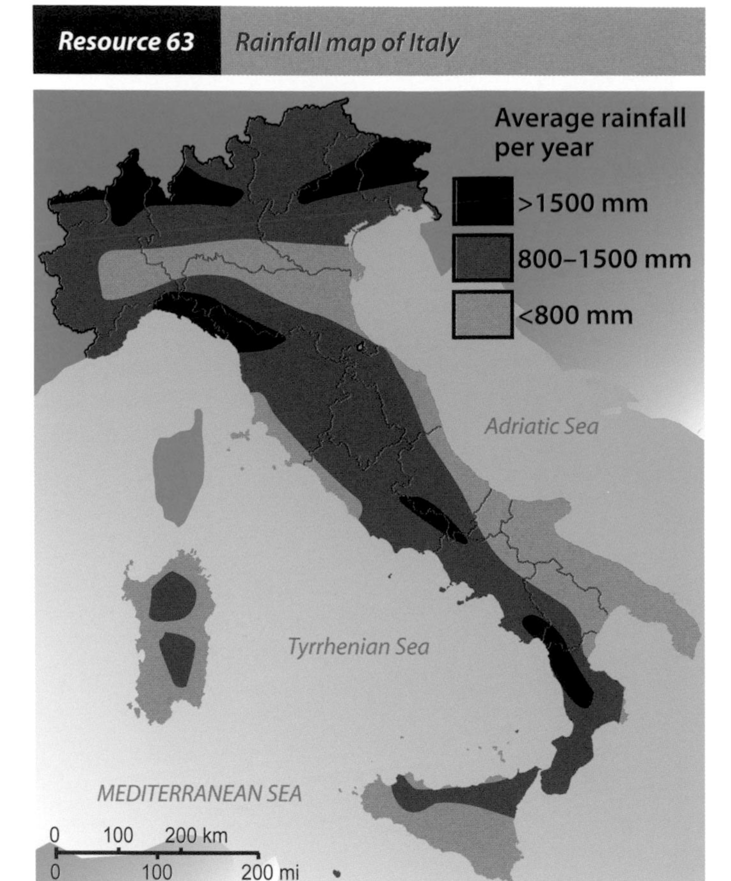

Agriculture

Agriculture has been the mainstay of life in this area for centuries but it has been characterised by poor organisation and low yielding crops. There are obvious physical difficulties in terms of soil fertility, high summer temperatures of 30 degrees centigrade, combined with drought conditions, but years of overgrazing of sheep and goats and deforestation have made conditions worse. Much of the land was in large estates or *latifundia* owned by landlords who rented or leased parts to tenant farmers. The landlords concentrated their interest in the more fertile, lower lying land with tenant farmers restricted to the higher and less fertile land. The tenant farmers practised an outdated peasant style of farming and lived in poor and overcrowded hilltop villages. In the poorest areas high birth rates and a tradition of subdividing land amongst children have resulted in farms becoming so small that they are not viable. Consequently, there are many landless labourers and high unemployment rates. For generations, many of these landless labourers moved to northern Italy to find work. Remittances from these migrants were an important income source for those left behind.

Post war developments

To many northern Italians the Mezzogiorno presented a very negative image and there seemed to be little chance of this region benefiting from the post Second World War economic prosperity that had occurred in the north. Until the 1950s much of the unemployment in the south was hidden by the numbers of men who joined the Italian army or migrated to the north. After the war was over and large numbers of men returned, with little chance of finding employment, it was clear that some form of government intervention was needed to assist the Mezzogiorno. To that end the Italian government set up the 'Cassa per il Mezzogiorno', a regional development authority for this region. The Cassa operated between 1950 to 1984 and its main work falls into three time periods.

1950–1957

Agricultural reform was the main achievement in this time period. The large estates or *latifundia* were broken up and allocated in plots of 5–25 hectares to the peasant farmers. Marshy areas were drained and cleared of malaria-carrying mosquitoes. Irrigation schemes were introduced to offset the difficulties caused by summer drought. In addition, improvements in infrastructure, including the building of a motorway, linking the Mezzogiorno and northern Italy, greatly improved accessibility. All homes in the area were connected to mains water supply during this period.

1957–74

This was the industrial development phase of the Cassa. The intention was to relocate large state-owned companies in the south and to attract private companies by tax incentives. The planners hoped that once large-scale and capital-intensive industry was established in the Mezzogiorno other spin-off industries would follow. A number of growth poles were established including Bari, Brindisi and Taranto. These towns were provided with all the infrastructure necessary to develop industry and they did attract some development such as Fiat at Naples and Palermo steel works at Taranto. However, the spin-off industries did not follow. In fact, the increased accessibility resulting from the new roads made it easier to bring components from the north. Increased accessibility also facilitated further out-migration and between 1950–1974 over four million people left the Mezzogiorno in search of better job opportunities elsewhere.

The Mezzogiorno has benefited from the Common Agricultural Policy but it remains a poor region with high out-migration.

1974–1984

The increase in oil prices in the mid 1970s affected many industrial nations but the impact was particularly severe in remote regions such as the Mezzogiorno. Increases in oil prices meant that much of the industry here was no longer competitive and resulted in a period of industrial decline. The Cassa now focused on smaller industrial developments which were more labour intensive as well as tourism.

Since 1984 the Italian government has continued to sponsor developments in the region, particularly those associated with tourism. The Mezzogiorno qualifies for Objective 1 funding from the EU and it has benefited from the Common Agricultural Policy. Living conditions have definitely improved as a result of all of these efforts but the Mezzogiorno remains a poor region with considerable out-migration. On average, some 80,000 economically active people migrate from this region annually.

Exercise

With reference to a country you have studied, describe and explain regional contrasts in development. (12)

3B ISSUES OF DEVELOPMENT

There are a number of issues that have had a dramatic effect on levels of development across countries and regions of the world. Some of the most important of these include: colonialism, neocolonialism, globalisation, aid, trade and debt.

Colonialism

Colonialism is where one country takes political control over another, usually as part of empire building. As far as development goes, the most important aspects of colonialism involved the building of empires by European nations as they took control of large parts of Africa and Asia. Frequently, the division of territories among the colonial powers meant carving out countries that bore no relationship to the existing tribal or economic areas. A good example of this is Gambia. This small country which lies along the banks of the River Gambia was formed after British traders set up a series of command posts along the river to control trade. Gambia has a narrow coastal section but is completely surrounded by the much larger Senegal, which was a former French territory. Both countries are now independent and relations between them, though generally peaceful, have been strained. In 2005 Gambia increased the price they charged Senegal for crossing their territory. Senegal retaliated by refusing to allow Gambia to cross Senegalese territory, and Gambia in turn prevented Senegalese-registered shipping and road transport from passing through their country. The issue was resolved peacefully, but not before intervention from Nigeria. This example highlights divisions that have existed since colonial times, even though Senegalese and Gambians belong to the same ethnic group.

Similarly, the creation of territories often resulted in tribal homelands being divided between two or more European nations. After independence, many countries have experienced bitter ethnic violence or civil war as in the partition of former British India into India and Pakistan in 1948, and even genocide as happened in Rwanda in 1995. Other examples of ethnic groups divided among several nations include the Kurds – clusters of whom exist in several neighbouring countries including Iraq and Turkey. Iraq was created by amalgamating several ethnic groups, and some see the division of Iraq along ethnic lines as a viable solution to the present problems in that country.

Colonialism also had a negative impact on the economy of the colonies. The colonies were seen largely as providers of raw materials to the western powers who then produced the higher-valued manufactured goods. Due to the fact that there was limited processing of the raw materials in the colonies themselves, no industry developed. After independence, the colonies had not only lost many of their raw materials but there was also no industrialisation. In some cases, even substantial numbers of the economically active population were taken as slaves. It is thought that Africa lost up to 11 million people during the slave trade.

In the colonies, new commercial crops were introduced in a system of monoculture, such as rubber and tea plantations. However, far from providing work for the natives, these large plantations were often farmed by workers brought from outside. The Ugandan Asians, who were expelled to England in the 1970s, were descendents of indentured workers brought from British India to work on the tea plantations in former British Uganda. Colonialism had left many countries in Africa with political and economic instability.

There are some positive outcomes of colonialism. Since the colonies were primarily concerned with exporting raw materials to the colonial power, most will have at least one well developed port city – Calcutta (India), Accra (Ghana) and Lagos (Nigeria). These cities were provided with the necessary infrastructure to carry out the export of goods. They often also served as an administration centre for the colonial ruler, and many have remained as capital cities post-independence. The city-state of Singapore, one of the wealthiest nations in South Asia, is a good example of the positive impacts that can follow from a colonial past. Singapore has built on the pattern of trade that developed during British rule to become one of the leading economies in the region. In many places colonial rulers initiated the building of extensive road and railway systems, albeit for their own ends, but the colonies did benefit from this infrastructure. The railway lines, which linked the gold and diamond mines in British territories in southern Africa to the coast, are important examples of this. Some former colonies have benefited from the educational system laid down in colonial times. Practically all colonial territories have now gained political independence, but most have retained some connection with their former rulers. In the case of Britain, the formation of the Commonwealth was designed to maintain contact and promote development and trade with former colonies. In addition, many economic migrants have come to their former colonial ruler in search of jobs, including the Afro-Caribbean to Britain and North Africans to France.

Aerial view of Singapore

Neo-colonialism (dependency)

As stated above, most colonial territories have gained political independence from their former rulers, mostly in the post Second World War period up to the 1960s. However, many are still dominated economically by MEDCs. This political independence with economic control from outside is referred to as neo-colonialism or dependency, and it is one of the most pressing issues facing many LEDCs. Many former colonies were left with very poorly developed economies. During colonial times their mineral wealth was exported for processing with the result that there was very little development of manufacturing industry. This was seen as vital if the countries were to develop. Many countries borrowed vast amounts of money from the World Bank, the International Monetary Fund or even from high street banks, all based in the MEDCs. These organisations were willing to lend this money because of the high rates of interest charged on such vast amounts of money. Some former colonies also received aid packages from the MEDCs and these too had high rates of interest. For many LEDCs, repaying these loans is almost impossible, but there seemed little alternative if industrialisation was to occur. Apart from the high interest rates, many of the loans also had other conditions attached. This might require the LEDC to purchase manufactured goods from the country that made the loan in the first place, or to provide them with primary goods at a fixed price. In other words, the LEDC colonies might have gained political independence, but their economy was tightly controlled by the MEDCs.

Describe how colonialism and neo-colonialism have influenced development.

Exercise

Globalisation

Globalisation has come to refer to the current interaction of most of the world's economies. In the world of today, national economies are no longer separate entities but rather part of a world or global economy. Developments in travel and communications mean that all parts of the world are accessible. In this way, globalisation is a recent process but the colonial system discussed earlier is actually a forerunner of globalisation. Following the end of the Second World War a number of global organisations were established. These were mostly concerned with promoting peace and developing trade (***Resource 64***).

Resource 64

Title	Full title	Date formed	Aim
WTO	World Trade Organisation (formerly known as GATT)	1948	Promote free trade
IMF	International Monetary Fund	1945	Promote international monetary cooperation and trade
UN	United Nations	1945	Promote peace and reduce conflict
IBRD	International Bank for Reconstruction and Development	1946	Provide funds and skills for economic development

Increasingly, the largest economies and the largest manufacturing organisations operate on a global scale. Transnational and multinational companies play a vital role in the globalisation process. (A transnational company (TNC) has its headquarters in a single MEDC and a number of branch factories in LEDCs, while a multinational company divides the running of the company across several countries.) These large companies provide employment for millions of people worldwide and some have incomes greater than the total GNI of many LEDCs. As a result, these organisations have a considerable influence on the world economy with some suggesting that they may also have a political influence as well. TNCs control the production and processing of most of our primary goods, including food, minerals and oil refining. They also control most of the world's vehicle and transport industries and the ever-growing high-tech computer and telecommunications industries. TNCs depend on global finance and banking and not surprisingly that sector of industry also falls under their control.

Resource 65 *The top non-financial multinational companies (ranked by foreign assets)*

Rank	2005	Country	Product
1	General Electrics	USA	Aero engines
2	Vodafone	UK	Telecommunications
3	General Motors	USA	Vehicles
4	BP	UK	Oil refining
5	Shell	Netherlands/UK	Oil refining
6	Exxon	USA	Oil refining
7	Toyota	Japan	Vehicles
8	Ford	USA	Vehicles
9	Total	France	Oil refining
10	Electricite de France	France	Electricity, gas and water
11	France Telecom	France	Telecommunications
12	Volkswagen	Germany	Vehicles
13	RWE	Germany	Electricity, gas and water
14	Chevron, Texaco	USA	Oil refining
15	E.ON	Germany	Electricity, gas and water
16	Suez	France	Electricity, gas and water
17	Deutsche Telekom	Germany	Telecommunications
18	Siemens	Germany	Electrical and electronics
19	Honda	Japan	Vehicles
20	Hutchinson Whampoa	Hong Kong	Diversified

Source: United Nations Conference on Trade and Investment report WIR/2007/TNCs

The globalisation process operates through these companies. The location of their businesses fall into three types of regions.

1. LEDCs that have stable governments and where labour costs are low.
2. Economic black spots in MEDCs such as Silicon Glen in Scotland, which also has universities close by for skilled labour.
3. Areas of international economic and research importance such as the M4 corridor in south-east England.

Advantages of globalisation

As with all of the issues affecting the level of development, globalisation has both positive and negative impacts. On the positive side, the relocation of part of the manufacturing process to the LEDCs has been largely welcomed because these companies offer the chance of employment to many who would otherwise be unemployed. There are at present over 45 million people in LEDCs employed directly by TNCs and many more in spin-off services. There will almost certainly be some technological development associated with the new factories, and this will enhance the skills level of the local population. Having TNCs in a country establishes links with the international economy and other spin-off industries, such as component suppliers for the car industry, may follow. The export of the finished product is often the only manufactured item to come from the LEDC, which usually only exports primary goods.

TNCs have also had a beneficial effect on some economically deprived areas in MEDCs. Because most industries are now flexible in their location, a number of TNCs have been attracted by financial incentives to some economic black spots. Silicon Glen in Scotland, an area of high unemployment following the decline of heavy industry, is a UK example of this. The M4 corridor in the south of England is yet another example of the positive impacts of globalisation. This area has many universities and has close proximity to London, its airports and ease of access to Europe, making it an attractive and prestigious location for international high-tech manufacturers. This region attracts the research and development side of manufacturing rather than the routine production associated with mass production in LEDCs.

Apart from the economic advantages, globalisation encourages communication between different countries and racial groups, and should promote mutual understanding and tolerance. Countries throughout the world can share the same sporting activities, music, films and television programmes, which have led some observers to discuss the development of a global society. The recent international conferences on climate change have emphasised the need for global action to respond to the challenges posed by global warming.

Disadvantages of globalisation

There are many negative aspects of this modern globalisation process. There are well-documented incidents of malpractice by these companies, including treatment of workers, wages and working conditions generally. Hundreds of millions of farmers and workers, many women, earn only one or two dollars per day while the TNCs become extremely wealthy. The combined annual incomes of Ford and General Motors are greater than the GDP of the whole of sub-Saharan Africa and in 2002, Nestlé recorded profits greater than Ghana's GDP for that year.

Many of the poorest nations, especially those in sub-Saharan Africa, have not been able to attract any TNCs and therefore have not experienced any economic benefits from globalisation. In sub-Saharan Africa there are close to 600 million people and in the mid 1990s they earned less from exports than the 3 million people who live in Singapore. In addition, sub-Saharan Africa received less than 2 billion dollars in foreign investment compared with the 12 billion dollars received by Malaysia with a population of 20 million. TNCs avoid any country where there is civil unrest or anything that could harm their interests. There is also the concern that decisions are taken at headquarters with no regard for the 'branch factory' or its workers. The development of skills in the LEDC location will not go beyond the current requirements of the company in that location. Actions taken in some of the largest economies have knock-on effects in all countries. At the time of writing, the downturn in the American economy is having serious repercussions on the British economy. There have also been environmental issues raised, including the location of dangerous chemical factories such as the American-owned Union Carbide chemical plant at Bhopal in India. A fire at this factory in 1984 resulted in 7,000 deaths in the immediate aftermath but the long term death toll exceeds 20,000 deaths and many more injuries. The American company paid only small amounts of compensation to the survivors, many of whom have been unable to work again and others have suffered from congenital birth defects as a result of the accident. Where financial incentives have been given to attract large companies, there is a concern in some cases that taxpayers' money could have been used in other aspects of development. Some question the TNC's commitment to the LEDCs, believing them to be motivated by increasing their profits with little regard for the well being of the workers in LEDCs.

Finally, many see globalisation as the spread of western values and cultures. Globalisation is also associated with material gains with little concern for environmental issues, although there have been some improvements in this area.

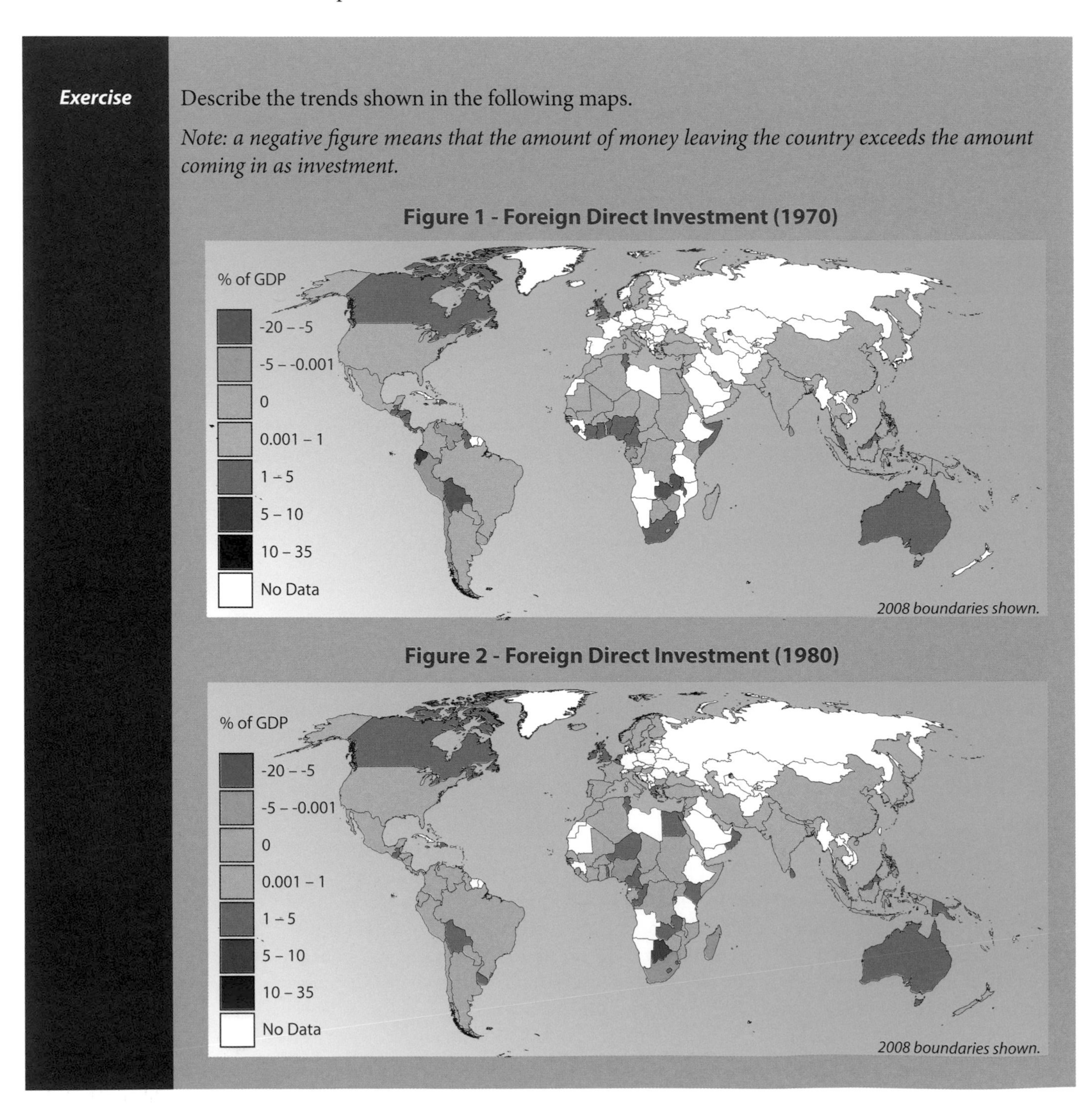

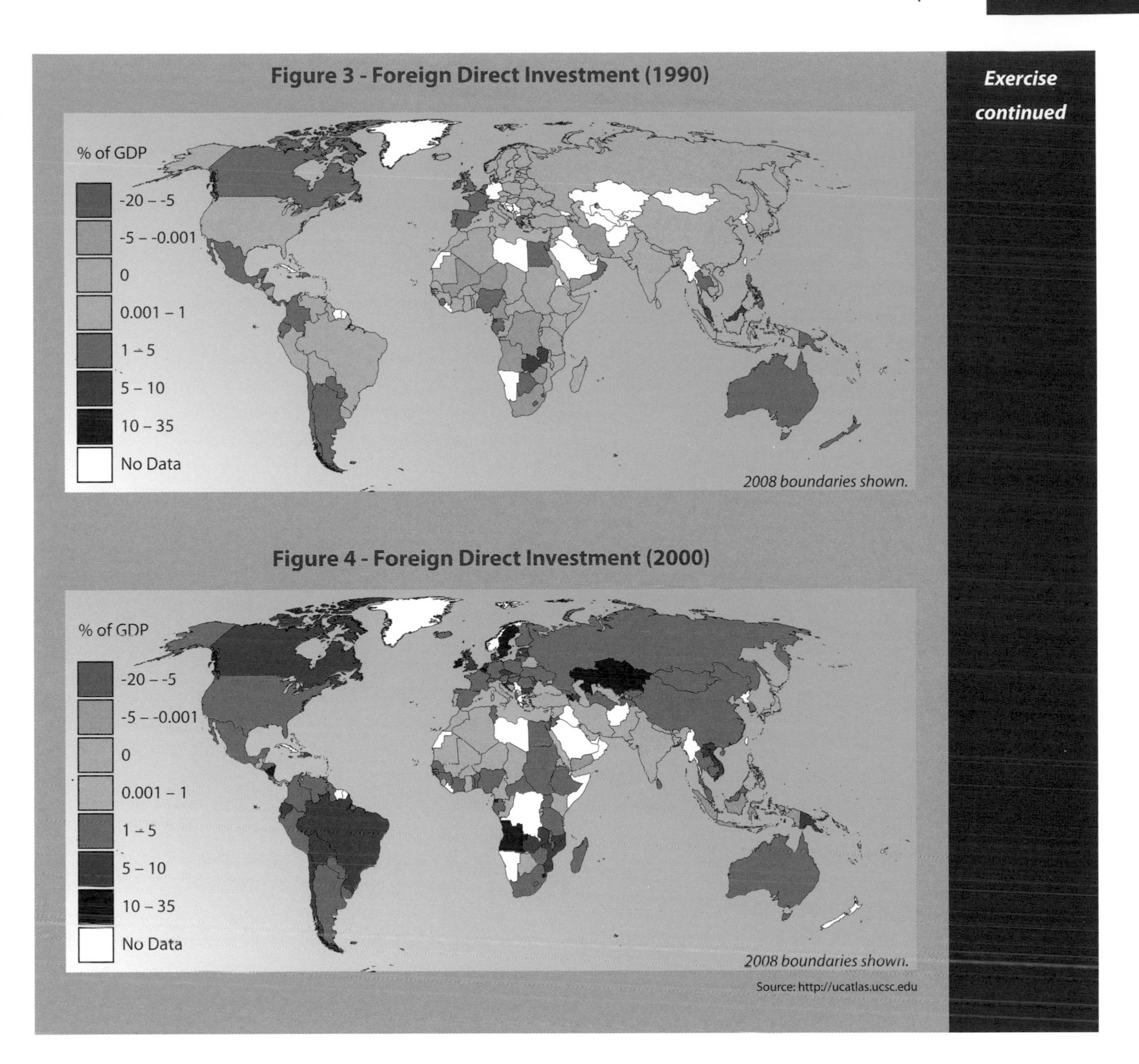

Aid

Aid is defined as the transfer of resources from richer MEDCs to the poorer, usually LEDCs. After the Second World War the US funded an aid package to help rebuild those European nations affected by the war. Aid can take the form of money, food, health care, education, technical advice or machinery. There are several types of aid.

1. Official aid

This comes directly from government sources. A distinction is often made between **bilateral aid,** which comes directly from one country to another, and **multilateral aid** which comes from several countries or organisations such as the World Bank. The UN has set targets for each country to donate official aid. This was set at 0.7 percent of the total GNI of each country but as **Resource 66** illustrates the target is often missed. In 2000 the United Nations held a Millennium Summit in New York where they set out targets to improve the quality of life in the LEDCs. These targets are known as the Millennium Development Goals (MDGs), which are set out overleaf, and are to be funded by official aid packages through multilateral aid. In 2005 the G8 met at Gleneagles in Scotland and agreed that an extra $25 billion was needed

Resource 66 *Aid as a percentage of GNI 2006*

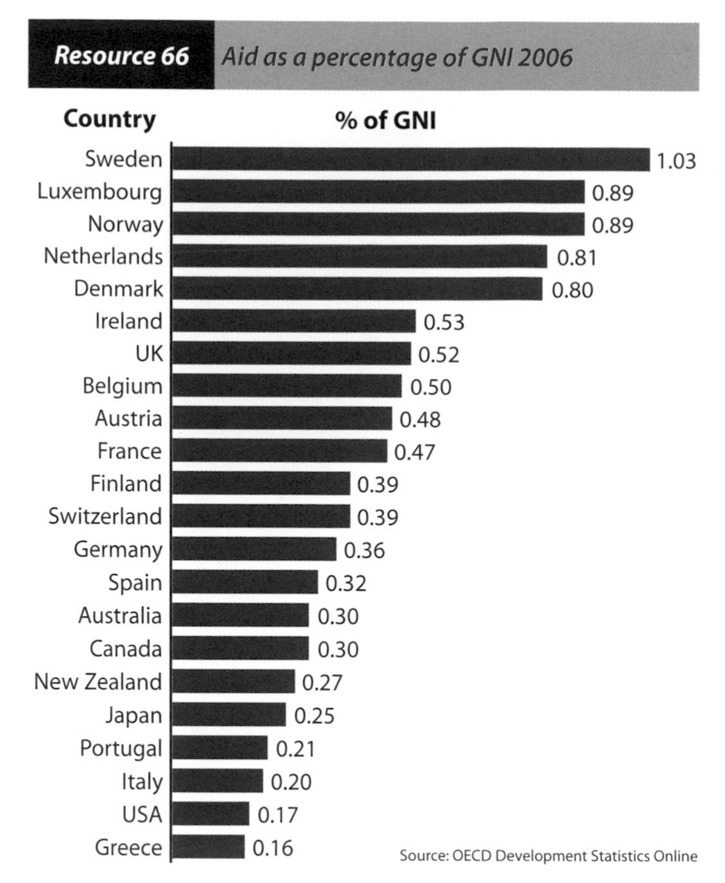

Source: OECD Development Statistics Online

annually until 2010 and double that amount between 2010 and 2015 if the MDGs were to be met. However, it seems that some countries in the G8 have not kept these pledges of additional money and even where the money has been available it has not always reached those who need it most.

2. Voluntary Aid

This comes from charitable organisations such as Oxfam, Trocaire and Action Aid. These are referred to as **Non-Governmental Organisations (NGOs)** and they rely entirely on voluntary contributions. These organisations provide emergency aid following a natural disaster as well as long term aid packages to provide clean water supply or irrigation schemes. Voluntary aid is a much smaller amount than official aid.

Most countries want to contribute towards official aid programmes, partly from a moral viewpoint, but it is also true that rich countries will benefit from any economic improvements in the LEDCs through increased markets for their goods. There is evidence to show that aid packages from whatever source do bring economic improvement to LEDCs. Some of the most dramatic examples are seen following a natural disaster such as the tsunami in Indonesia in 2004. Aid packages have also contributed to vaccination programmes, which have resulted in a decrease in infant mortality. Education programmes have been established in many countries and the numbers attending primary school have increased significantly in recent years. In Africa as a whole, enrolment in primary education has risen from 57% in 1999 to 70% in 2005. Documentary evidence provided by the UN, charting progress of the MDGs, has shown many positive developments from aid packages but it seems unlikely that all of the targets will be met, especially in the poorest region of all – sub-Saharan Africa.

The Millennium Development Goals

Goal 1: Eradicate extreme poverty and hunger

Goal 2: Achieve universal primary education

Goal 3: Promote gender equality and empower women

Goal 4: Reduce child mortality

Goal 5: Improve maternal health

Goal 6: Combat HIV/AIDS, malaria and other diseases

Goal 7: Ensure environmental sustainability

Goal 8: Develop a global partnership for development

Source: www.undp.org/mdg/basics.shtml
Note: further information regarding the millennium development goals, including indicators used to monitor progress for each goal, is available from the above website.

Official aid in the form of money does incur interest but at much lower rates than normal loans. In this way, aid can also prevent countries from getting into serious debt and some countries have used aid for large-scale development projects. Aid works best in countries where there is effective democratic government and political stability. Unfortunately, in many LEDCs these are frequently absent.

There has been considerable criticism of aid as a means of encouraging development in LEDCs. Too often much of the financial aid is wasted, either through corruption, bad management, poorly thought-out schemes or spent on arms deals. One of the best-documented examples is that of former Zaire (Democratic Republic of Congo) where millions of pounds of aid money was stolen by the corrupt ruler in the 1970s and 1980s and never reached the poor for whom it was intended. Much bilateral aid is 'tied aid', meaning that the aid has conditions placed upon it. It is estimated that tied aid accounts for about 40% of all aid donations to LEDCs. This could mean that the recipient country has to buy manufactured goods from the donor country, or the recipient may have to agree to terms of trade beneficial to the donor. Britain and other countries have been accused of linking aid packages with arms sales in LEDCs. Whatever the conditions, this tied aid does restrict the recipient country. Multilateral aid from organisations such as the World Bank is often criticised for only providing aid to those countries which support western style politics. Israel receives vast amounts of technical aid from the United States and many view this as being politically motivated. The US has also received criticism for its support for Pakistan (see below). Many large-scale projects funded by the World Bank have had serious environmental repercussions (see below). Others claim that aid creates a dependency culture and prevents initiative.

Examples of Aid

World Bank accused of razing Congo forests

The World Bank encouraged foreign companies to destructively log the world's second largest forest, endangering the lives of thousands of Congolese Pygmies, according to a report by an independent inspection panel. The report also accuses the bank of misleading Congo's government about the value of its forests and of breaking its own rules.

Congo's rainforests are the second largest in the world after the Amazon, locking nearly 8% of the planet's carbon and having some of its richest biodiversity. Nearly 40 million people depend on the forests for medicines, shelter, timber and food.

It is particularly embarrassing for the British government, which is a development partner of the bank and its third largest financial contributor. It encouraged the bank to intervene in the Congo forests with export-driven industrial logging and has earmarked £50m for further Congo basin forestry aid.

But although the bank is legally committed to protecting the environment, and trying to alleviate poverty, the panel found that the policies it imposed on the Congo were having the opposite social and environmental effects.

Source: The Guardian, 4/10/07

United States Criticised Over Aid to Pakistan

Some Senate Democrats challenged the Bush administration's support for Pakistan's President Musharraf and questioned the 10 billion dollars in U.S. assistance given to Pakistan since 2001 in light of the treatment of opposition groups in Pakistan.

But a government official defended the aid, saying that the country's success as a nation is essential to U.S. security. And he said the United States has played a positive role in moving Pakistan towards democratic change:

"Part of the reason why we are in a period of transition to democracy right now, part of the reason why there are political leaders in Pakistan contesting elections that will be held in a month, part of the reason why the state of emergency once imposed is going to be lifted soon is because of the role of the United States," he said.

Much of the U.S. aid to Pakistan since the September 11, 2001, terrorist attacks on the United States has been spent to reimburse Islamabad for its assistance in the war on terrorism. After President

Musharraf imposed the state of emergency on November 3, the State Department began a review of U.S. assistance. The US now claims that $200 million in U.S. aid, previously dispensed by the Pakistani treasury, would now be spent directly by U.S. agencies for health and education programs.

Source: Voice of America News, D Tate 7/12/07

Bangladesh seeks food aid after cyclone

DHAKA, Nov 28, 2007 (AFP) - Bangladesh has sought half a million tonnes of food aid, an official said Wednesday, as hundreds of cyclone survivors demonstrated to demand more relief.

At least 3,400 people died in the November 15 cyclone. Some 1,700 are still missing and more than 360,000 are homeless and in need of supplies.

The World Food Programme has promised 71,000 tonnes and the Indian government 50,000 tonnes. USAID, the US government's international development arm, had also pledged 10 million dollars worth of food.

Source: Agence France-Presse 28/11/07

Exercise

1. Research the Millennium Development Goals on the following website: www.undp.org/mdg/. Use that information in conjunction with the information on Aid (page 169 onward) to discuss the role of aid in development.
2. (a) Study the table below, which shows changes in infant mortality rates. Discuss the pattern of change from 1990 to 2005.

   (b) How could aid packages have helped bring these changes about?

Deaths of children before reaching the age of one, per 1000 live births

	1990	2005
World	65	52
Developing Regions	71	57
Northern Africa	66	30
Sub-Saharan Africa	110	99
Latin America and the Caribbean	43	26
Eastern Asia	37	23
Southern Asia	87	62
South-Eastern Asia	53	31
Western Asia	53	45
Oceania	59	47
Commonwealth of Independent States	39	33
Developed Regions	10	5
Transition countries of South-Eastern Europe	25	14

3. One type of aid package is associated with small-scale projects that are labour intensive, use local skills and resources, and are not harmful to the environment. Such projects, often sponsored by NGOs, are said to use Appropriate Technology. They often involve supplying a village pump or well that will provide clean, fresh water for the inhabitants of the village.

   Use one of the following websites to research one of these schemes:

   http://journeytoforever.org/at.html

   http://www.atasia.org.uk

4. Discuss the advantages and disadvantages of these schemes compared to the larger schemes.
5. Identify and discuss two problems with aid as a process of development for LEDCs.

Additional Reference

Action Aid Kenya: a case study of an NGO – GEOFILE April 2004

Exercise continued

Trade

In an ideal world trade between countries would operate to reduce inequalities in development. Surpluses produced in one country could be exchanged for goods in demand from another. With increased transportation capabilities, we in the west have access to products on a global scale, and this increased accessibility has facilitated the global shift in manufacturing discussed under globalisation. However, the reality is that trade is controlled to a large extent by the MEDCs and by the G8 in particular. (G8 countries are UK, USA, France, Germany, Italy, Japan, Canada and Russia.) The population of the top five MEDCs is similar to that of the bottom 49 LEDCs yet they have 100 times more trade than the poorest countries.

Disparity in trade

There are many reasons for this disparity in trade between LEDCs and MEDCs. One of the main reasons is associated with patterns of interdependence that originated in colonial times. During the colonial phase, western nations laid down a foundation of trade between MEDCs and LEDCs that still exists today. The LEDC colony was a source of raw materials for the MEDC. In post colonial times many LEDCs have continued to export raw materials, either in the form of minerals or agricultural products. These were exported in an unprocessed state to be processed or used in manufacturing industries in the MEDC. Raw materials are generally low in value or at least lower in value than a processed item. Again partly as a result of colonialism, the LEDCs have little in the way of manufacturing industry and they are forced to import higher-valued manufactured goods from the MEDC. In addition, many LEDCs have a limited range of primary products for export. This creates a vulnerable situation for the LEDCs because:

1. The price of primary products fluctuates greatly and in recent times there has been a sustained fall in their price.
2. Over dependence on one product makes a country vulnerable to changes in demand or value of that product.
3. Crop failure or a finite resource being worked out will seriously reduce their export potential.
4. Competition from other producers can force down the price of a product, resulting in a drop in export revenues.

The end result of these factors is that most LEDCs earn less for their exports than what they pay for imports. This situation is referred to as **trade deficit**. Many MEDCs have a **trade surplus** – ie their exports are more valuable than their imports. Some LEDCs find themselves locked into trade agreements with MEDCs, often as a result of tied aid. This usually means that the MEDC will provide aid in return for a market from the LEDC for their manufactured goods. There is much well-documented evidence to show this happening on a wide scale. One example relates to the UK government supplying aid to Malaysia in return for the purchase of weapons and tanks from the UK.

Study **Resources 67** and **68.** Compare and contrast the trade flows to and from Western Europe with those for Africa.

Resource 67 *Trade flows to and from Western Europe 2000*

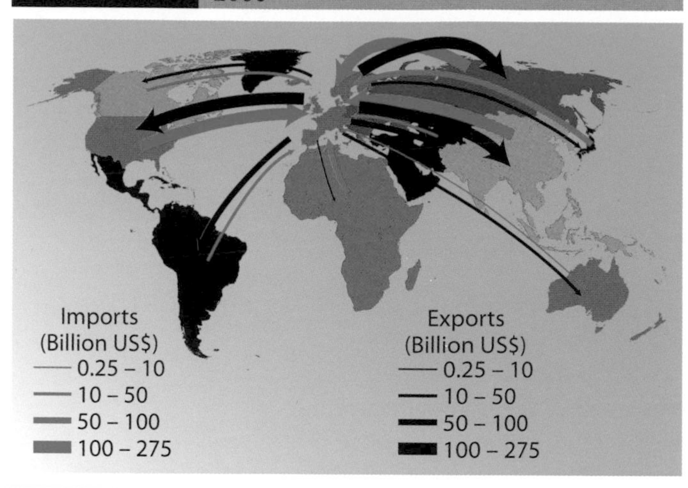

Resource 68 *Trade flows to and from Africa 2000*

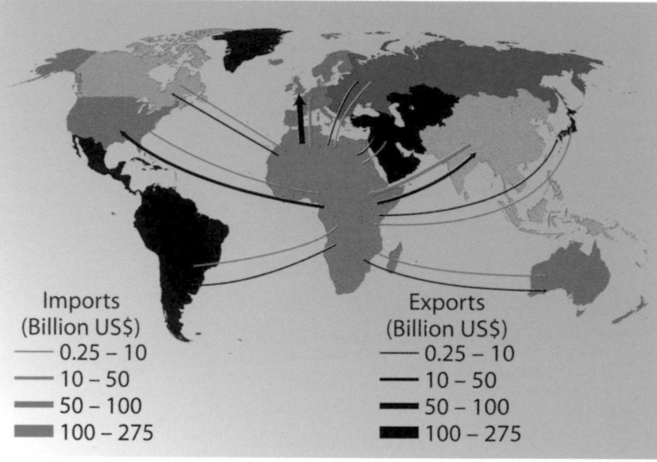

Source: http://ucatlas.ucsc.edu/trade/tradeflows.htm

Trade blocs

The formation of trade blocs (a group of countries that join together to protect trade amongst themselves) operates to protect western nations, mostly to the detriment of the poorer countries. The European Union is an example of a trade bloc. Trade blocs' main aims are:

1. To establish some form of regional control regarding trade of nations within that region.
2. To use tariffs to protect intra-regional trade from non-member states.
3. To exert some form of political control regarding international trade.

These trade blocs protect their own trade by imposing quotas (limits) on the amount of imports or by imposing tariffs or taxes on imports from non-member states. Prior to entering the EU, Britain had a number of trade agreements with the Commonwealth but on joining the EU these had to be terminated. This caused considerable hardship to many of our former trade partners. Trade blocs can also subsidise production of goods in their own countries to prevent reliance on imports, and as a group they can also boycott trade with other countries, usually on political grounds. At present the EU operates a boycott on trade with Zimbabwe and in the past Libya and Iraq have faced similar actions.

Global organisations

Global organisations also appear to operate to the advantage of the richer nations. The International Monetary Fund (IMF) was established after the Second World War to facilitate global cooperation on monetary matters and international trade. One of the aims of the IMF was to give loans to poorer countries to help reduce the development gap. However, the MEDCs have control over where and when loans will be made as well as setting out the conditions of the loan. Many argue that this organisation works for the interests of the MEDCs more than the LEDCs. The World Bank – which provides aid following natural or man-made disasters – faces similar criticism. In recent times the World Trade Organisation (WTO) has received the most criticism for its attitude to the LEDCs. Dominated by the G8, the WTO has come to symbolise to many the domination of world trade by large companies, particularly the TNCs. Wherever the WTO meets, there are always anti-globalisation protests. These protesters and many LEDCs accuse the WTO of protecting trade and profits for the MEDCs and presenting a situation whereby the LEDCs are forever locked into a cycle of deprivation and debt. The policies of the WTO which are most criticised are:

1. Price fixing for raw materials and agricultural products that seem to benefit the MEDCs.
2. Allowing surplus agricultural products from MEDCs to be sold at low price. This means that LEDCs have to lower their prices if they wish to compete.
3. Allowing pharmaceutical companies to charge high prices for medicines that are beyond many in LEDCs.
4. Trade blocs such as the EU as well as the USA are allowed to have protectionist policies to safeguard their own agricultural and industrial products.

Some MEDC manufacturers establish companies in overseas countries in order to overcome import tariffs and restrictions. An example of this are the Japanese car manufacturers who have set up factories in Britain in an attempt to gain entry to the European markets.

There is one example where the MEDCs have been at a disadvantage regarding trade. This refers to the global trade in oil on which so much of the west's development relies. In the 1970s the oil-producing countries joined together as the Oil Producing and Exporting Countries (OPEC). They meet at regular intervals to control the quantity and price of oil. The increase in oil prices in the 1970s had a serious impact on the western economies, causing prices of manufactured goods to increase. This also had a knock-on effect on the LEDCs who were engaged in industrialisation programmes. They found that they not only had to borrow more money if they wished to continue their industrialisation programme but that the interest rates had increased.

Additional References

http://ucatlas.ucsc.edu/ – An excellent website that provides up to date and interactive maps, graphs, tables and commentary on many aspects of development.

Geofile no. 513 Transnational Corporations

Geofile no. 541 Globalisation of food production

www.undp.org/mdg/ – Millennium Development Goals Report 2006

Exercise

1. *Question from CCEA June 2003*
   Study **Figure 1**, showing population and GDP per capita for some regions in China and **Figure 2** showing some of the main physical characteristics of China.
   (a) Using **Figures 1** and **2** describe and account for the variations in GDP per capita in China. (6)
   (b) Discuss **one** effect of globalisation. (3)

All of China: Population 1166 million, mean GDP per capita $470.

Figure 1

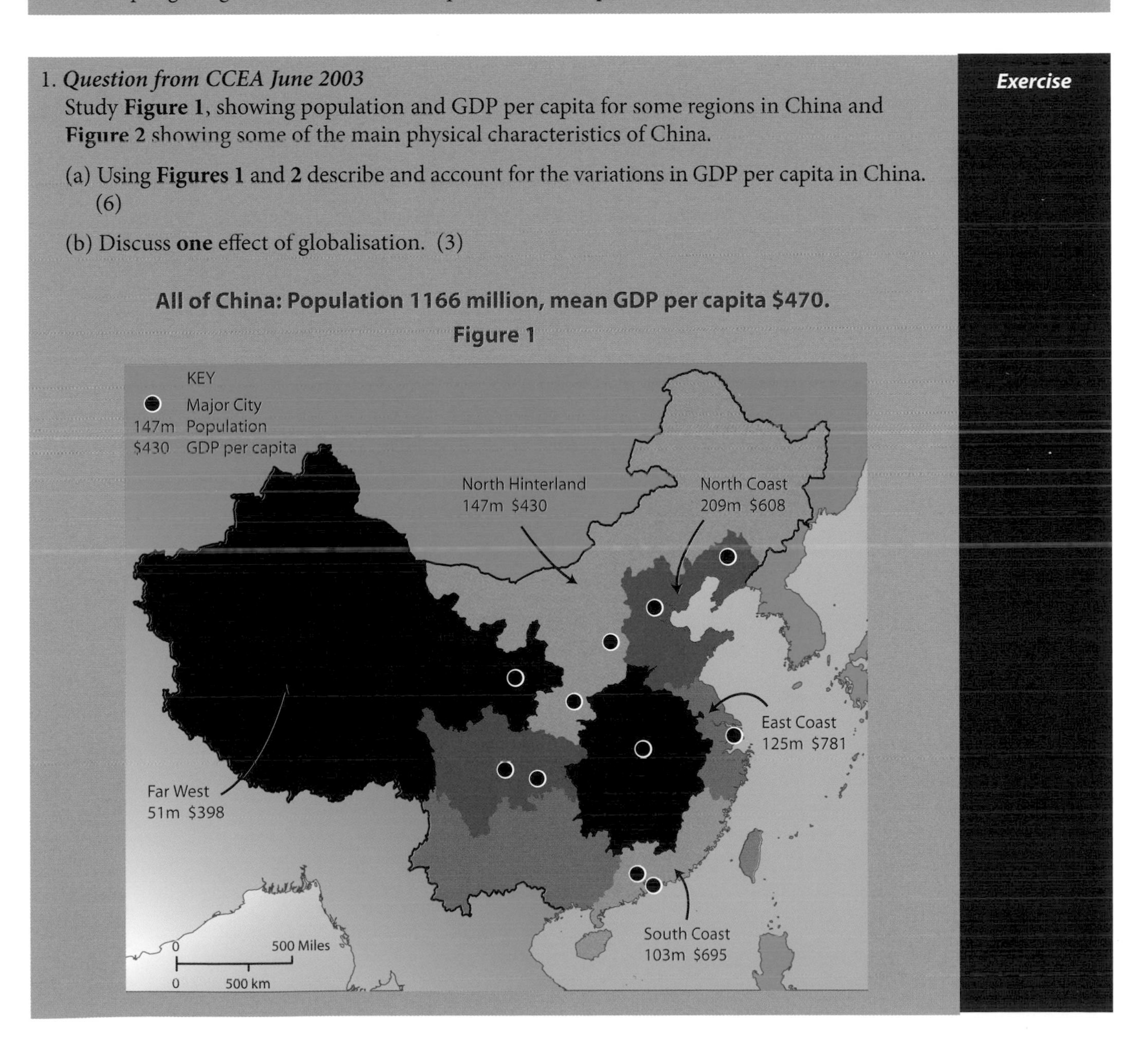

Exercise continued

Figure 2

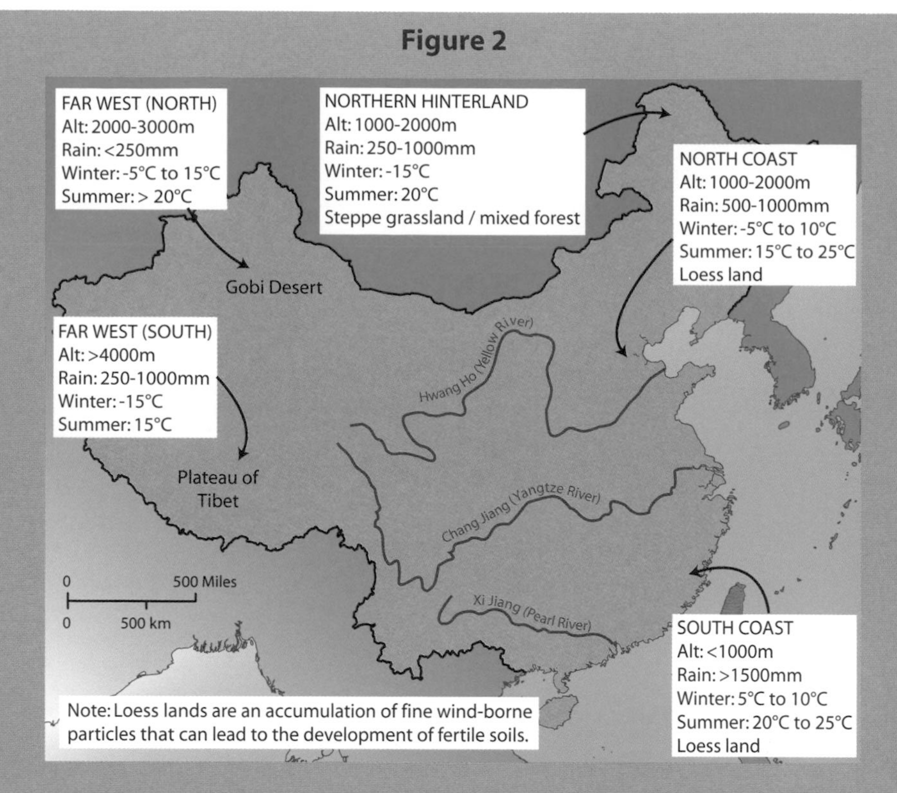

2. *Question from CCEA June 2004*
Study **Figures 3** and **4**, showing European expansion into Africa in 1870 and 1914.

(a) Identify the European country that colonised Angola. (1)

(b) Discuss **two** effects colonialism has had on the development of countries such as those in Africa. (6) (3 marks for each effect)

Figure 3 – Africa in 1870

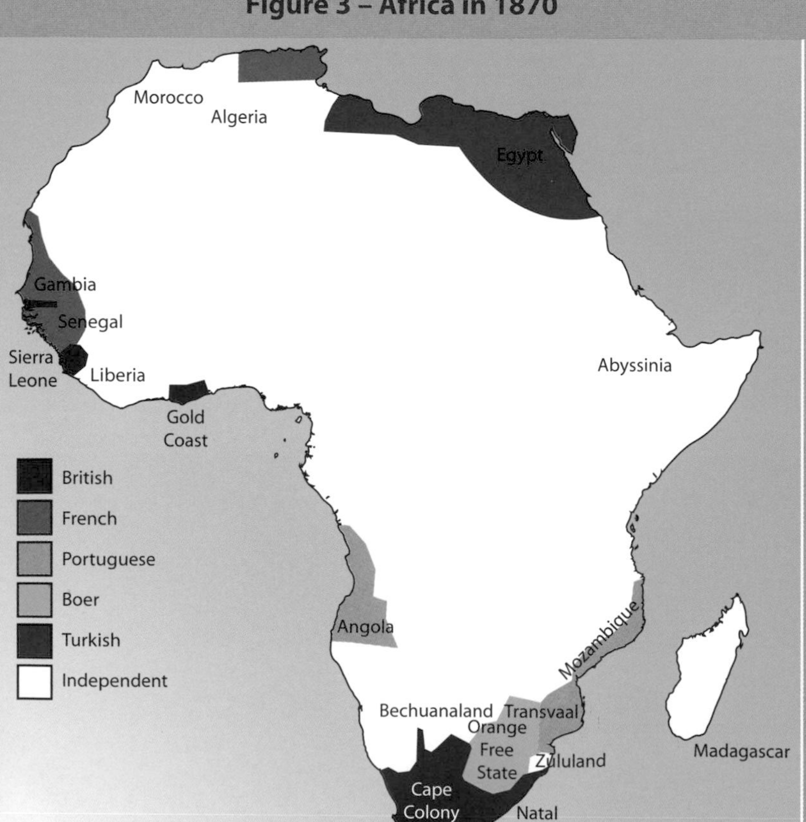

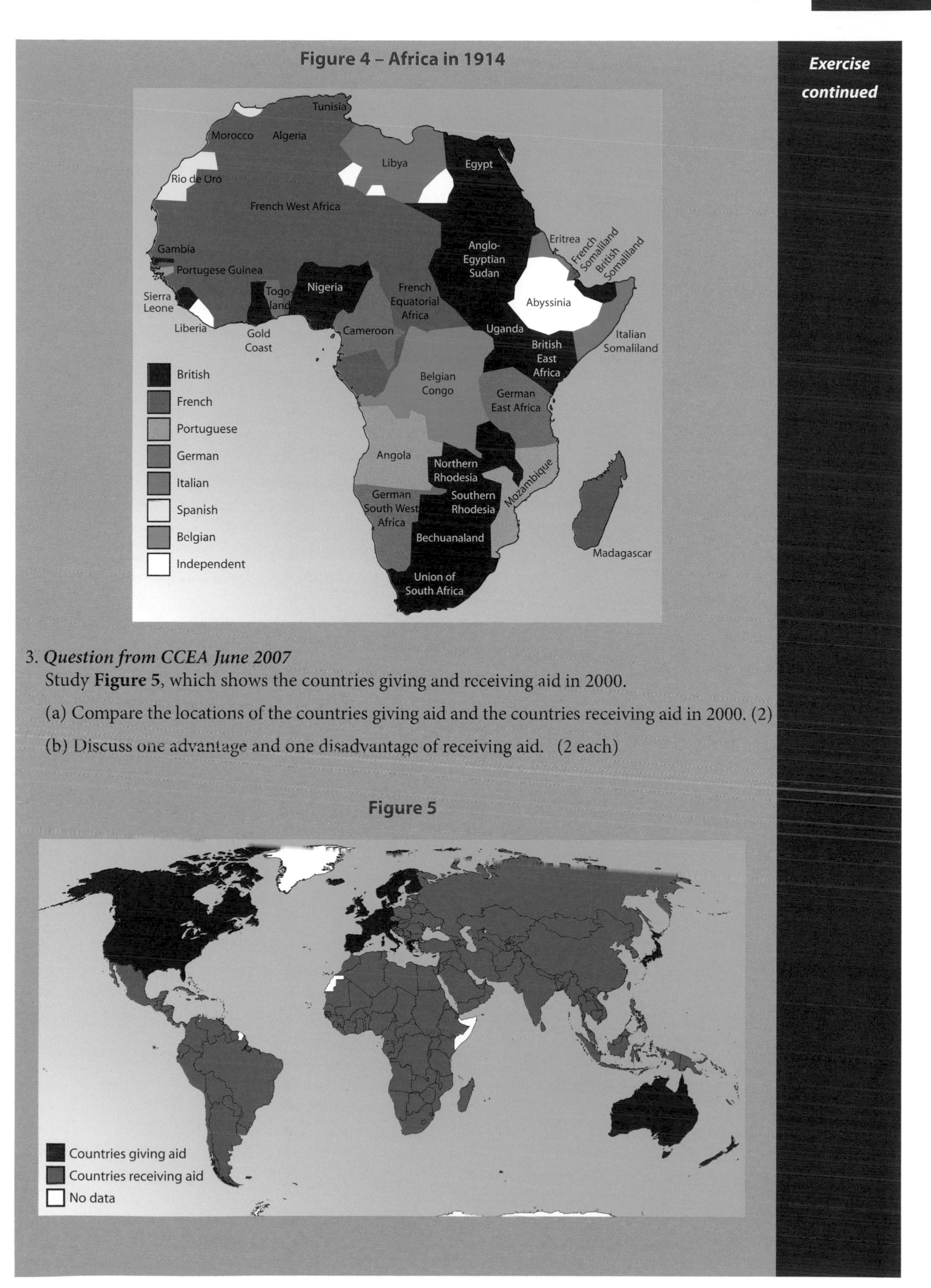

3. *Question from CCEA June 2007*

Study **Figure 5**, which shows the countries giving and receiving aid in 2000.

(a) Compare the locations of the countries giving aid and the countries receiving aid in 2000. (2)

(b) Discuss one advantage and one disadvantage of receiving aid. (2 each)

Debt

One thing all LEDCs share in common is the fact that they all are in debt to some extent to either MEDCs or to one or other of the international monetary organisations. The amount of debt has increased in the last thirty years. Initially, the debt burden was a major issue in South America but in the last decade the problem has escalated in Africa and particularly in Sub-Saharan Africa. Some countries are so indebted that there seems no possibility that they will ever be able to repay the debt. The reasons for this increase in the debt burden are varied and complex but the impacts on development projects are simple – countries have to sacrifice health and education programmes in order to repay the interest on the loan. It is estimated that about 21 million children will die each year and many more will not receive proper education or health care because of debt problems. This is because conditions, known as Structural Adjustment Programmes (SAPs), have been placed upon the LEDCs that have had to take out additional loans. SAPs usually meant LEDCs had to increase their exports and commodities in order to pay off loans. This often meant there was less money spent on food, health and education.

Reasons for the growth of the debt burden

1. Many LEDCs borrowed heavily from the 1960s onward in an attempt to develop industrialisation. At this stage the MEDCs and the World Bank were willing to lend at relatively low rates of interest. The loans were usually to finance capital-intensive projects such as power stations, often using western technology and expertise. In the 1970s, partly due to a downturn in the global economy, these interest rates were increased.
2. The increase in the price of oil had a further serious impact on the LEDCs. Many LEDCs were heavily dependent on imported oil, and some were forced to borrow still more to maintain their level of progress.
3. Trade problems also affected the LEDCs. In the west the cost of manufactured products increased but the price of primary goods or commodities fell. As LEDCs exported mostly commodities and imported manufactured goods, their balance of trade deteriorated, pushing them further into debt.
4. In some cases the loans were spent unwisely on large capital schemes and the LEDCs did not have an adequately trained workforce to make these schemes operational.

Recent developments

The debt problem is an emotive issue for many in the west. There are those who believe that some assistance is needed for the countries with the worst debts, while others argue that the LEDCs needed to act more prudently with the loans. In the late 1990s many LEDCs received some help with their mounting debt problems. These included additional loans at much lower rates of interest but most were tied to strict conditions such as the SAPs mentioned earlier. In 2000 the setting up of the Millennium Development Goals and the Gleneagles summit 2005, along with the Make Poverty History Campaign, have heightened awareness of the problem. Prior to the Gleneagles Summit, the G8 agreed that if nothing changed, it would take Africa 150 years to reach the poverty reduction target set out in the Millennium Development Goals. The proposed measures that were set out to improve the situation in Africa included an extension of the debt relief scheme known as the Heavily Indebted Poor Country Initiative (HIPC), see opposite. The poorest 18 countries are to have their debts completely written off by the IMF and World Bank as well as by individual countries, while the debt relief scheme is being extended more widely. Africa's most populous country, Nigeria, which had not qualified before, convinced the G8 that it had done enough to combat corruption to earn the relief of some of its bad debts. These measures would cost the west about $30 billion. (For comparison Microsoft Corporation makes $20 million profit daily.) The G8 also pledged to double the amount of aid given to Africa by 2010. However, in spite of these and many other promising statements there is still a long way to go towards solving the debt problem for many LEDCS.

Debt Relief Under the Heavily Indebted Poor Countries (HIPC) Initiative

The HIPC Initiative is a debt reduction scheme set up in 1996 for heavily indebted poor countries that satisfy economic and political criteria laid down by the IMF and World Bank. In 2005, to help accelerate progress towards the United Nations Millennium Development Goals (MDGs), the HIPC Initiative was widened by the Multilateral Debt Relief Initiative (MDRI). The initiative allows 100% debt relief.

How countries have benefited from the HIPC Initiative

Before the HIPC Initiative, eligible countries were, on average, spending slightly more on debt service than on health and education combined. Now, they have increased markedly their expenditures on health, education and other social services and, on average, such spending is about five times the amount of debt-service payments. It has taken time and effort to ensure that money is redirected to help the poor in ways that will best reduce poverty. Other countries face challenges to meet the criteria to get the debt relief.

Exercise

1 'LEDCs should repay all of the loans otherwise they will not be encouraged to follow prudent economic policies in the future'. Discuss the extent to which you agree or disagree with this statement.

2 Describe and explain how LEDCs have been affected by debt.

Map of Ghana ***Resource 69***

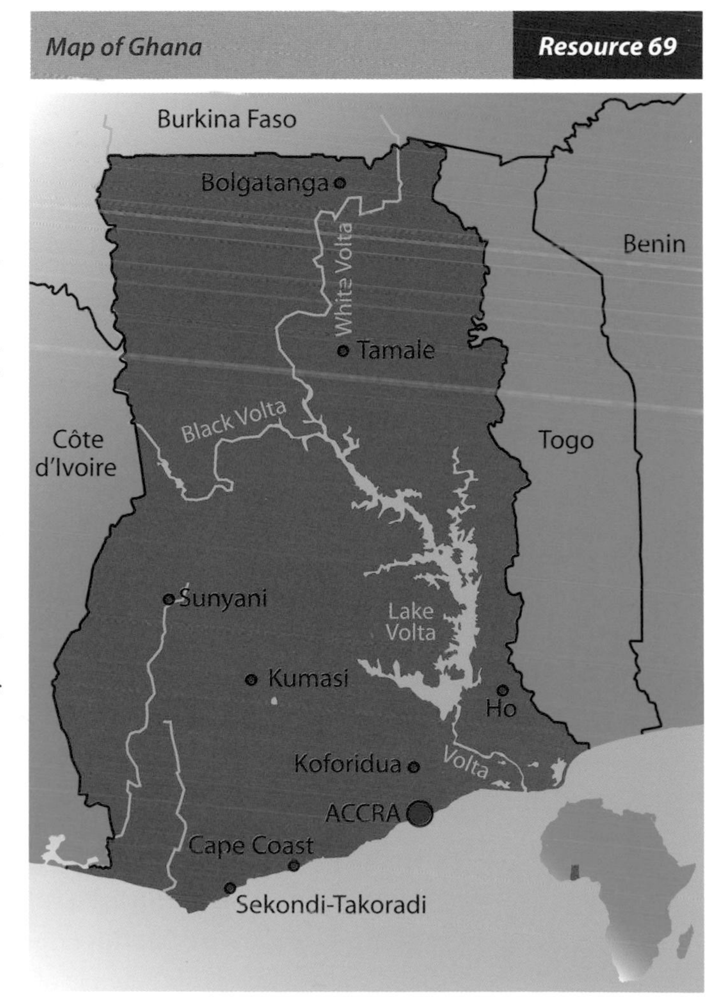

Case Study of Ghana – a LEDC

Location: Ghana is located in Western Africa and is part of the region referred to as Sub-Saharan Africa. It shares a border with Togo, in the east, Burkina Faso in the north and Cote d'Ivoire in the west. Ghana has an Atlantic coast in the south.

Climate: There are three distinct climatic zones – a hot wet region in the south west, a warm and dry south eastern region, and the remainder of the country to the north and east is hot and dry with only seasonal rainfall.

Relief: Ghana is mostly low lying but there is an area of higher land of less than 800 m in the south central region. There are two large rivers – the Black Volta and the White Volta which join and flow into Lake Volta before entering the Atlantic Ocean. A large delta has formed at the mouth of the Volta.

Natural Resources: Ghana has a varied resource base including gold, industrial diamonds, bauxite, manganese, silver, various tree products and oil (recently discovered).

Present day Ghana, formerly the Gold Coast, was the first sub-Saharan African country to be colonized by Europeans. Britain took control of the Gold Coast in 1874 and established the port city of Accra as the administrative capital. In 1884, at the Berlin Conference, the European powers divided much of Africa amongst the various colonial powers. The British section became Ghana. The years of British administration of the Gold Coast did bring significant progress in social, economic, and educational development. Communications were greatly improved, especially in the southern region. Telecommunication and postal services were initiated as well.

There are promising developments in cocoa processing in Ghana.

New crops were also introduced and gained widespread acceptance. Cacao trees, introduced in 1878, brought the first cash crop to the farmers of the interior. By the 1950s the Gold Coast was exporting more than half of the world's cocoa supply.

The colony's earnings increased further from the export of timber and gold with most of the wealth going to the colonial government. There were few settlements developed by the Britain apart from Accra. The economy of Ghana remained mostly subsistence agriculture and mining. Ghana became independent in 1957 and although there have been political problems and claims of corruption, the country has been relatively peaceful. The country operates a constitutional democracy and there is universal suffrage for all over eighteen. The legal system is based loosely on the British system.

Economy

As Ghana was the first country to gain independence from Britain there was no real understanding of the problems that a newly independent country would face. The economy had been largely geared towards the export of raw materials, and one of the first tasks of the new government was to establish a manufacturing sector. This necessitated increasing supplies of electricity. One of the most controversial schemes was the construction of the multi-purpose scheme on the River Volta built between 1961–1965. This involved the building of a large dam – the Akosombo Dam – south of the confluence of the Black and White Volta. This scheme was to supply electricity, some of which would be used in the processing of local bauxite as well as supplying irrigation water. Half of the total cost of the project came from bilateral aid from Britain and the USA as well as multilateral aid and loans (40 million pounds) from the World Bank. The government provided the remaining cost. The dam did benefit some industrial activity associated with lake transportation, fishing, tourism and increased farming activities. However, critics of the scheme claim that Ghana provided the financial cost of the dam but actually received very few benefits because:

1. The construction of the dam and the flooding of the land to create Lake Volta resulted in 80,000 people losing their land and having to be resettled elsewhere.
2. The American-owned aluminium companies Kaiser Aluminium and Volta Aluminium Company, who received generous financial incentives to locate in Ghana including low cost electricity, imported bauxite rather than develop mining in Ghana.
3. Ghana also entered into agreements with neighbouring countries to export electricity resulting in less electricity for local Ghanaians. In fact, Ghana only benefits from 20% of the power generated.
4. Soil fertility along the Volta has decreased because there is no addition of silt from flood-waters and costly artificial fertilizers have had to be used.

Many have claimed that this large-scale project shows MEDCs controlling the economy of Ghana and see this as an example of neocolonialism.

This example illustrates a negative impact of globalisation in the early years of the country's independence. In recent times there are more positive impacts associated with cocoa production; Ghana has recently agreed a deal with a Chinese company to provide them with cocoa beans in exchange for the building of a HEP in eastern Ghana. The large US company, Cargill, has started building a cocoa processing plant at Tema in the south east. This should provide more job opportunities for locals and result in increased income as processed cocoa is worth more than the cocoa beans. Furthermore, the recent discovery of oil offshore should heighten foreign interest in Ghana.

Ghana's trade is typical of most LEDCs with over reliance on raw materials and foodstuffs for its main exports and importing higher-priced manufactured goods and energy resources. The country has experienced a trade deficit since independence and although there are promising developments in cocoa processing and the potential for oil drilling, the situation is unlikely to change in the near future (***Resource 70***).

Two of the major problems that any LEDC has to deal with are repayment of loans and debt. To be competitive in the modern world, Ghana has to develop industry but in order to do this the country is dependent on international financial and technical assistance. Since independence, Ghana has received over $5 billion in loans from the World Bank. To date, Ghana's domestic economy still revolves around subsistence agriculture (***Resource 71***).

Ghana's trade deficit **Resource 70**

	1986	1996	2005	2006
Total exports (US$ millions)	749	1810	2802	3685
Cocoa	503	552	908	1002
Timber	44	147	227	207
Manufacturing	70	137	194	323
Total imports (US$ millions)	806	2523	5762	7264
Food	121	378	623	637
Fuel & Energy	135	284	862	868
Machinery	343	1212	1309	1553

Structure of the economy **Resource 71**

% of GDP	1986	1996	2005	2006
Agriculture	47.8	39.0	37.5	37.2
Industry	17.2	23.6	25.1	25.4
Manufacturing	11.1	8.6	1.6	1.3
Services	35.1	37.5	37.4	37.3

Exercise

1. Calculate the trade balance (difference between import costs and export revenue) in Ghana for each of the years shown in **Resource 70**. Describe the changes in Ghana's trade balance from 1986–2006.

2. Choose an appropriate graph to show how the values of Ghana's imports and exports have changed from 1986–2006.

3.(a) How do the changes in the value of cocoa contrast with the changes in the price of fuel and energy?

(b) What does your answer to (a) above, suggest about the problems LEDCs often have regarding trade?

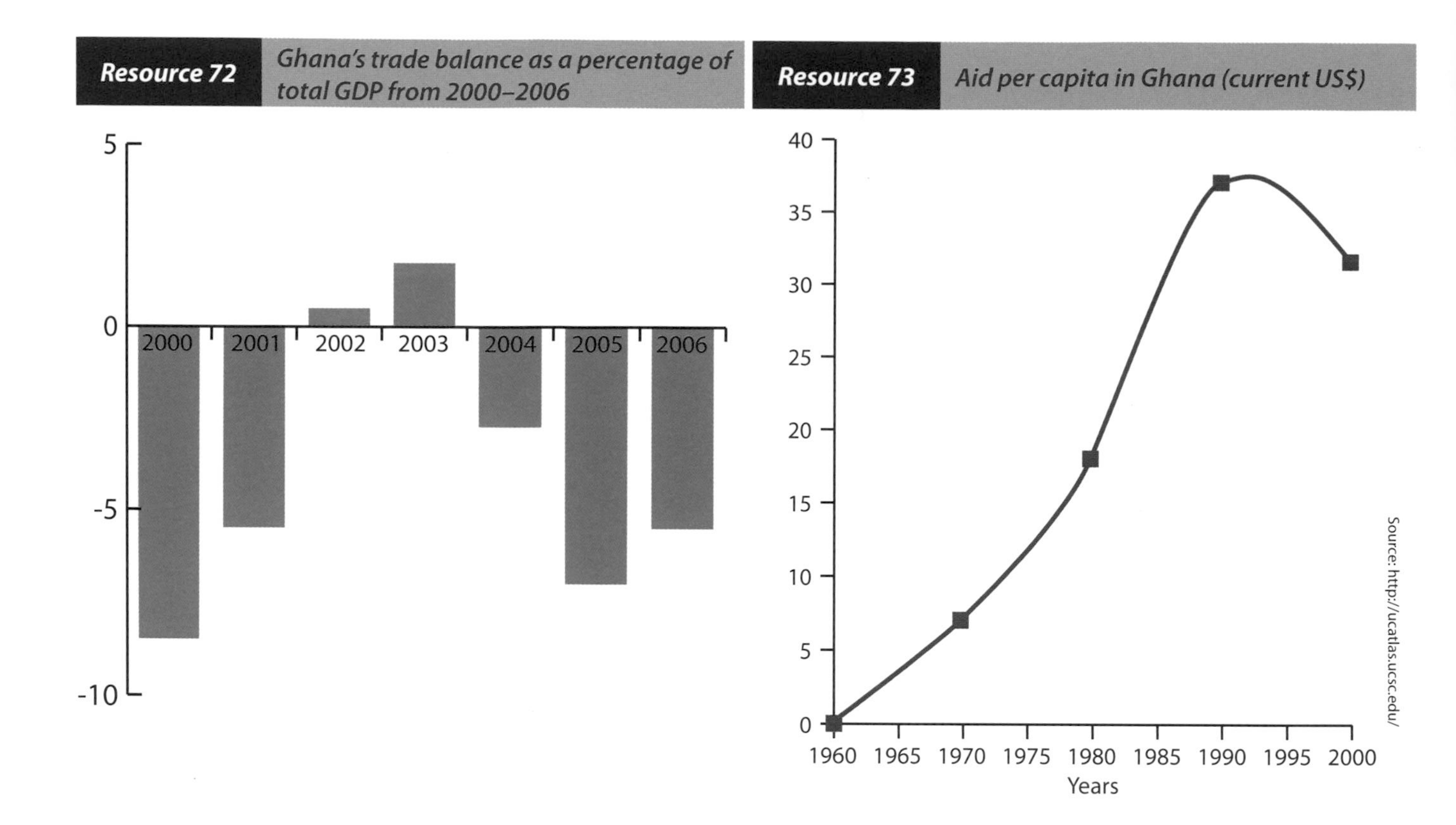

It is evident from the information shown in **Resources 70–73** that Ghana had serious problems to overcome if the country was to escape from an increasing dependence on aid and high interest rates. However, there have been a number of promising developments and several projects have gone some way towards improving the quality of life for many. In 2002 Ghana fulfilled the necessary economic reforms to qualify for debt relief under the Heavily Indebted Poor Country Initiative (HIPC) and this has meant that almost $6 billion of debt has been cancelled. While Ghana had to adopt strict economic reforms to qualify for this debt relief, most feel it has been worthwhile. A study of the information below gives an insight into some of the improvements that have resulted from aid.

- Today, Ghana is one of the best performing economies in Africa.
- Overall poverty has declined from 52 percent in 1992 to 28 percent in 2006, and Ghana is on course to exceed the 2015 MDG of halving its poverty.
- Real GDP growth averaged 5 percent from 1983–2005. Since 2005 it has risen to 6 percent.
- Over the last two decades the World Bank has provided finance for 8,000 classroom blocks and more than 35 million textbooks, and helped improve adult literacy, teacher training, primary and tertiary education. The UK government, through its International Development Agency, is providing £105 million between 2005–2015 to provide free primary school education for all children in Ghana. Ninety four percent of all children are currently enrolled in primary schools.
- A $20 million HIV/AIDS Programme (2005–2011) is aiding Ghana's anti-HIV/AIDS efforts and an Aids Response Fund has helped to achieve a reduction in the HIV prevalence rate among pregnant women.
- An extensive water project will provide clean water to many rural areas in the south and east.
- A number of NGOs, including Oxfam, are providing small-scale development projects such as building wells and affordable housing in some of the more remote areas in Northern Ghana.

- A Fair Trade scheme involving setting up farmers' cooperatives in over 1,200 villages has been initiated. These cooperatives deal directly with the main cocoa marketing companies and retain more of the revenue for development projects in their local communities.
- The Human Development Index (HDI) is currently 0.553 which gives Ghana a rank of 135 out of 177 and above the average for Sub-Saharan Africa. However, the Gender Development Index (GDI) is 0.549, illustrating a slightly lower level of development for women.

Ghana has made improvements in recent decades but there are still many areas of weakness in the economy. The country is still reliant on aid packages, although the heavy debt burden has been removed. Manufacturing industry is relatively undeveloped and the reliance on imported fuel resources is a major problem. Social development still shows an imbalance between males and females, especially in areas such as literacy and mortality rates. However, at the recent celebrations of the country's fiftieth anniversary as an independent state there was hope that the improvements that have come may have laid the foundations for sustained development in the future.

Tribeswoman grinding grain. A Fair Trade scheme operates in 1200 villages in Ghana, supporting development projects for local communities.

Exercise

Use all of the information presented in this case study of Ghana to describe the country's economic problems. Make use of figures in your answer.

EXAMINATION TECHNIQUE

Maximising Your Potential

Ok, so you have read the book and you have completed the homeworks; you may even have a well organised file of notes and handouts. A really good grade is within your grasp – or is it? To ensure that you maximise your potential and achieve a good grade in the AS Level Geography examination you need to develop, and put into practice, good examination technique.

What is examination technique?

Examination technique refers to the skills involved in taking an examination and includes:

- Paying attention to time
- Following instructions
- Understanding the demands of the question
- Recognising the scope of the question
- Developing answers to achieve higher marks

The Basics

Requirements

It may seem obvious but you need to know the requirements for each examination paper you will sit. The following summarises the main requirements for CCEA's AS Level Geography.

Assessment Unit	Length of examination	Structure of examination paper		
		Section A	Section B	Section C
Assessment Unit AS 1 Physical Geography	1 hour 30 minutes	A compulsory, multi-part question assessing fieldwork skills	Three compulsory short structured questions	Three extended response style questions – students choose any two
Assessment Unit AS 2 Human Geography	1 hour 30 minutes	A compulsory, multi-part question assessing data handling skills and techniques	Three compulsory short structured questions	Three extended response style questions – students choose any two
			In Sections B and C, questions focus mostly on one of the three main elements in the unit content. However, parts of questions may be used to assess the relationships between the elements.	

Terminology

It is not the intention here to look at study skills in detail; there are many excellent books already on the market which will help you to develop good study skills. Before you even start to think about examination technique there is one important point to remember – you will be expected to know, understand and use appropriate geographical terminology. There is a clear link between marks awarded and the use of appropriate terminology – here's what the Chief Examiner has to say: *"answers written in the correct terminology in good depth and detail will get high marks"*.

> **Hot Tip!**
> Use a copy of the current specification to identify key terms and concepts. Then draw up your own glossary/definition (you can download your own copy of the specification from www.ccea.org.uk).

Checklist

Terms and concepts used on recent AS Geography exam papers:		
Drainage basin	Abiotic components	Neo-colonialism
Open system	Ferrel cell	Random sampling
Basin size	Counter-urbanisation	Systematic sampling
Basin geology	Population structure	Stratified sampling
Ecosystem	Colonialism	

Questions

Many students seem to think that when the moment comes to turn over the page and start the exam there's nothing more that can be done. Wrong! This is exactly when you need to have really sharp examination technique.

Let's start with the question – you need to have a clear understanding of what is required and what the scope of the question is. The best way to do that is to systematically deconstruct the question. Start by identifying the command word, for example are you being asked to 'describe' or 'explain'? It is useful to circle, highlight or underline the command word and other key phrases.

Let's take an example.

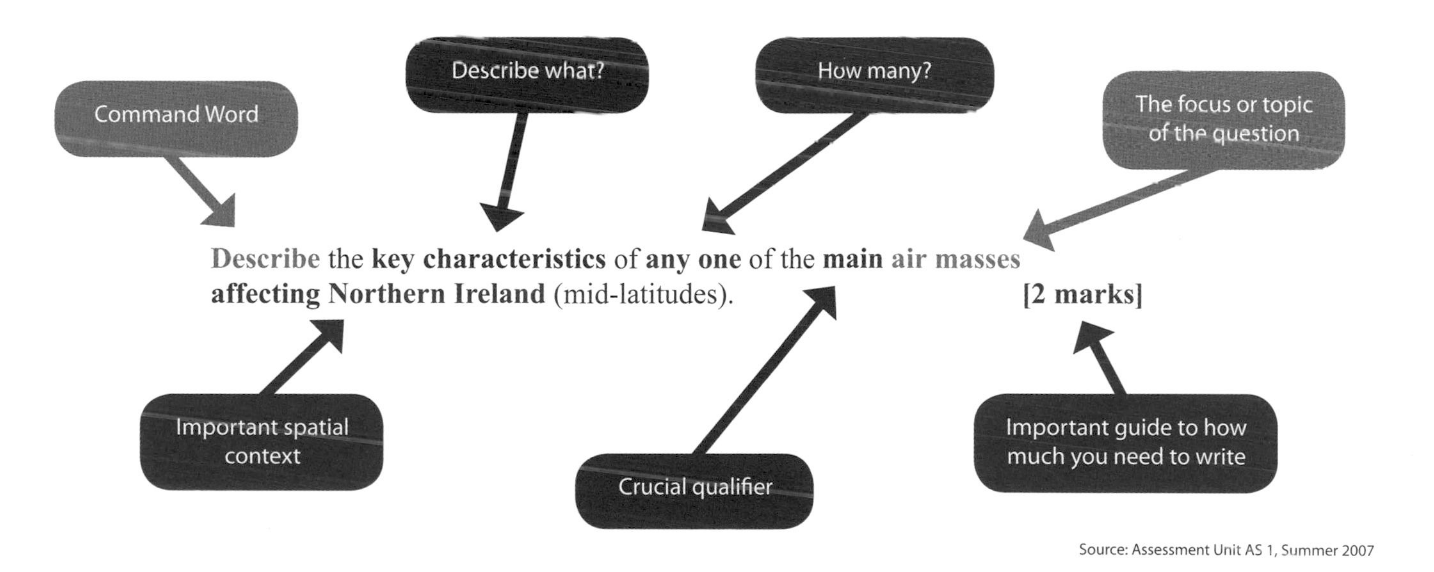

Source: Assessment Unit AS 1, Summer 2007

Plan of attack:

- Identify the command word: **describe**. Make sure you understand what is meant by commonly used command words.
- What is the **subject or focus** of the question, ie what are you being asked to describe? In this case the topic is **air masses**.
- The main focus or topic is often **qualified**, that is, it is often narrowed down by specifying

elements or aspects; in this case you are being asked to focus on any one of the **main** air masses. Note that the Examiner is drawing your attention to the words '**any one**' which were emboldened on the examination paper.

- Sometimes the question includes a spatial context – in this case it is the key characteristics of any one air mass **affecting Northern Ireland**.
- The number of marks available and, for Section B questions, the number of lines provided for your answer will give you a guide to how much you need to write as your answer.

This may seem like it will take a long time. In reality, with practice, you will find that it takes only a few seconds and as the Chief Examiner says *"the few seconds it takes to reflect on precisely what the question requires are worth spending"*.

Hot Tip!

It never says 'write all you know about' a topic or a case study; questions always require you to select from your knowledge. In short – ***answer the question exactly as it appears on the paper!***

Commonly used command words in AS Geography

Command Word	Meaning
Annotate	Add labels to summarise main features and or processes as required.
Compare	What are the main differences and similarities?
Contrast	What are the main differences?
Define	State the meaning of the term.
Describe	Give a detailed account; if a resource has been provided make sure you include details from the resource.
Discuss	In relation to a topic, this means both describe and explain. This command word may also be used in relation to a statement – in which case you should put forward arguments for and against (agree and disagree). A balanced answer is required.
Examine	A general instruction: both describe and explain.
Explain	Give reasons why.
Identify	This means choose or select.
Outline	This requires description and some interpretation but not in any detail.

Using Resources provided on the Exam Paper

A feature of Section B of the CCEA AS Geography examination papers is that you may be asked to make use of resources provided by the Examiner. You should make full use of the material provided; it may contain enough information to provide a decent answer. You might be asked to:

- Describe and/or explain patterns on a map;
- Describe and/or explain trends on a graph;
- Describe and/or explain issues from a text source.

When using a map remember to quote place names and to use the main points of the compass – avoid descriptions such as *'in the bottom right hand corner ...'*. If the resource is a graph then use the axes to quote figures to support your points as well as, if appropriate, to identify anomalies. Text sources are less frequently used at AS Level; if you are describing issues from a text source it is a good idea to make use of carefully selected, short quotations.

Let's look at an example.

Study **Resource 1**, which shows the population distribution of Brazil and climate graphs for locations A, B and C.

Resource 1

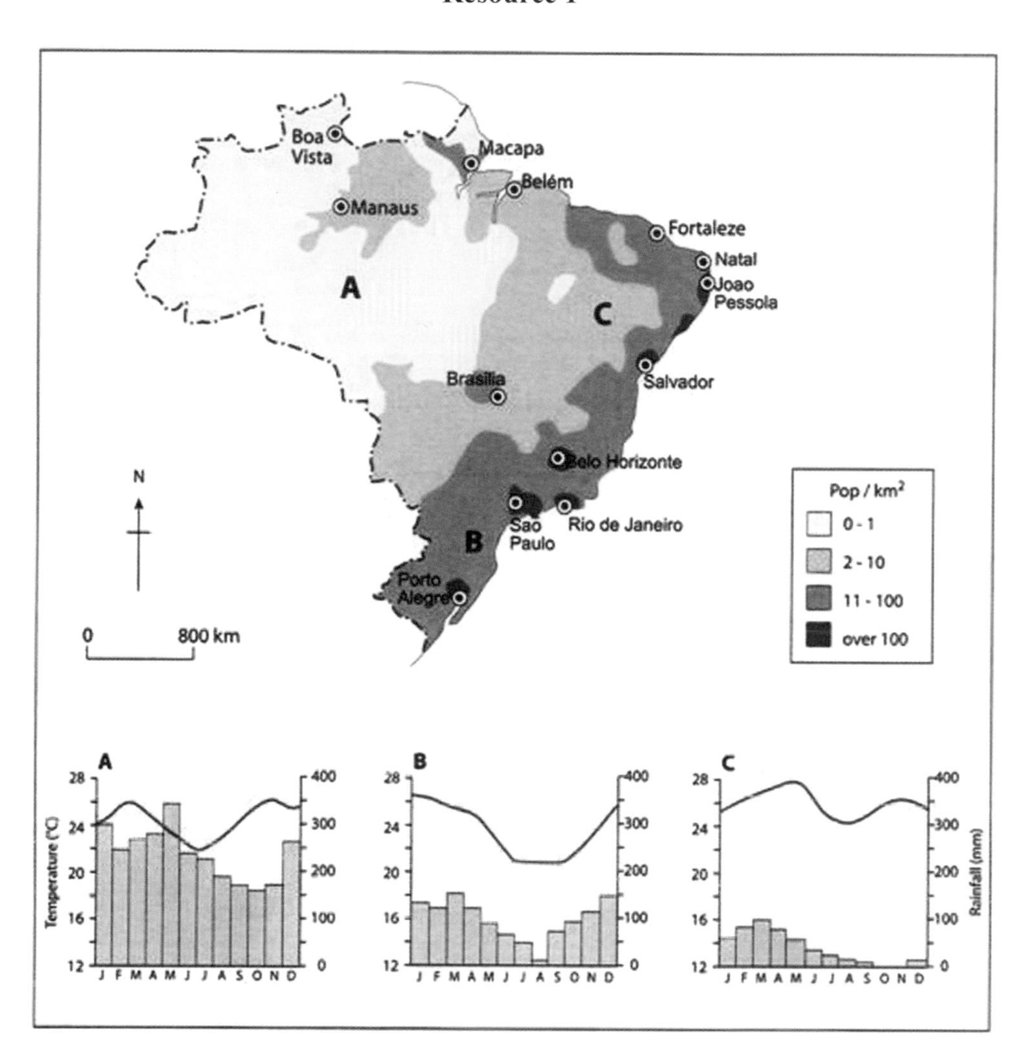

For any **two** of the locations A, B, C, describe and explain, using **Resource 1**, how their contrasting population densities have been influenced by climate.

[6 marks]

Source: Assessment Unit AS 2, Summer 2007

First make sure that you read the opening instruction carefully. It may contain important information that will be useful to you as you answer the question. Note that the Examiner has emboldened the word '**two**' as well as the phrase '**Resource 1**'. This is to draw your attention to the fact that you have to deal with two locations in your answer as well as making clear reference to information provided in the resource. You will be expected to use figures from the climate graphs to support your answer. If you do not include information from the resource in your answer you will not be awarded full marks for your answer.

Hot Tip!

Make full use of the resource materials provided otherwise you may lose marks.

Use of Case Study Material

Throughout the CCEA AS Level Geography, you will be expected to draw on case studies to illustrate your answer. This is particularly so in Section C of both assessment units although you may also be asked to make reference to case study material in your answer to questions in Section B.

Such questions are usually marked using a levels mark scheme and you will need to provide specific place-related information including facts, figures and dates as appropriate, to ensure that you are awarded a Level 3 mark.

Let's look at an example.

> With reference to a national case study, describe and explain the regional variations in development. [12 marks]

Level 3 ([9]-[12])

Candidate provides a thorough and detailed description of the regional variations in development with specific figures and places mentioned. These differences are explained thoroughly showing sound understanding.

Level 2 ([5]-[8])

Candidate provides a general but accurate answer but there is less detail and depth throughout or one aspect only is dealt with in a superficial manner. The answer may be limited to a description of regional variations in their chosen case study.

Level 1 ([1]-[4])

Candidate provides a limited answer which is lacking in detail and depth on all aspects or there may be incorrect information. Reference to case study material may be limited, inaccurate or omitted altogether.

Source: Assessment Unit AS 2, Specimen Paper

In order to be awarded a Level 3 mark you will need to provide accurate facts and figures from your chosen case study. It is important that the case study is at the correct scale – in this case a national scale case study is required. If you are in any doubt about your case study you should refer to your teacher or the specification for guidance.

Hot Tip!
Case studies have to be detailed, with facts and figures.

Planning

You may think that planning is taking up valuable time but the few minutes it takes to jot down a plan will be worthwhile and may result in your answer being awarded high marks. Essay questions in Section C on both AS examination papers, as well as longer sections of structured questions in Section B, require planning before you start to write your answer. Planning your answer will help you to:

- focus on the question exactly as it appears on the examination paper
- recall and select the material required ; this will help you to avoid the tendency to write all you know about the topic
- ensure your case study (if applicable) is appropriate and includes detailed facts and figures
- give your answer a logical sequence

Getting Help

There is no doubt that candidates do best when they know what to expect. In the first instance you should seek advice, help and guidance from your teacher. You may also find useful information within the Geography microsite on the CCEA website: www.ccea.org.uk where you will find:

- The current GCE geography specification
- Specimen papers and mark schemes
- GCE Geography Student Guide
- Chief Examiner's reports
- PowerPoint presentations:
  - AS Geography: A Student Guide
  - Techniques in Geography: A Student Guide

Past papers and mark schemes will also be useful but they come with a warning – they were written for the previous specification and you need to use them carefully, taking the time to check that the question relates to the current specification.

GLOSSARY

Physical Geography

Abrasion: the scraping, scouring and wearing away of the bed and bank of rivers by rock fragments carried along by the river flow.

Afforestation: the deliberate planting of trees.

Alluvium: the term used to describe any sediment deposited by a river.

Annual hydrograph: a graph which shows the variation in a river's discharge over one year. In Europe the water year normally runs from October to September.

Arcuate delta: a delta shaped like the letter 'delta' in the Greek alphabet or a wedge of cake. The Nile delta in Egypt is one example.

Atmosphere: the layer of transparent gases that surrounds the earth held by gravity.

Attrition: the wearing down of river load material as it collides and rubs against other pieces of load.

Base flow: the part of a river's discharge that is contributed by the groundwater store.

Barsha floods: the normal and beneficial annual inundations of the Ganges delta in Bangladesh.

Bangladesh: a low-lying country that is largely the deltaic product of the Ganges, Brahmaputra and Meghna rivers as they flow into the Bay of Bengal.

Billabong: an Australian name for an **oxbow lake** or meander cut-off.

Bird's foot delta: a delta in which the distributaries extend their channels and banks into the sea at the mouth of a river such as the Mississippi in the Gulf of Mexico.

Braiding: the dividing of a river into several interweaving channels, normally the result of a loss of energy or overloading with sediment.

Bonna floods: the exceptional and damaging annual inundations of the Ganges delta and floodplain in Bangladesh.

Channel catch: the term to describe rain that falls directly into river channels.

Confluence: the junction of two rivers.

Corrasion: an alternative term for **abrasion.**

Corrosion: an alternative term for **solution** erosion by rivers.

Deferred junction: where a tributary flows downstream alongside another river before its **confluence**. Also known as a **yazoo.**

Deforestation: the clearance of trees from an area.

Delta: the depositional landform built up from alluvium deposited at a river mouth.

Discharge: the volume of water passing a point in a given time. It is represented by the equation $Q = A \times V$, where Q is discharge, A is channel cross-section area and V is velocity. Discharge is measured in cumecs, cubic metres per second (m^3/sec).

Distributary: a river that branches from the main channel, commonly found in deltas.

Drainage basin: the land area drained by a river and its tributaries.

Drainage density: an index of the average length of river channel per unit area of a drainage basin (km/km^2).

Erosion: the wearing away of the land by natural processes, including ice, flowing water and wind.

Evaporation: the change of water from its liquid to its gaseous form. Along with transpiration it is an output from the drainage basin system termed evapotranspiration.

Flashy hydrograph: a water graph with steep limbs and a short time lag associated with the risk of river flooding.

Flocculation: the joining together of clay particles when fresh water meets salt water, eg at a river's mouth.

Flood (storm) hydrograph: the term for a graph representing a river's response to a single rain event.

Gorge: a vertical sided river valley often formed by a retreating waterfall.

Hjulström curve: a graph that shows the velocity at which differing sizes of particle are eroded, transported or deposited by water flow (rivers).

Hydraulic action: a process of river bed and bank erosion; it involves the energy of the flowing water itself.

Hydrograph: a line graph or changes in river discharge over time.

Impermeable: a soil or rock type that does not let water pass into or through it.

Infiltration: the process of water soaking into the soil from the surface.

Interception: the process in which precipitation falls onto vegetation rather than directly to the ground.

Kharif: the name given to the summer wet farming season in Bangladesh.

Levees: raised river banks. They may be formed by deposition of larger alluvial particles during floods but are often artificially raised to prevent such floods.

Load: the term used to describe any sediment carried by a river.

Meander: a bend in a river.

Mort lake: another term for a **oxbow lake** or cut-off meander.

Overland flow: water run-off over the land surface but not in a channel.

Oxbow lake: the remnant section of unused channel when a river changes its course, cutting off a meander.

Peak discharge: the period of highest river water flow during a storm or flood episode.

Peak rainfall: the period of most intense precipitation during a storm.

Percolation: the downward movement of water from the soil into the deeper stores such as ground water.

Permeable: a soil or rock type that allows water to pass into and through it.

Plunge pool: the deep section of river bed immediately below a waterfall.

Point bar deposits: the alluvial sediments left on the inner bank of a meander by slow moving water. They form the slip-off slope.

Precipitation: the name for any form of water leaving the atmosphere, including rain, hail, sleet, snow, dew, fog and frost.

Rabi: the name given to the winter dry farming season in Bangladesh.

Rainfall floods: inundations of land due to direct heavy rainfall exceeding the infiltration capacity and not rivers simply overflowing.

Recession limb: the declining slope following peak discharge on a hydrograph.

Riffles and pools: these are the regular pattern of shallow and deep sections of channel that form in river beds.

Rising limb: the increasing curve on a hydrograph leading to peak discharge.

River floods: inundation of land as the result of a river channel overflowing.

Saltation: the process by which material is transported in a bouncing motion such as bed load moving downstream in a river or sand across a beach.

Snowmelt: the release of water when winter snows melt in spring, often causing flooding.

Solution: the term used for both an erosion process and the transportation of material chemically dissolved in water.

Stores: in a system these are the places to and from which material and energy is transferred.

Storm flow: describes all the river discharge resulting from a period of rain. It is graphically represented as the area under the curve in a hydrograph above base flow.

Suspension: the river transport process where light particles are held and carried in the flowing water.

Systems theory: a framework used to describe and analyse the interaction of parts of the physical and human world. Examples include ecosystems, drainage basins, factory production and farms.

Thalweg: the line of maximum velocity in a river channel.

Traction: a river transport process involving bed load material rolling downstream.

Transfers: exchanges of material and/or energy between stores in a system.

Transpiration: the loss of moisture from vegetation during photosynthesis.

Transport: in fluvial studies, this is the movement of material downstream.

Tributary: a river or stream that joins and adds (*contributes*) water to another channel.

Throughflow: is the term used to describe water moving downhill within the soil.

Water table: the height of water stored as groundwater within rocks beneath the surface of a drainage basin.

Waterfall: a steep or vertical section of channel where the river falls freely.

Watershed: the line that marks the boundary between drainage basins.

Yazoo: a term used for a **deferred junction** on a river. The name comes from an example on the Mississippi floodplain.

Human Geography

Adult literacy rate: the percentage of adults who can read and write.

Aid: the transfer of resources from richer MEDCs to the poorer, usually, LEDCs.

Areas of Outstanding Natural Beauty (AONBs): these are usually smaller in area than National Parks. There are 36 in all, covering about 15 per cent of England. Their management is the responsibility of the local council.

Bilateral aid: aid that comes directly from one country to another.

Colonialism: where one country takes political control over another, usually as part of empire building.

Composite measures of development: those measures that use several indicators to provide an index of development which can then be used to rank countries.

Counterurbanisation: a movement of urban workers from a city to rural towns and villages within commuting distance of the city.

Crude Birth Rate* (CBR): the number of live births in a year per 1,000 of the mid-year population.

Crude Death Rate* (CDR): the number of deaths in a year per 1,000 of the mid-year population.

**If the difference between these two measures is positive there is an increase in population and conversely if the difference is negative, there is a population decrease.*

Debt: the amount of money owed, usually by a LEDC, to another country or organisation.

Dependency ratios: the ratio of economically active population to the economically inactive or dependent population.

Aged Dependency is calculated:

$$\frac{\text{Total number over 65}}{\text{Total number 15–64}} \times 100$$

Youth Dependency is calculated:

$$\frac{\text{Total number 0–14}}{\text{Total number 15–64}} \times 100$$

Enrolment in primary, secondary and tertiary level education: the percentage enrolment in each stage of education.

G8: a group of the world's wealthiest countries – UK, USA, Canada, Japan, France, Germany, Italy and Russia.

Gentrification: a process where areas of old housing in the inner city have been bought by middle class buyers or developers with sufficient funds to modernise them. As a result of this the value of the house increases and the area becomes more fashionable.

Globalisation: the current interaction of most of the world's economies. In the world of today national economies are no longer separate entities but rather part of a world or global economy.

Greenfield sites: rural land that is being developed for non-rural use.

Gross National Income per capita (GNI pc): GNIpc, formerly known as GNPpc, is the total value of goods and services produced in a country plus taxes and income from abroad in one year divided by the total population. In order to allow international comparisons to be made it is always expressed in US dollars.

Heavily Indebted Poor Countries (HIPC) Initiative: is a debt reduction scheme set up in 1996 for heavily indebted poor countries that satisfy economic and political criteria laid down by the IMF and World Bank.

Human Development Index (HDI): measures the average figures for life expectancy, enrolment in primary, secondary and tertiary level education and Purchasing Power Parity (PPP). The values of HDI fall in the range 0–1 and countries are ranked according to their calculated value.

Infant Mortality Rate: the number of live children per thousand who die within the first year of life.

Informal settlements: housing built without planning permission on any available piece of land in LEDCs, usually without any basic infrastructure. These settlements are built using whatever materials are available – corrugated iron, timber, even plastic sheeting. The people living in these informal settlements have no legal right to occupy the land and local authorities do forcibly remove them on occasions.

Life Expectancy at birth: the number of years a person is expected to live calculated at the time of their birth.

Migration Balance: the balance between in-migration and out-migration.

Multilateral aid: aid that comes from several countries or organisations such as the World Bank.

Multinational Company: divides the running of the company across several countries.

National Census: a count of all of the population and those social and economic characteristics that can easily be counted on a specific date and usually every ten years.

Natural population change: the balance between birth rates and death rates.

National Parks: a form of protected land. The first National Parks were set up during the 1950s to manage conservation and recreation. There are currently 14 National Parks in the UK. Each Park has a National Park Authority whose responsibilities include conservation of the natural beauty, wildlife and cultural heritage of the Park while at the same time improving opportunities for public understanding and enjoyment of the Park.

Neo-colonialism or dependency: where a former colony has political independence but is still tied economically to its colonial ruler either through aid packages or trade agreements.

Official aid: aid that comes directly from government sources.

Physical Quality of Life Index (PQLI): uses three indicators – life expectancy at the age of one, infant mortality and literacy – each having equal weighting to give an index of development. Values of PQLI fall within the range of 0–100 and countries are ranked according to their calculated value.

Primate City: a city that is disproportionately large compared to other cities in a country. It is a situation that usually occurs where development is uneven but not exclusively in LEDCs.

Purchasing Power Parity (PPP): this takes account of the real purchasing power of a given amount of money in different countries and as such is a more reliable measure than GNIpc for LEDCs.

Reurbanisation: the movement of high-income groups into high status accommodation in the inner cities.

Rural-urban fringe: the zone where urban and rural land use meet.

Sites of Special Scientific Interest (SSSIs and ASSIs in Northern Ireland)): these are areas that have special wildlife or rare flora. There is a list of restrictions in force in SSSIs and applications for development within a SSI must be passed by the SSSI regulators. Most are in private ownership.

Suburbanisation: a process that refers to the decentralisation of people, services and industry to the edge of the existing urban area (urban sprawl).

Structural Adjustment Programmes (SAPs): conditions that have been placed upon those LEDCs which have had to take out additional loans.

Trade: the exchange of goods between one country and another.

Transnational Company: has its headquarters in a single MEDC and a number of branch factories in LEDCs.

Vital Registration: the official recording of all births, including stillbirths, adoptions, marriage and civil partnerships, and deaths. In Scotland there is also information on divorce.

Voluntary Aid: comes from charitable organisations such as Oxfam, Trocaire and Action Aid. These are referred to as Non-Governmental Organisations (NGOs) and they rely entirely on voluntary contributions.

Geography for CCEA

Martin Thom and
Eileen Armstrong

AS LEVEL

ISBN: 978 1 904242 90 1

First Edition
Third Impression

Layout and design: Colourpoint Books

Printed by: GPS Colour Graphics Ltd

Colourpoint Educational
An imprint of Colourpoint Creative Ltd
Colourpoint House
Jubilee Business Park
21 Jubilee Road
Newtownards
County Down
Northern Ireland
BT23 4YH

Tel: 028 9182 6339
Fax: 028 9182 1900
E-mail: info@colourpoint.co.uk
Web site: www.colourpointeducational.com

The Authors

Martin Thom is a graduate of QUB and teaches Geography at Sullivan Upper School, Holywood. He has worked with CCEA for over 20 years and published a number of A-levels texts and articles.

The author would like to thank the Geography staff at Sullivan Upper and to the many teachers who replied to the questionnaire that helped shape the thinking behind this book.

Eileen Armstrong has a BA (Hons) in Geography from QUB. She began her teaching career in St Dominic's Grammar School, Belfast and is currently teaching in Sullivan Upper School, Holywood. In addition she has widespread experience with CCEA as an examiner. She became Principal Examiner for A-level Human Geography in 1996 and was a co-writer of the AS and A2 level specification which comes into force in September 2008.

The author acknowledges the help and support of her family during the writing of this book. She also wishes to acknowledge her colleagues in Sullivan Upper School for many helpful discussions.

Picture credits

All photographs by the authors except for the following which are included with kind permission of the copyright holders:

istockphoto: 83, 85, 87, 95, 99, 103, 107, 111, 123, 131, 139, 142, 153, 155, 163, 165, 180, 183

Wesley Johnston: 14, 159

Page 130: Ordnance Survey map reproduced with permission from Peak District National Park Authority